USER ASPECTS

OF PHASE DIAGRAMS

ORGANISING COMMITTEE

Dr F. H. Hayes
(University of Manchester/UMIST, UK)
Conference Chairman

Dr T. G. Chart
(National Physical Laboratory, UK)

Dr G. Effenberg
(Max Planck Institute, Stuttgart, Germany)

Professor R. Ferro
(Universita de Genova, Italy)

Professor B. Legendre
(Université de Paris-Sud, France)

Dr M. Merz
(Joint Research Centre, Petten, The Netherlands)

Professor A. P. Miodownik
(University of Surrey, UK)

Professor A. Prince
(Brunel University, UK)

Dr N. Saunders
(University of Surrey, UK)

Dr J. R. Taylor
(Johnson-Matthey Technology Centre, UK)

User Aspects of Phase Diagrams

Proceedings of the International Conference

held at the Joint Research Centre,
Petten, The Netherlands
25-27th June 1990

Organised by
The Institute of Metals
and the
Commission of the European Communities,
Joint Research Centre,
Institute for Advanced Materials, Petten

THE INSTITUTE OF METALS
1991

Book 501

Published by

The Institute of Metals
1 Carlton House Terrace
London SW1Y 5DB

and

The Institute of Metals
North American Publications Center
Old Post Road, Brookfield, VT 05036, USA

British Library Cataloguing in Publication Data

Applied for

ISBN 0 901462 91 8

Made and printed in Great Britain by
Bourne Press Ltd, Bournemouth

CONTENTS

PREFACE

The papers contained in this volume were presented at the "User Aspects of Phase Diagrams" Conference held at the Joint Research Centre, Petten, Netherlands from 25-27th June 1990. This meeting was jointly organised by the Institute of Metals, London and the Commission of the European Communities, Joint Research Centre, Institute for Advanced Materials, Petten, in collaboration with the following organisations: ASM International, Associazione Italiana di Metallurgia, Deutsche Gesellschaft für Metallkunde, Indian Institute of Metals, Max-Planck-Institut für Metallforschung, U.S. National Institute of Standards and Technology, Società Chimica Italiana, Société Française de Chimie, Société Française de Metallurgie and the Verien Deutscher Eisenhüttenleute. The main theme of the meeting was to focus attention on the wide range of industrial applicability of phase diagrams in the extraction, production, processing and use of materials ranging from advanced ceramics, magnetic materials, superconductors, composites and semiconductors to conventional hard metals, steels and nonferrous alloys. A total of nine sessions over three days were held of which seven were devoted to papers on different classes of materials, one session to on-line demonstrations of modern software, thermochemical databanks and computer calculations and one poster session with twenty contributions.

As the content of the papers presented at Petten clearly demonstrates, phase diagrams of various kinds are widely used by Metallurgists, Materials Scientists and Design Engineers in major industries to locate conditions where particular phases will be stable or metastable, where particular properties will be exhibited, where particular reactions or processes become feasible and to understand and predict solidification sequences. Phase diagrams of interest can range from the relatively simple pressure-temperature plots for pure elements to complicated multicomponent, multiphase systems where several variables may have to be fixed before a graphical representation even becomes possible. Over the past fifteen to twenty years the impact of computer calculation of multicomponent phase diagrams via thermodynamic data stored in databanks combined with advances in experimental techniques has been dramatic and has begun to provide the type of detailed knowledge concerning phase relationships in complex systems and real engineering materials required by process engineers. Increased academic activity in this area has been characterised by an unusually high degree of international cooperation through for example the Calphad group, the SGTE group, the ASM/NBS binary phase diagram assessment programme coordinated through the Alloy Phase Diagram International Commission and the European COST 507 Project which currently involves fourteen countries. This cooperation has led to rapid developments in the underlying thermodynamics, in the modelling of phases, in creation of databases and associated computer software. Several new high quality scientific journals now exist which provide a forum for reporting assessments, new research results and for discussion of new experimental and theoretical developments. This resurgence of interest in phase diagram research will continue to play its full role in the future by its contribution to satisfying the ever increasing demands made by the industrial nations for new and improved engineering materials.

Finally I would like to thank the members of the Conference Organising Committee and the Institute of Metals staff involved for their hard work in contributing to the success of the conference. I would also like to express my gratitude to Dr M. Merz, Dr M. de Groot and Dr Ben Seysener of JRC, Petten, for the help that they provided at the Conference Centre, to Mr de Haan for his help in setting up the computer demonstrations and to Professor E. Hondros FRS for hosting the meeting and for giving the opening address.

F. H. Hayes
Materials Science Centre
University of Manchester and UMIST

1

PHASE DIAGRAMS AS A TOOL FOR THE PRODUCTION AND DEVELOPMENT OF CEMENTED CARBIDES AND STEELS

B. UHRENIUS

AB Sandvik Coromant R & D, Materials and Processes, Cemented Carbide, Stockholm, Sweden

ABSTRACT

Data bases and programs for the calculation of phase diagrams have become valuable tools in the research and development of cemented carbides and steels. The number of applications of this technique for cemented carbide systems have increased lately due to the fact that some basic systems have now been analysed and entered into such data bases.

In the following, some cases are reviewed where the knowledge of phase diagrams, especially calculated diagrams play an important role in the development work. These cases refer to the accurate control of the carbon balance necessary for a successful sintering of some WC-Co alloys, as well as the information needed for the sintering of Fe-Ni based carbides.

The development work on steels within the Sandvik Steel company is further ahead when it comes to using phase diagrams and especially calculated diagrams as a tool for the development of new grades. This work is now facilitated by the great number of assessed systems for iron based alloys and the reliable data bases available today. A few examples concerning steels are also reviewed where the use of calculated diagrams has been successful.

INTRODUCTION

The Sandvik group is active in two major business areas, steels and cemented carbides. The manifoldness of products produced for various applications in these business areas means that today the Sandvik company is a Materials Science based company. The heart of the company is located in Sandviken where it started as a steel producer more than 125 years ago, and the research and development as well as the production of steels are still located in Sandviken. The cemented carbide activities, however, are more scattered and the research and development of such materials is located in Stockholm.

Today most commercial alloys are very complex, and that holds for steels as well as for cemented carbides. Many steels contain about 10 components. The constitutional diagrams of such alloys are almost impossible to describe in detail and there is always a strong need for simplification. In many text books the phase diagrams presented are related to binary or ternary systems and it is solely in very special areas that one can find higher order systems.

Modern computer techniques and thermochemical programs for the calculation of phase equilibria and phase diagrams thus have offered very important tools for the research and development of steels and cemented carbides.

Of great importance in Sweden is the "THERMO-CALC" data base [1] at the Institute of Technology in Stockholm. The so called "POLY-program" and the data bases linked to that system are used within the Sandvik organization for the research and development of steels and cemented carbides.

However, the successful use of such computer programs is very much dependent on the accuracy and the amount of information available in the data bases. The number of alloy systems investigated related to steels or iron base systems is much more comprehensive than those based on cobalt or other metals of importance to the cemented carbide producer.

The calculation of phase diagrams is not only a great help when dealing with multi-component systems but offers better means to present the results. One can make isothermal plots for the particular temperature of interest and vertical sections for any alloy composition. Calculations of activities of the individual components can be made as well as presentations in mole fraction, weight percentage or whatever measure one may wish. In the following, all phase diagrams presented are obtained by using the THERMO-CALC-system and the POLY-program. I would say that the use of these programs has become a very important step forward at least in the field of steel metallurgy and I am convinced that it will become more powerful also for applications related to cemented carbides when more of the alloy systems involved have been analysed and assessed.

APPLICATIONS TO CEMENTED CARBIDES

Traditionally cemented carbides are based on the Co-W-C system. The early investigations of this system were made by Takeda[2] as early as in the 1930s and by Rautala and Norton[3] in the 1950s. Much of the information available today is based on their work but also on work by Pollock and Stadelmaier[4], Gurland[5], Ettmayer and Suchentrunk[6] and others. All these investigations were carried through by using classical methods and it was not until in the early seventies that people started to combine these classical metallographic methods with thermodynamic calculations. Some contributions in the area of cemented carbides have been made by the group at the Institute of Technology in Stockholm in co-operation with Sandvik. An attempt to analyse the Co-W-C-system in thermodynamic terms was made by Leif Åkesson[7] but it was not until the data could be handled in a more efficient way by using the THERMO-CALC-system that a consistent description was presented by Armando Fernández-Guillermet[8].

However, tradition and experience is a strong tool too and I would say that most cemented carbide producers can make high quality cemented carbides without the detailed knowledge of the phase diagrams. This is particularly true for the Co-W-C system, which is a very straightforward ideal system for cemented carbides and consequently there is little need for a more detailed description of the constitutional diagram of this system.

However, in order to develop more sophisticated grades or to fully exploit the possibilities of the system, I would say that by using phase diagram calculations the work could be considerably facilitated.

The production of sintered carbides based on the Co-W-C-system always involves a very careful control of the carbon content of the alloys.

This production involves a mixing of tungsten carbide and cobalt and a processing which means heating to high temperatures, 1400-1500°C, sintering and cooling usually without allowing any formation of graphite, or "eta carbides" (carbides formed by both cobalt and tungsten). The range of compositions within which the carbon content might be allowed to vary is not very wide. Too high a carbon content would cause precipitation of graphite whereas in those cases when the carbon content is too low the eta carbide precipitates either during sintering or during cooling.

In addition, a variation of the carbon content of the alloy also causes a variation of the amount of alloying of the cobalt-phase by tungsten. This will have a strong influence on the properties of the sintered body especially the mechanical properties.

The proper control of the carbon content is not only influenced by the amount of carbon or carbides added from the beginning but it is also essential owing to the carbon losses which are caused by reactions with absorbed oxygen or hydrogen containing gases added during the presintering or sintering stages. The production usually involves some correction of the carbon balance by adding soot or tungsten metal powder to the powder mixture in order to reach the desired carbon balance.

Figure 1 shows a section through the Co-W-C system at 1400°C which is a common sintering temperature. This diagram is calculated on basis of a thermodynamic analysis of the classical metallographic information as well as the thermochemical information of the Co-W-C system and the lower systems involved.

From this figure it is obvious that WC is in stable equilibrium with liquid cobalt only in a rather narrow region of composition. During cooling the solubility of tungsten carbide in the liquid will decrease. In order to follow the changes during cooling, a number of isothermal sections could be calculated showing the precipitation of more tungsten carbide, the freezing of the liquid into solid cobalt and later on also the formation of the ternary graphite eutectic. However, such a procedure can more easily be shown by calculating vertical sections through the ternary system for the particular alloy chosen.

Figure 2a shows such a section showing the constitution of an alloy with a molar ratio of tungsten to cobalt equal to 2.7. This figure clearly shows what happens if the carbon content is increased or decreased. Consequently, one could easily follow what occurs at each temperature during cooling. However, from such a diagram one can neither evaluate the amount of the individual phases nor the composition of the various phases. The diagram simply shows what phases are present at a certain temperature for a certain carbon content. However, in the data base or in a result file there is all the information one might need to make such plots.

Further, Fig. 2b shows what might be possible to plot from such a calculation. That figure shows the amount of WC, liquid and FCC-Co present at any temperature during sintering and solidification. In the same diagram the carbon activities of some alloys with different carbon content are included for comparison. These are obtained by slightly increasing or decreasing the carbon content.

From the diagram it is obvious that the stoichiometric alloy will not cause any precipitation of graphite during solidification. The carbon activity will only reach graphite saturation after solidification has been completed. Due to slow progress of the graphite precipitation in the solid state, graphite will probably not occur in the microstructure of that alloy.

However, an increased carbon content by 0.05% by weight, as shown in Fig. 2b, will cause graphite precipitation during the freezing of the liquid.

The above diagrams relate to the plain Co-W-C-system, a system which is still the only one for many applications of cemented carbides such as rock drilling applications and certain wear parts. For cutting tools the systems used are more complicated. In addition to cobalt, carbon and tungsten, many alloys involve the carbides of titanium, tantalum, niobium and molybdenum. In some grades cobalt is replaced by nickel to some extent.

The assessment of the systems related to these components is under progress. Some of them are already stored in a local data base at Sandvik or in data bases available at the Institute of Technology in Stockholm.

However, much of the information needed is still missing and a lot of work is needed before all these systems can be dealt with in an efficient way.

During the past fifteen years some efforts have been made on systems in order to find replacement for cobalt as a binder.

Depending on the world situation or marketing strategies, the prices of cobalt and tungsten regularly fluctuate and this has a strong impact on the demand for substitutes for cobalt and tungsten. From time to time there has thus been a strong demand for knowledge of other systems for the production of cemented carbides. Many attempts at replacing cobalt by iron and nickel or combinations thereof have been made. Frequently there have been attempts made at replacing tungsten by molybdenum.

In the beginning of the eighties a French/Swedish research project was started in order to strengthen the picture of the Fe-Ni-W-C system, and the background was to find information in order to develop new alloys based on the Fe-Ni-W-C system rather than the Co-W-C system.

The high temperature part of the work was carried out at the Ecole Polytechnique in Grenoble[9] where experiments were made involving the liquid phase whereas some heat treatments and equilibrium studies at lower temperatures were made at the Institute of Technology in Stockholm. The results obtained were later analysed in thermodynamic terms[10].

This piece of information was added to the information already available for the Co-W-C system and renewed analyses of all these systems were made[11]. Therefore today there is a powerful coverage of the entire Co-Fe-Ni-W-C system.

However, the Co-WC alloys are usually referred to as being based on an unusually happy marriage between two phases, and the combinations of these two phases offer properties which are not easy to obtain by using iron or nickel or combinations thereof. The reasons behind are not only found in the phase diagram but also in the mechanical strength and the good wetting between cobalt and tungsten carbide. A successful commercial utilization is still lingering.

Thanks to the fact that calculations can be made it is now easier to find out how the various alloy contents influence the structure at equilibrium and a commercial realization should now be closer at hand.

Figure 3a shows a vertical section through the Fe-Ni-W-C system taken from the work by A. Fernández-Guillermet[12]. The calculation was made for an alloy containing 20% binder by weight and having an iron to nickel ratio equal to 3:1. It can be seen from that figure that an alloy of the stoichiometric mixture of WC and the pure metals would precipitate eta phase during sintering at 1350°C. Even though the alloy after cooling would arrive at a stable two-phase field of solely FeNi-binder and tungsten carbide at temperatures below 1300°C, the precipitation of eta phase formed at sintering temperature would not dissolve but stay in a metastable form at lower temperatures. It is obvious from the calculated section that it would be better to increase the carbon content of the alloy to between 5.0 - 5.10% carbon in order to avoid either eta phase precipitation or the graphite precipitation which would occur if the carbon content exceeds the 5.10%.

In Fig. 3b[10] a temperature projection of the areas in the Fe-Ni-W-C system is shown which would give graphite or eta phase in equilibrium with the liquid phase during sintering or the successive cooling. The zone in the middle shows the most favourable compositions where neither of these two phases will precipitate. From the figure it is obvious that stoichiometric mixtures of WC and the pure metals will give favourable structures if the mixtures of Ni/Ni+Fe fall in the region between 45 to 65%. When using a data base to calculate phase diagrams the choice of alloys could thus be facilitated and one could easily restrict the choice of alloys to those which are most favourable from the compositional point of view. Similar descriptions of the entire Fe-Ni-Co-W-C system can now be obtained on the basis of thermodynamic calculations. An experimental investigation which covers the entire five-component system would be very tedious and time consuming.

In Fig. 4, obtained through the work by A. Fernández-Guillermet[11], a section is chosen showing an alloy which has 10% binder containing 5% iron and equal parts of nickel and cobalt. According to those calculations it would be possible to sinter such an alloy using WC and the pure metals without any additional amounts of soot or tungsten.

Attempts at producing cemented carbides based on iron and nickel and other hard constituents than WC were made as early as in the thirties with varying success. Today a growing number of alloys containing nickel-cobalt or even iron-nickel-cobalt binders are produced. Many of these alloys contain other hard constituents than WC and to a great extent they are based on TiC or Ti(C,N). Many of them are referred to as cermets but they are rather a different kind of cemented carbides than the traditional ones.

Some of the most important applications for cemented carbides are cutting tools for metal cutting. Such cemented carbides contain relatively large amounts of TiC, TaC, NbC in addition to Co and WC. All these carbides have a cubic structure of FCC-type. At sintering temperatures these carbides dissolve tungsten to a great extent. A solid solution of FCC-(Ti,W)C is thus formed. This solid solution is most often referred to as "gamma phase". The amount of tungsten in the gamma phase is dependent on the sintering procedure but also to some extent on the mixture of raw materials used. The diffusion of the metallic elements in these carbides is rather sluggish. Even though the grain size of the carbide is very small, usually in the size of 2-5μm, equilibrium conditions are very seldom realized. However, the knowledge of the equilibrium conditions is of great importance in order to understand the development of the structure.

A major part of the research efforts on phase diagrams is now dealing with the mapping of the titanium containing systems. In addition to this nitrogen, has now become an important alloying element. A great deal of the refractory systems including titanium, tungsten, tantalum and niobium were investigated by Rudy[13] at an early stage and our analysis is to a great extent based on his work.

A research program is now run by Sandvik in co-operation with Chalmer's Institute of Technology in Gothenburg and the Institute of Technology in Stockholm. The aim of this project is to develop better information on the equilibrium conditions in the C-Co-N-Ti-W system.

Figure 5 shows the equilibrium conditions in the Ti-W-C system at 1500°C[14]. From this figure it is evident that the cubic carbide, the gamma phase, contains about 20 atomic-% tungsten in equilibrium with the hexagonal tungsten carbide and graphite at that temperature.

The predominant carbides in cemented carbides i.e. the hexagonal WC and the cubic gamma phase dissolve very little cobalt, iron and nickel. Thus the equilibrium conditions prevailing are not very much influenced by the addition of these elements unless new phases for instance the eta-carbides, (Co,Fe,Ni,W), C, are formed. These carbides are, however, usually avoided or not formed during sintering of cemented carbides due to the relatively high carbon activities prevailing during sintering conditions.

The addition of nitrogen makes the picture more complex. Nitrogen like carbon forms an FCC-phase with titanium. This cubic FCC-TiN-phase is isomorphous with TiC and there is a complete miscibility between the two phases. However, tungsten has a much higher affinity to carbon than to nitrogen and when tungsten is added the miscibility is decreased.

Titanium and niobium are strong nitride formers whereas tungsten, tantalum and molybdenum more easily react with carbon. It was shown by Rudy[15] that the Mo-Ti-C-N system has a miscibility gap between a molybdenum enriched FCC-carbide and a titanium enriched nitride. It is likely that tungsten behaves in a way similar to that of molybdenum. Preliminary results of equilibrium studies performed at 1750°C show that this is the case.

Commercial alloys always contain small amounts of impurities. In cemented carbides there are always minor quantities of the elements silicon, calcium and aluminium in the order of 10 to 100 ppm. These small amounts are often easily dissolved in the liquid cobalt phase during sintering. However, oxygen and sulphur are also present in the raw materials or in the furnace atmosphere during sintering. Calcium forms a very stable sulphide and when oxygen is present calcium might enter into silicates together with aluminium. Precipitates of calcium sulphide, aluminium oxide and some silicates are thus often found on the surfaces of cemented carbides or in the interior of sintered pieces. The presence of such phases might influence the mechanical properties, especially of large wear parts where a bigger volume suffers stressed conditions.

The calculation of the conditions which might cause precipitation of such phases is shown in Fig. 6. From this it is obvious that even very small amounts of oxygen or low oxygen potentials might result in the formation of the oxides of aluminium, calcium or silicon. As soon as silicon oxide is formed, silicates of these elements might be precipitated[16, 17]. In this figure the oxygen potential prevailing during sintering in a graphite lined sintering furnace has been indicated. The corresponding sulphur potential is usually not known.

Owing to the composition of the cemented carbides and the potential of volatile species inside the sintering furnaces the precipitation of oxides, silicates or sulphides could be explained. Such calculations might be useful to predict to what extent these elements could be allowed in cemented carbides.

Cemented carbides for wear parts applications are frequently exposed to corrosive environments. The binder phase of such cemented carbides is often alloyed by chromium to increase their resistance against corrosion. However, chromium is also a strong carbide former and the solubility of chromium in the binder is thus very much limited by the formation of carbides.

In order to find out to what extent chromium can be used as an alloying element in cemented carbide binders, the THERMO-CALC system was used[18]. The results of a few calculations involving an alloy containing 6% binder are shown in Fig. 7. This figure shows the liquidus projections of the areas which on cooling from the WC+liquid state would result in FCC-cobalt and WC only. The first region is limited by precipitation of graphite, chromium carbides and the $(Co,W)_6C$-carbide. The other area shows what would happen if the remaining liquid transformed into FCC-Co and WC at temperatures corresponding to completed solidification.

The area which could be used to reach a maximum of chromium alloying is indicated by the shaded area in the figure. This is true with the assumption that no further precipitation of chromium-rich phases will occur during the cooling in the solid state.

When using cemented carbides for cutting tool applications the temperature in the cutting edge often reaches very high levels in the order of 1000-1200 °C. Thus the high temperature properties, mechanical as well as chemical, of the cemented carbides are of great importance. When developing new grades it would be of value to know the temperatures prevailing inside the insert and in the vicinity of the cutting edge.

One method to measure these temperatures used by Dearnley and Trent[19] was based on the fabrication of certain alloyed inserts. The binder of these inserts consisted of a low silicon alloyed iron-carbon matrix. On heating to the high temperatures prevailing during operation such a binder would transform to austenite which on cooling would form martensite. That martensitic structure could be differentiated from the areas where no transformation from ferrite to austenite had taken place. Even though such inserts do not show a performance in the level of cobalt based inserts, the temperatures reached at the cutting edge would presumably be similar to those of ordinary cemented carbides. It is thus believed that a certain knowledge could be gained by analysing the structure of such Fe(Si)-WC grades to support the development of ordinary cobalt based grades.

In order to optimize the composition of such Fe-Si-C binders a number of calculations were made in the Fe-Ni-Co-Si-W-C system[20]. The aim was to find alloys which would give a well-defined transformation temperature from ferrite to austenite and where the transformation would take place in the interesting temperature range of 900-1200°C. One alloy will thus only give information about one temperature isotherm and a number of alloys must be used in order to develop a complete picture of the temperature distribution in the cutting edge and its vicinity.

The result of calculations in the Fe-Si-W-C system for an alloy containing approx. 5% binder is shown in Fig. 8. This calculation was performed by allowing a relationship between the silicon and carbon content in order to maintain graphite saturation. The results indicate that such alloys can be used to detect transformations up to 1150°C. However, it is obvious that the interval of transformation is getting broader at higher alloy contents and it is much less well-defined at the high silicon contents. Experiments based on these calculations are now under progress and it is yet too early to say whether the calculations will be successful or not.

APPLICATIONS TO STEEL

Hacksaw blades are often produced in the form of bimetal products consisting of a high speed steel as cutting material and a low-alloy steel in the rear part. These two steels are welded together by laser or electron beam techniques. The weld itself then obtains an average composition of the two steels and its structure consists of very hard martensite and the strip usually has to be subject to a heat treatment to make setting or rolling possible. Any heat treatment involves extra costs and an investigation was undertaken to find out whether an alternative method could be used to avoid the post-weld heat treatment.

This method could involve alloying during welding in order to reach a composition of the weld which would make the heat treatment redundant. This might be possible if a weld hardness of approx. HV200 was obtained. It would be even easier if the steel was coated with some alloying element to facilitate the welding procedure.

Instead of making expensive experiments a number of calculations based on various additions were made. The simulations involved the addition of either steels or pure metals. By combining equilibrium calculations with assumptions of the average composition reached and by using available information to calculate Ms-temperatures and the amount of retained austenite, untransformed ferrite and the amount of carbides formed in the post-weld structure, the hardness of the weld could be calculated.

The most efficient alloying element or combination of elements considered was pure chromium and the result of a calculation with chromium additions is shown in Fig. 9 where the chromium is supposed to be added in the form of a thin coating on the steel strip before welding. This figure shows how the final hardness of the weld is depending on the thickness of the chromium coating.

The hardness is drastically reduced as the chromium content increases. This is due to the dissolution effect by the chromium addition and to the fact that chromium lowers the Ms temperature of the steel. It can be seen that a hardness of HV 200 is reached by a layer of approx. 25μm of chromium. The minimum of the hardness shown in the figure is due to the formation of ferrite when alloying by chromium.

During heat treatment, carbides precipitate in many spring steels. These carbides might cause embrittlement of the steel. After annealing coils of wires of such steels there is always some inhomogeneity in the temperature distribution depending on the cooling conditions. The susceptibility to carbide precipitation during cooling very much depends on the steel grade but also varies from one batch to another depending on the temperature distribution within the coil and on the variation of the chemical composition allowed for the the particular grade.

In order to show how minor variations in the composition of different steel grades influence the structure and the tendency to carbide precipitation some phase diagram calculations were made. From these diagrams the variation in composition could be transformed into a temperature interval for the onset of carbide precipitation. Figure 10 shows how the variation of the carbon content of two steel grades influences the critical temperature for the onset of carbide precipitation. Besides showing the importance of keeping the carbon content at its lower limit a ranking of the susceptibility to carbide precipitation of such grades can be made.

Another problem involves the possibility of obtaining an even distribution of the individual phases after the annealing of low alloyed steels. The variation of the annealing temperature and the variation of the composition allowed for each steel strongly influence the amount of austenite or ferrite obtained after annealing. Phase diagram calculations were thus made for a number of steels. The result of a calculation for a steel containing 0.35% Mn, 1.2% Si, 0.5% Cr, 0.3% Mo and 0.15% V is shown in Fig. 11. Figure 11a shows the stable phases present as a function of temperature and carbon content. The influence of temperature and carbon content on the amount of austenite is shown in Figure 11b. In principle this figure shows that a temperature change of only 3-4°C will have the same influence as if the average composition of the steel was changed by 0.15% carbon.

By making two calculations one allowing for the maximum amount of ferrite-forming elements, the other for the maximum of austenite-forming elements according to the specifications of the grade, the extremes of the amount of austenite or ferrite were achieved and the graph in Fig. 11c was obtained. From this figure it is obvious that the amount of austenite might vary from 0 to 91% from one charge to another depending on the alloy composition and the annealing temperature. In order to avoid such variations the annealing temperature must be kept well below 740°C.

Within the Sandvik steel company, the development of new grades in the last few years has been much more efficient owing to the use of the data bases mentioned above. One very striking example of a successful development of a new steel is the development of the Sandvik SAF 2304 grade which is a duplex ferrite austenitic stainless steel. From the very beginning the aim of the development of this steel was clearly defined in terms of corrosion resistance and mechanical properties. These goals were translated into terms of structure and composition of the individual phases.

The THERMO-CALC-system used became a very efficient tool and the number of experimental melts could be minimized and the costs were consequently reduced. In addition the time needed for this particular project was much shorter.

The examples mentioned above quite clearly show that the calculated phase diagram itself is a very efficient tool which offers flexibility and can easily be adopted to the particular grades or compositions involved. Most commercial grades are of such a complex nature that it is almost impossible to draw phase diagrams for these alloys without using programs for the calculation of the diagrams needed.

Calculations based on thermodynamic models also offer possibilities of deriving quantities of thermodynamic nature for instance activities, vapour pressures and chemical potentials which only with difficulty could be obtained by using other methods.

There is, however, a lot of work needed before these computer programs and the related data bases could be used in the most efficient way. Apart from the iron based systems, the data bases contain too little information, e.g. regarding systems of importance for the production of hard materials, cemented carbides and ceramic grades.

ACKNOWLEDGEMENTS

I would like to express my sincere thanks to the director of the Sandvik Steel Research, Mr Henrik Widmark and Dr Thomas Thorvaldsson for allowing me to use results obtained within the frame work of their research. I would also like to express many thanks to Dr Bo Jansson at the Swedish Institute for Metals Research and to the members of the joint program for cemented carbides at the said institute for allowing me to use the results obtained. Furthermore, I would like to thank Dr Armando Fernández-Guillermet presently at CONICET, Centro Atomico Bariloche, Argentina for many stimulating discussions and for his permission to let me use some of his results. Finally I would like to express my thanks to Sandvik Coromant and to the Swedish National Board for Technical Development which to a great extent financed the joint research program.

REFERENCES

1. B. Sundman, B. Jansson and J-O. Andersson, Calphad 9 (1985) 153.
2. S. Takeda, Sci.Rep.Tohoku Univ., Honda Anniversary Vol., (1936) 864.
3. P. Rautala and J.T. Norton, Trans AIME, 194 (1952) 1045.
4. C.B. Pollock and H.H. Stadelmaier, Met Trans, 1 (1970) 767.
5. J. Gurland, Trans AIME, 200 (1954) 285.
6. P. Ettmayer and R. Suchentrunk, Monatsch. Chem., 101 (1970) 1098.
7. L. Åkesson, Ph.D. Thesis, Royal Institute of Technology, Stockholm, Sweden, 1982
8. A. Fernández-Guillermet, Met Trans, 20A (1989) 935.
9. A. Gabriel, "Replacement du cobalt dans les carbures cementes", Ph.D Thesis, L'Institut National Polytechnique de Grenoble, Grenoble 1984, France.
10. A. Fernández-Guillermet, Z. Metallkunde, 78 (1987) 165.
11. A. Fernández-Guillermet, "The Co-Fe-Ni-W-C phase diagram: A thermodynamic description and calculated sections for Co-Fe-Ni-bonded cemented WC-tools, TRITA-MAC 0374, Royal Inst. of Technology, Stockholm, Sweden 1988.
12. A. Fernández-Guillermet, J. Refractory and Hard Metals, 6 (1987) 24.
13. E. Rudy, J. Less-Common Metals, 33 (1973) 245.
14. B. Uhrenius, Calphad, 8 (1984) 101.
15. E. Rudy, J. Less-Common Metals, 33 (1973) 43.
16. B. Uhrenius, L. Åkesson and M. Mikus, High Temperatures — High Pressures, 18 (1986) 337.
17. B. Uhrenius, H. Brandrup-Wognsen, U. Gustavsson, A. Nordgren, B. Lehtinen and H. Manninen, The 12th International Plansee Seminar, Reutte, Austria, 1989.
18. B. Jansson, "Thermodynamic calculations for Co-Cr bonded WC-tools", Rep. IM-2386, Swedish Institute for Metals Research, Stockholm, Sweden 1988.
19. P.A. Dearnley and E.M. Trent, Met. Technol., 9 (1982) 60.

8

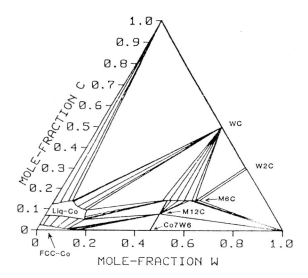

Fig. 1. Isothermal section of the Co-W-C system at 1400°C.

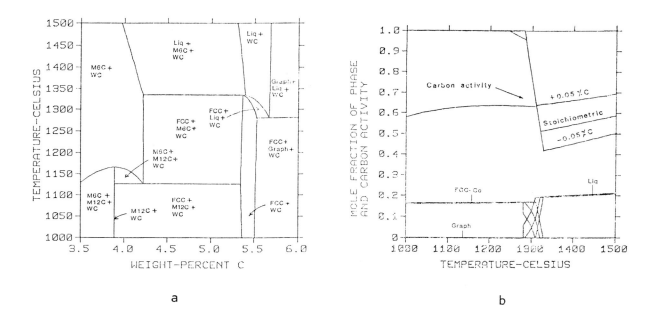

a

b

Fig. 2(a). Calculated section of the Co-W-C system at a molar ratio of W/Co = 2.71. (b). Calculated mole fractions and carbon activities of Co-W-C alloys having a molar ratio of W/Co = 2.71 and different carbon contents.

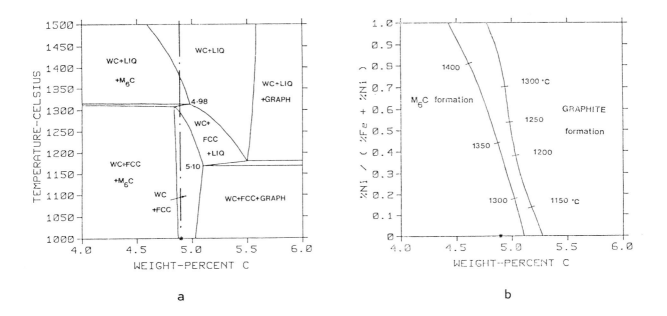

Fig. 3(a). A calculated section of the Fe-Ni-W-C system at Fe+Ni = 20% by weight and %Fe/%Ni = 3. The stoichiometric WC+15%Fe+5%Ni composition is shown by the broken line.
(b). Calculated curves showing the composition of a WC+liquid mixture in equilibrium with FCC+M$_6$C (left) or FCC+graphite (right). The amount of Fe+Ni = 20% by weight.

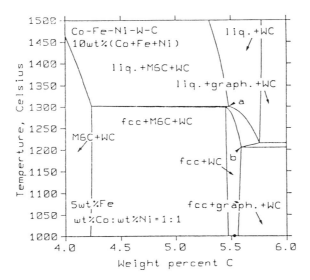

Fig. 4. Calculated section of the Co-Fe-Ni-W-C system at 5% Fe, 2.5% Ni and 2.5% Co by weight. The stoichiometric mixture of WC and the pure metals is indicated by the dot at 5.52% C. Alloys with a carbon content between "a" and "b" solidify by forming WC+FCC only.

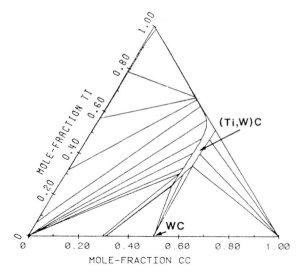

Fig. 5. Calculated isothermal section of the Ti-W-C system at 1500°C.

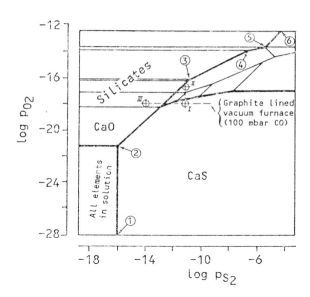

Fig. 6. Predominance area diagram calculated for the Ca-Al-Si-S-O system at 1400°C. The activities of the components are a(Ca) = 2 10^{-4}, a(Al) = 8 10^{-6} and a(Si) = 2.5 10^{-6} which approx. correspond to 100 ppm of these elements in a graphite saturated WC-6%Co alloy.

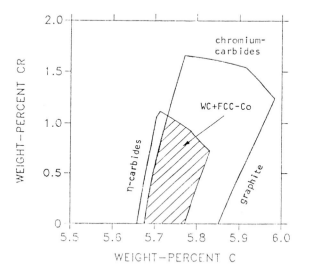

Fig. 7. Calculated regions of the Co-Cr-W-C system at Co+Cr = 6% by weight showing the composition of the WC+liquid mixture in equilibrium with different carbides and the FCC-phase. Within the shaded area solidification is completed without formation of other carbides than WC.

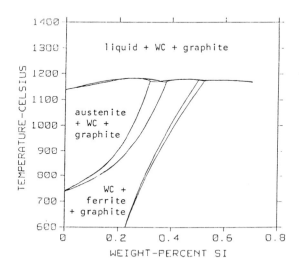

Fig. 8. A calculated section of the Fe-Si-W-C system at 88.7% W and %C = 6.70+3/7x%Si by weight.

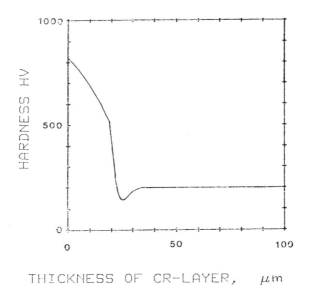

Fig. 9. The calculated hardness after quenching of a weld composed of equal amounts of a high speed steel (M2) and a low alloy steel. Small additions of chromium are added as a thin coating on one of the steels before welding.

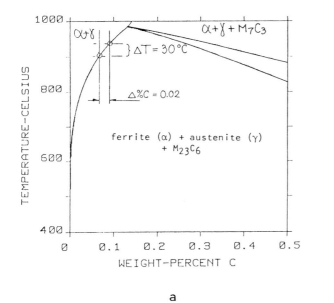

a

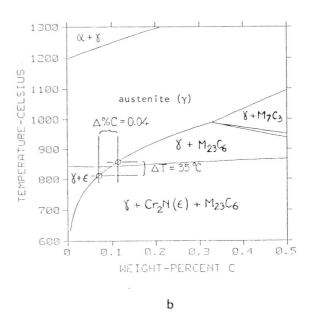

b

Fig. 10(a). Calculated phase diagram for a spring steel showing the tendency to carbide precipitation due to variations in the carbon content of the steel. (b). Calculated phase diagram for a nitrogen alloyed spring steel showing the tendency to carbide and nitride precipitation due to variations in the carbon content of the steel.

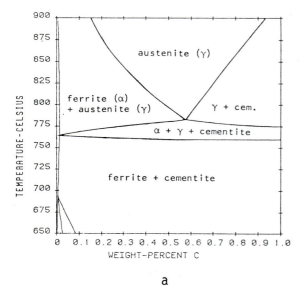

a

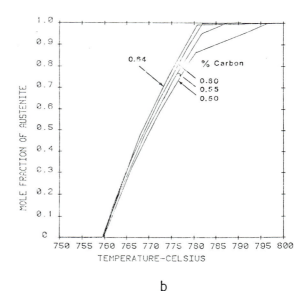

b

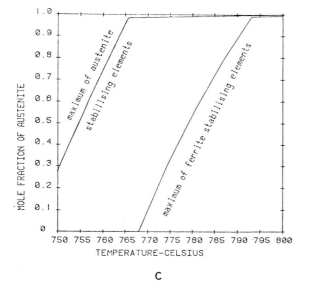

c

Fig. 11(a). Calculated phase diagram for a steel containing 0.35% Mn, 1.2% Si, 0,5% Cr, 0.5% Mo and 0.15% V. (b). The calculated amount of austenite in the steel of Fig. 11(a) depending on the carbon content and annealing temperature.

(c). The amount of austenite obtained in the steel (Fig. 11a) at different annealing temperatures and by allowing a maximum of austenite and ferrite-forming elements within the specifications of the grade.

2
THERMODYNAMIC ANALYSIS OF SINTERING PROCESSES FOR THE PRODUCTION OF CEMENTED CARBIDES

M.H. RAND AND B. UHRENIUS[*]

Harwell Laboratory, Didcot, Oxon, England
*Sandvik-Coromant, Stockholm, Sweden

ABSTRACT

A thermodynamic data set for the eleven-component system: C, Co, O, Ti, W, Y, Zr, Ca, Si, Al, S has been assembled to study the reactions and equilibria occurring during the sintering of Co-Ti-W-C cemented carbides on yttria stabilized zirconia (YSZ) in the presence of calcium alumino-silicate and calcium sulphide impurities.

Results are presented for the following equilibria at 1450°C:

YSZ + graphite + (Co, W, Ti, C)
CaO-SiO_2-Al_2O_3 + graphite + CaS
YSZ + (Co, W, Ti, C) + CaO-SiO_2-Al_2O_3 + CaS + graphite.

In the absence of impurities, CO<g> and Co<g> are the only gas species with pressures greater than 1×10^{-6} bar. In the presence of impurities a liquid oxide phase can be formed, and appreciable pressures ($\sim 3 \times 10^{-3}$ bar) of SiO<g> and SiS<g> occur. These could form deposits on the cemented carbides on cooling, leading to possible contamination with Si and S.

1. INTRODUCTION

Cemented carbide products (carbides of W, Ti, Ta and Nb, dispersed in cobalt metal) are sintered to shape by being heated on graphite trays, (frequently coated with yttria-stabilized zirconia) to 1400-1500°C in vacuum. Since the demands for product control are very stringent, an analysis of the furnace atmosphere during sintering has been carried out to throw light on possible variabilities in the process.

Small amounts of calcium alumino-silicates and calcium sulphide have been identified as impurities in sintered carbides, so these materials have been added to the system studied.

The paper includes calculations for 1450°C in sub-systems of the eleven-component system Co-W-Ti-C-Y-Zr-O-Ca-Si-Al-S.

Attention is focussed mainly on two aspects:

* Interactions in the Zr-Y-C-O system in the presence of (Co, W, Ti, C)
* Phase relations and activities in the CaO-SiO_2-Al_2O_3 system, with additions of graphite and CaS.

Some final calculations are presented in the full eleven-component system.

2. OVERVIEW OF THE CHEMICAL SYSTEM

The analysis concentrates on various parts of the CaO-SiO_2-Al_2O_3-CaS-(Y_2O_3-ZrO_2)-C-Co-(Ti,W) system. Although many calculations involve only subsets of this system, an overall data set for the eleven component system was first accumulated. This file contains thermodynamic data for 57 phases, the gas phase containing 52 species. The data are taken from the following sources:

(i) Solution phases in the Co-W-Ti-C system: data file provided by Sandvik-Coromant.

(ii) Most condensed phases and gases: data taken from the SGTE[†] Substance Data Bank, supplemented where necessary from the compilations published by the Institute of High Temperatures, Moscow (78GLU/GUR, 79GLU/GUR, 81GLU/GUR, 82GLU/GUR).

(iii) For two phases, yttria-stabilized zirconia and yttrium-zirconium oxycarbide, new simplified descriptions were developed.

(iv) For the CaO-SiO$_2$-Al$_2$O$_3$ system a recent data set developed by KTH, Stockholm has been used (see for example 88HAL, 88HIL/SUN). This data set was the best available when the calculations were carried out, but has since been slightly modified in view of assessments in quaternary systems (89BAR/DIN). However, the data set is known to give a satisfactory representation of the CaO-SiO$_2$-Al$_2$O$_3$ system at the temperatures of interest in the current project. It utilizes the KTH ionic 'sublattice' model with varying site ratios, with the following species:

$$(Ca^{+2}, Al^{+3}) (SiO_2, SiO_3^{-2}, SiO_4^{-4}, O^{-2})$$

In addition to the liquid phase, 22 solid phases relating to the pure CaO-SiO$_2$-Al$_2$O$_3$ system were introduced.

All calculations were performed using the MTDATA software developed by the Division of Materials Metrology at the National Physical Laboratory, England. This is an integrated suite of programs for data-fitting, tabulation of thermodynamic functions for individual substances and reactions, for binary and ternary phase diagram calculations, and importantly for this project, a sophisticated, user-friendly and reliable program (MULTIPHASE) for solving the multicomponent, multiphase equilibrium problem, for a wide range of models for solution phases.

3. DATA ASSESSMENT

The only data assessment carried out was for the yttria-zirconia solid solution and the zirconium oxycarbide phase, which was assumed to dissolve yttrium monocarbide. Since these two phases at a defined carbon activity essentially control the carbon monoxide pressure, it was thought necessary to have at least an approximate treatment for these phases.

3.1 Yttria-stabilized Zirconia

This phase was taken to be an ideal solution of YO$_{1.5}$ and ZrO$_2$; the data for these components were taken from source (ii).

3.2 Yttrium-zirconium oxycarbide

Since YO$_{1.5}$ is appreciably more stable than ZrO$_2$, and ZrC more stable than yttrium carbides, carbon reduction of (Y, Zr)O$_{2-x}$ will leave most of the yttrium in the oxide phase, with very little in the oxycarbide phase. Most attention was therefore placed on providing a simple description of the Zr(C,O) phase.

Figure 1, taken from the work of Ouensaga and Dodé (76OUE/DOD) shows that at 1555°C, Zr(C,O) can exist as a single phase material over a range of compositions. On the carbon-rich side, the phase in equilibrium with ZrO$_{2-x}$ and graphite has a composition of approximately ZrC$_{0.83}$O$_{0.06}$; on the zirconium-rich side, the oxygen content can increase to give an oxycarbide of composition ZrC$_{0.73}$O$_{0.14}$.

A proper description of this oxycarbide phase would thus undoubtedly involve a sublattice model of the type Zr:C,O,Va. However, the available funding did not permit the assessment of the data in this form. Instead a simple model of a regular (in fact ideal) solution between a stoichiometric ZrC phase and an oxycarbide of composition ZrC$_{0.7}$O$_{0.14}$ was used. The data for the ZrC component was taken from the SGTE data base. Those for ZrC$_{0.7}$O$_{0.14}$ were developed in the following manner:

C$_p$(T), S(298.15K) were taken to be equal to the sum of 0.7 ZrC + 0.07 ZrO$_2$ + 0.23 Zr; the C$_p$ values were fitted to a simple expression of the form $A + BT + CT^2 + DT^{-2}$.

Δ_fH(298.15K) was adjusted to give Δ_fG(1828K) = -225.4 kJ/mol as given by 76OUE/DOD.

It was anticipated that a binary interaction term between ZrC and ZrC$_{0.70}$O$_{0.14}$ might be required in order to

---

† Scientific Group Thermodata Europe, a consortium of seven European laboratories.

reproduce the correct composition of the oxycarbide phase and pressure of CO(g) in the ZrO_2-C-Zr(OC) phase field. In fact, an ideal solution between the components gives, at 1828K,

Composition of oxycarbide phase = $ZrC_{0.84}C_{0.073}$

$p(CO) = 0.409$ atm $= 310$ Torr

compared to the experimental values (given by 76OUE/DOD) of $ZrC_{0.83}O_{0.06}$ and $p(CO) \sim 300$ Torr. Agreement with the p(CO) values at lower temperatures is also satisfactory (at 1723K, p(CO,calc) = 0.078 atm, p(CO,expt) = 0.093 atm. The oxycarbide compositions were measured by 76OUE/DOD only at 1828K. A simple ideal solution of ZrC and $ZrC_{0.7}O_{0.14}$ was thus taken for the Zr-C-O system.

As noted above, in the range of calculations encountered for the present study, there is unlikely to be much yttrium carbide formation. Again, therefore, an approximate treatment of the 'YC'-phase and its dissolution in the Zr: C,O phase is all that is justified.

The Y-C system is quite complex, but at 1673-1773K, there are three phases: YC_{2-x}, a phase variously written Y_3C_4 or Y_5C_6, and a gamma phase with a range of homogeneity extending from c. 30 to 45 at.% carbon. Above 1645°C, the gamma and YC_2 phases form a single solution phase, probably of the type $Y:C,C_2,Va$. Although Ran (87RAN) has recently assessed this system, he does not seem to have included the important work by Storms (71STO) on the stabilities of these carbides. We have therefore made an estimate of the stabilities of YC_2 and of metastable 'YC' based mainly on Storms' values. The experimental values used to assemble the set of data for YC_2 and 'YC' and the assessed values are given in Table 1.

The Y, Zr : C,O phase was taken to be an ideal solution of the three components ZrC, $ZrC_{0.7}O_{0.14}$, YC.

4. THERMODYNAMIC CALCULATIONS

4.1 Yttria-stabilized Zirconia + Cemented Carbides

Since yttria-stabilized zirconia(YSZ) and graphite are not thermodynamically compatible at high temperatures, a series of calculations was carried out to characterize the parameters of the system, particularly the pressure of CO<g> and the composition of the oxycarbide phase in the various phase regions.

The conditions selected were:

Mass of YSZ	: c. 3 kg
Free volume	: 3 m³
Temperature	: 1400, 1450 and 1500°C
Yttrium content	: 0, 10 and 20%

The phase diagram was scanned through the range of homogeneity of the oxycarbide phase by changing the amount of carbon in the system. This passes from the three-phase field: YSZ + graphite + oxycarbide to the diphasic field: YSZ + oxycarbide. Since the simple model we have used for the oxycarbide is not strictly valid other than in oxygen-saturated regions, we have not pursued the calculations into the Zr + oxycarbide phase field. The component amounts were chosen to given conditions typical of the sintering runs, giving a low ratio of oxycarbide to YSZ.

Since in the data-set used there is no interaction between the (Y, Zr)-containing phases and the cemented carbides [i.e. the (Y-Zr)-containing phases cannot dissolve (Co,Ti,W) and vice-versa], and the gas contains essentially only CO<g> and CO_2<g>, the important gas pressures are exactly the same in the Zr-Y-C-O system as in the more complete Co-Ti-W-Y-Zr-C-O system.

4.1.1 General Conclusions

(a) Since Y_2O_3 is appreciably more stable than ZrO_2, and yttrium carbides are less stable than ZrC, it is clear that in the formation of oxycarbide, there will be segregation of the yttrium and zirconium, with Y/Zr<oxide> greater than Y/Zr <oxycarbide>. In fact the calculations show that essentially none of the yttrium goes into the oxycarbide phase. This, however may overstate the segregation, since the simple model for the oxycarbide (see Section 3.2) assumes only ideal mixing of 'YC' with ZrC and $ZrC_{0.7}O_{0.14}$. In fact, in this simple model there could be negative interactions between YC and the assumed $ZrC_{0.7}O_{0.14}$ component, leading to higher yttrium solubilities in this phase, but they will always be small.

(b) The only appreciable metal-bearing gas species is Co<g> from the cemented carbide — all other such species are < 1 x 10⁻¹⁰ bar at 1450°C.

(c) The yttrium activities are very low (c.1 x 10⁻⁶), even in the carbon-saturated region.

4.1.2 Detailed Results

Figures 2 and 3 show respectively the pressures of all species with log(p/bar) > -8, and the phase amounts of all species, when c. 250 kg of a typical cemented carbide is heated with 3 kg of YSZ in the presence of a variable amount of graphite. More detailed tables of minor gas pressures and compositions of the cemented carbide, YSZ and oxycarbide phases are of course available, but are not included.

4.2 Calculations including $CaO-SiO_2-Al_2O_3$ (C-S-A) and CaS impurities

4.2.1 Ternary C-S-A System

Figure 4 shows the calculated section at 1450°C, and Table 2 gives the activities of each of the three components, referred to the solid phases, at the 26 points indicated on Fig. 5. At 1450°C, there are no miscibility gaps in the system, though of course, one appears at the SiO_2-rich corner at higher temperatures.

4.2.2 Reactions with Graphite and CaS

In the presence of graphite, SiO_2, even at less than unit activity, will react with graphite via the reactions

$$SiO_2 + 3\ C<gr> = SiC<g> + 2\ CO<g>$$
$$SiO_2 + C<gr> = SiO(g) + CO<g>$$

which can be a source of CO<g> which will interact with the equilibria discussed in Section 4.1 — see below. The SiO<g> and CO<g> pressures will of course be a function of the volume of the system in some situations. Gaseous silicon sulphides are also relatively stable species. The following conditions were taken, corresponding to an impurity level of c.0.1 wt% of both the equimolar phase ($CaO.SiO_2.Al_2O_3$) and CaS.

Free volume	: 3 m³
Temperature	: 1450°C
Mol(CaO)	: 1.0
Mol(SiO_2)	: 1.0
Mol(Al_2O_3)	: 1.0
Mol(CaS)	: 3.5

Figure 6 gives the partial pressures of the important gaseous species and Fig. 7 the molar amounts of the condensed phases, both as a function of the carbon content. It is seen that both SiO<g> and SiS<g> are present at significant pressures (c. 1 x 10³ bar), giving a possible mechanism for contamination of the cemented carbides with both silicon and sulphur. As the amount of carbon reacting increases, more silicon carbide (SiC) is formed at the expense of SiO_2, and the ternary diagram (Fig. 4) is traversed downwards from the original (liquid + gehlenite + CA6) phase region to the (gehlenite + CA6 + CA2) field. Thus there is no liquid predicted to be present in the carbon-saturated case, for the $CaO-SiO_2-Al_2O_3$-C-CaS system. We shall see in Section 4.3 however, that this is not true, at least in some cases, when YSZ is also present.

The concomitant decrease in silica activity is shown in Fig. 8. The carbides of aluminium and calcium are not formed, reflecting the greater stabilities of their oxides over silica.

4.3 Reactions in the Full System

We are now in a position to do some calculations in the full 11-component system : Co-W-Ti-C-Zr-Y-O-CaO-SiO₂-Al₂O₃-S. For ease of presentation, they are shown as graphs — an artificial abscissa of amount of C is used, chosen so that all the equilibria lie in the same phase field, thus giving essentially horizontal lines, except for the amount of graphite present. Figure 9 shows the partial pressures and Fig. 10 the amounts of the various phases for the equilibration of a typical cemented carbide, contaminated by c. 0.1 wt.% of an equimolar solution of $CaO-SiO_2-Al_2O_3$ plus c.0.1 wt. % of CaS, with a YSZ containing 15 at.% Y (i.e. Y/(Y+Zr) = 0.15).

They show that the following phases occur at equilibrium in the carbon-saturated state:

Cemented carbides	: Liquid, TiC and WC
Zirconia supports	: YSZ + oxycarbide
Impurities	: Oxide melt + CaS + SiC

Graphite	: Graphite
Gas	: CO,SiS,SiO ; (rest < 1ppm)

It will be seen that for all the grades, a liquid oxide is now present even in the carbon-saturated condition, unlike the simpler system CaO-SiO$_2$-Al$_2$O$_3$-C-CaS discussed in Section 4.2.3. This difference has not been investigated further, but presumably arises from the buffering of the CO<g> pressure by the reaction of YSZ with graphite, at a level which leaves the activity of SiO$_2$ high enough for the liquid oxide still to be present.

CONCLUSIONS

The following brief conclusions arise from the relatively small number of detailed calculations which have been carried out with the data-set that has been assembled for this study. Further detailed calculations are certainly warranted to define any significant differences between grades of cemented carbides and the assumed composition and level of oxide contamination.

5.1 Zr-Y-C-O System

The YSZ layers can react slowly with the underlying graphite to give an oxycarbide; the carbon monoxide pressures at which this occurs and the compositions of the oxycarbide formed have been calculated. The currrent model suggests that virtually all the yttrium remains in the YSZ phase. This may overstate the segregration of yttrium, but this effect undoubtedly occurs.

5.2 CaO-SiO$_2$-Al$_2$O$_3$ System

The isothermal section for this important impurity system at 1723K has been calculated, together with the component activities around the liquid region.

5.3 C-S-A + Carbon + CaS

The coexistence of C-S-A impurities with graphite and CaS can give rise to subsidiary reactions producing both CO<g> — which can affect the reaction between YSZ and graphite — and Si-containing gases, particularly SiS and SiO, which could lead to contamination of the cemented carbides, and perhaps deposition of traces of sulphides and oxides on cooling.

5.4 Eleven-component Calculation

Calculations in the full eleven-component system at 1450°C for a typical cemented-carbide shows there is a possibility of the reactions of graphite with YSZ and SiO$_2$, which both produce CO<g>, can influence one another. In the calculations shown here, the YSZ + graphite reaction buffers the CO<g> pressure and less SiO$_2$ is reduced than in the absence of YSZ. However the contrary situation, where p(CO<g>) is controlled by the reaction of (dissolved) SiO$_2$ and graphite presumably cannot be ruled out (with a different oxide impurity composition). Although the nominal amount of YSZ is very much greater than that of SiO$_2$, the effective amount of YSZ (i.e. that able to react with graphite) may be very much less.

6. FURTHER WORK

It is hoped that the present calculations give a better understanding of the relevant equilibria involved in the sintering of cemented carbides in graphite furnaces, in the presence of yttria-stabilized zirconia and impurities of calcium alumino-silicates and CaS.

Apart from the need for more extensive calculations using this data-set, the following points arise naturally from the discussions above:

1. The current funding has permitted the development of only a very simple model for the important (Zr-Y) oxycarbide phase. For further work it would be very desirable to develop a proper sublattice model, probably of the type:

(Y,Zr)(C,O,Va)

Clearly microprobe studies of the black materials formed on the zirconia supports, if possible to establish the C,O and Y levels, would yield very useful information for this modelling.

2. The current analysis includes only a crude assessment of the thermodynamics of the Y-C system. Although one such analysis has recently been published (87RAN) it seems not to have included an important study by Storms. A re-assessment of this system is required.

ACKNOWLEDGEMENTS

It is a pleasure to acknowledge the considerable (and enjoyable) help received from colleagues from the National Physical Laboratory, Teddington in the preparation of the data sets used in this work and valuable discussions on the technical aspects of the sintering processes with Sandvik staff (Helene Brandup-Wognsen and Jan Qvick*). This work was funded by Sandvik-Coromant, Stockholm, Sweden.

* Now with Seco Tools A.B., Sweden.

REFERENCES

65DeM/GUI G.De Maria, M. Guido, L. Malaspina and B. Pesce, J. Chem. Phys. 43, 4449 (1965).

71STO E. K. Storms, High Temp.Sci. 3, 99 (1971).

76OUE/DOD A. H. Ouensanga and M. Dode, J. Nucl. Mater. 59, 49 (1976).

78GLU/GUR V. P. Glushko, L. V. Gurvich, G. A. Bergman, I. V. Veits, V. A. Medvedev,G. A. Khachkuruzov, V. S. Jungman, 'Termodinicheskia Svoistva Individual' nykh Veshchtv', Tom I, Nauka, Moskva (1978).

79GLU/GUR idem, Tom II (1979).

81GLU/GUR idem, Tom III (1981).

82GLU/GUR idem, Tom IV (1982).

87RAN Q. Ran, Thesis, University of Stuttgart, W. Germany (1987).

88HAL B. Hallstedt, 'An Assessment of the $CaO-Al_2O_3$ System' KTH Report TRITA-MAC 380 (1988).

88HIL/SUN M. Hillert, B. Sundman and Xizhen Wang, 'An Assessment of the $CaO-SiO_2$ System', KTH Report TRITA-MAC 381 (1988).

89BAR/DIN T. I. Barry, A. T. Dinsdale, M. Hillert, B. Sundman, Private communication (1989).

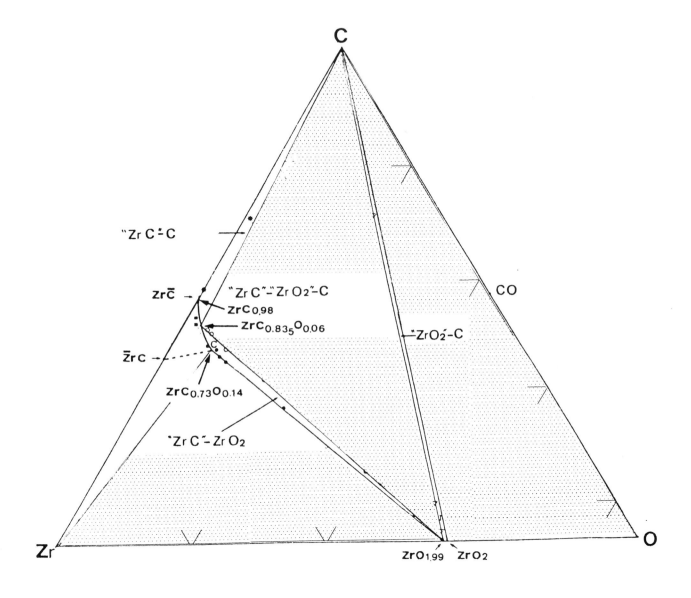

Fig. 1 Zr-C-O System at 1828 K.

20

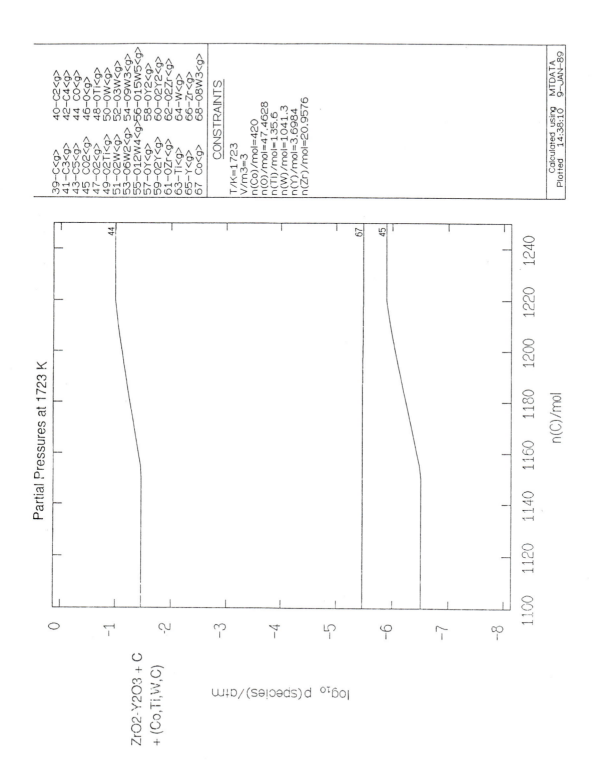

Fig. 2 ZrO$_2$-Y$_2$O$_3$ + C + (Co, Ti, W, C) partial pressures at 1723 K.

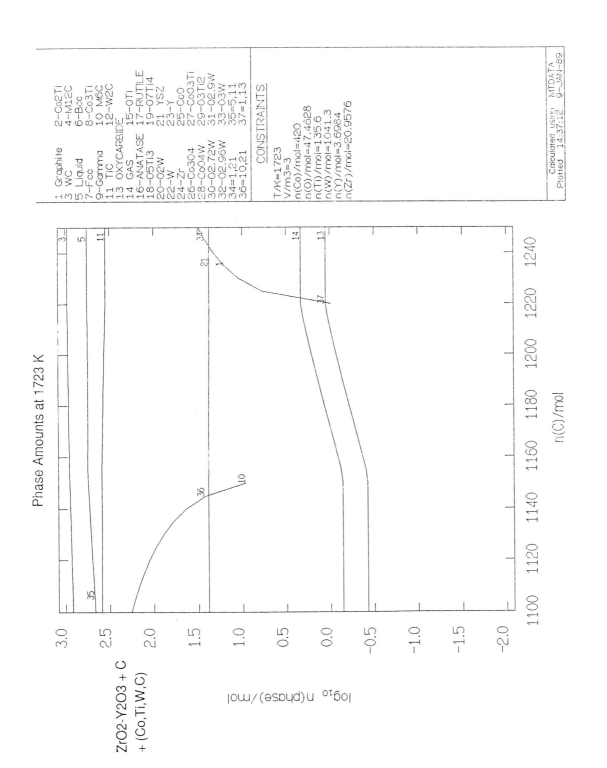

Fig. 3 ZrO_2-Y_2O_3 + C + (Co, Ti, W, C) phase amounts at 1723 K.

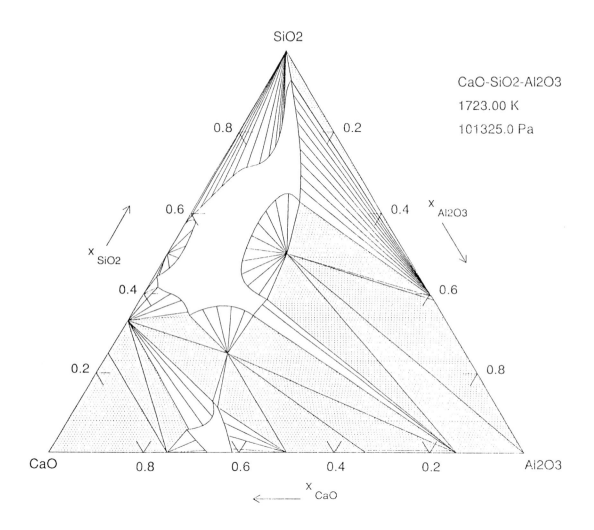

SiO2

CaO-SiO2-Al2O3
1723.00 K
101325.0 Pa

0.8   0.2

x_{SiO2} 0.6   0.4 x_{Al2O3}

0.4   0.6

0.2   0.8

CaO   0.8   0.6   0.4   0.2   Al2O3

x_{CaO}

Fig. 4 CaO-SiO2-Al$_2$O$_3$ 1723.00 K, 101325.0 Pa.

23

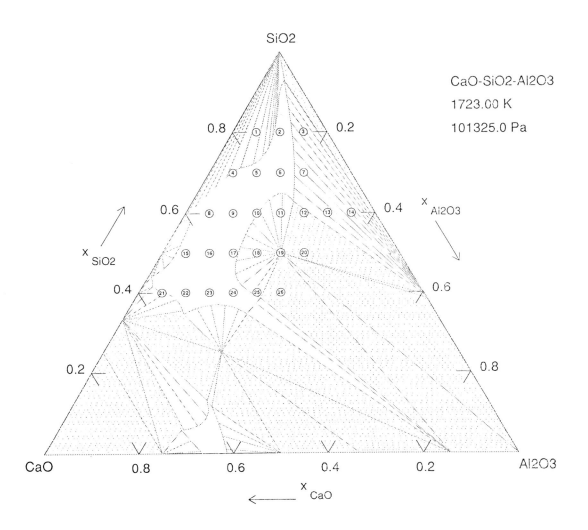

Fig. 5 Location of points for which component activities are tabulated.

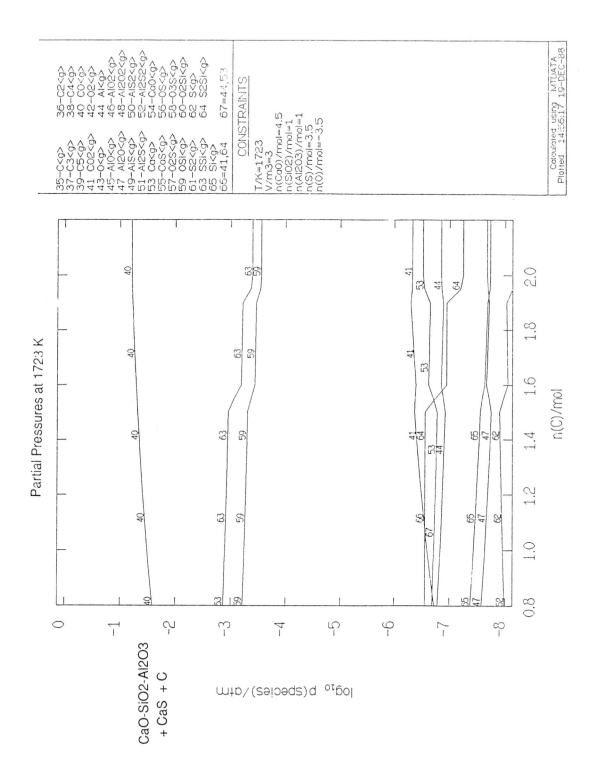

Fig. 6 CaO-SiO₂-Al₂O₃ + CaS + C Partial pressures at 1723 K.

25

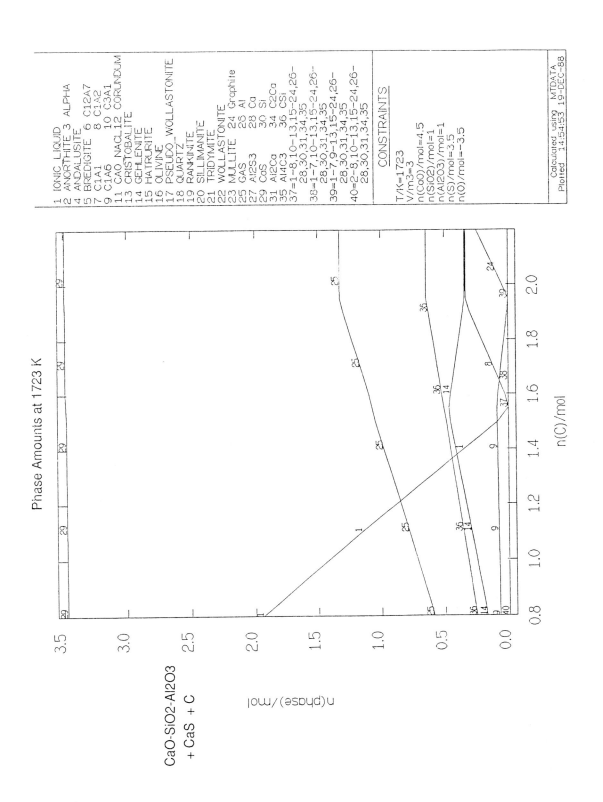

Fig. 7 CaO-SiO2-Al₂O₃ + CaS + C phase amounts at 1723 K.

26

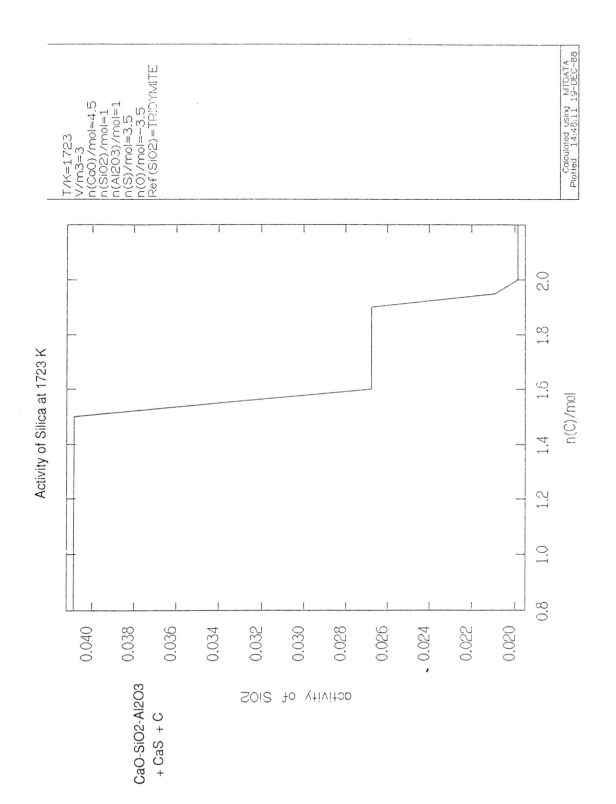

Fig. 8 CaO-SiO2-Al₂O₃ + CaS + C Activity of silica at 1723 K.

27

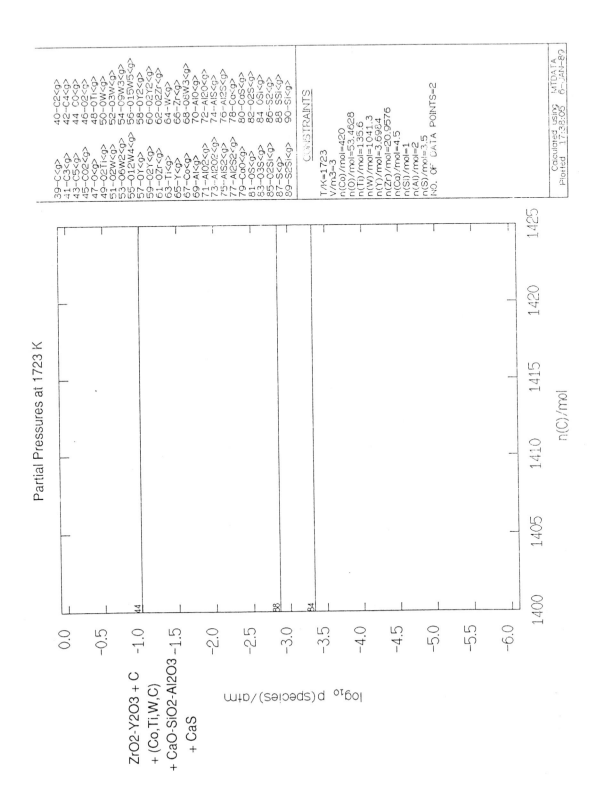

Fig. 9 ZrO2-Y2O3 + C + (Co, Ti, W, C) + CaO-SiO2-Al2O3 + CaS . Partial pressures at 1723 K.

28

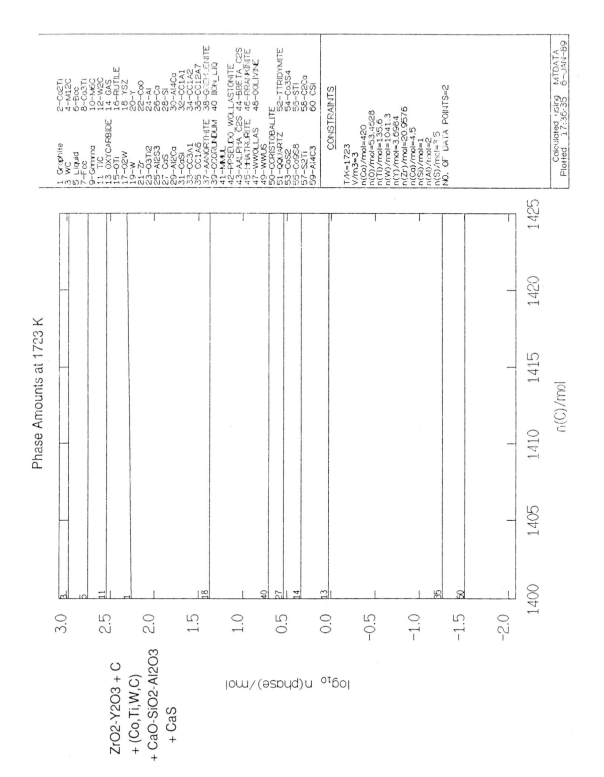

Fig. 10 ZrO₂-Y₂O₃ + C + (Co, Ti, W, C) + CaO-SiO₂-Al₂O₃ + CaS. Phase amounts at 1723 K.

Table 1 YC_2 (s) experimental data (all from 71STO)

T/K	Quantity	Value	Fitted Value
1700	$\Delta_f G$	−143.2	−144.2 kJ/mol
2100	$\Delta_f G$	−153.5	−152.8 kJ/mol
298.15	$\Delta_f H$	−116.3*	−116.3 kJ/mol
1900	$\Delta_f S$	+19.9*	19.7
1800	$\Delta_f S$	25.1	25.4

* Using C_p values from 65 DeM/GUI.

Assessed Data

$\Delta_f H(298.15K)$ $= -116.32$ kJ mol^{-1}
$S(298.15K)$ $= 49.0$ J mol^{-1} K^{-1}
$C_p(T)/$J mol^{-1} K^{-1} $= 75.647 + 13.096 \times 10^{-3}T - 17.071 \times 10^5 T^{-2}$
$\qquad$ (298.15 to 2100K)

'YC'(s) Experimental or Estimated Data

T/K	Quantity	Value	
1700	$\Delta_f G$	−102.9 kJ mol	(Fig. 6 of 71STO)
298.15	$\Delta_f S$	−0.10	(2/3 of $\Delta_f S$ for YC_2)
T	C_p	≈ Same as ZrC	

Assessed Data

$\Delta_f H(298.15K)$ $= -96.9$ kJ mol^{-1}
$S(298.15K)$ $= 46.7$ J mol^{-1} K^{-1}
$C_p(T)/$J mol^{-1} K^{-1} $= 48.954 + 6.5945 \times 10^{-3}T - 1.11431 \times 10^{-6}T^2$
$\qquad - 1.5514 \times 10^6 T^{-2}$ (298.15 to 2000K)

The Y,Zr:C,O phase was taken to be an ideal solution of the three components ZrC, $ZrC_{0.7}O_{0.14}$,YC.

Table 2 Activities in CaO-SiO$_2$-Al$_2$O$_3$ systems at 1723 K

Reference States for Activities are:
CaO   — Fcc (halite) phase
SiO$_2$ — Tridymite
Al$_2$O$_3$ — Corundum

Pt.	Mole fraction			Phases Present	Activities		
	CaO	SiO$_2$	Al$_2$O$_3$		CaO	SiO$_2$	Al$_2$O$_3$
1	0.15	0.8	0.05	trid + liq	5.7×10^{-4}	1.0	4.6×10^{-3}
2	0.10	0.8	0.10	liq	1.1×10^{-4}	0.96	7.8×10^{-2}
3	0.05	0.8	0.15	mull + liq	6.0×10^{-6}	0.98	0.61
4	0.25	0.7	0.05	liq	7.5×10^{-4}	0.98	2.0×10^{-3}
5	0.20	0.7	0.10	liq	4.9×10^{-4}	0.91	1.5×10^{-2}
6	0.15	0.7	0.15	liq	1.2×10^{-4}	0.80	0.21
7	0.10	0.7	0.20	mull + liq	2.1×10^{-5}	0.84	0.68
8	0.35	0.6	0.05	liq	2.5×10^{-3}	0.51	8.6×10^{-3}
9	0.30	0.6	0.10	liq	2.0×10^{-3}	0.52	1.5×10^{-2}
10	0.25	0.6	0.15	liq	8.7×10^{-4}	0.51	0.10
11	0.20	0.6	0.20	an + liq	1.6×10^{-4}	0.66	0.34
12	0.15	0.6	0.25	an + mull + liq	7.7×10^{-5}	0.61	0.85
13	0.10	0.6	0.30	mull + liq	4.3×10^{-5}	0.72	0.76
14	0.05	0.6	0.35	mull + liq	1.3×10^{-5}	0.90	0.65
15	0.45	0.5	0.05	liq	5.3×10^{-3}	0.28	1.5×10^{-2}
16	0.40	0.5	0.10	liq	4.6×10^{-3}	0.29	2.2×10^{-2}
17	0.35	0.5	0.15	liq	3.3×10^{-3}	0.25	9.5×10^{-2}
18	0.30	0.5	0.20	an + liq	3.0×10^{-3}	0.25	0.13
19	0.25	0.5	0.25	an	8.7×10^{-4}	0.45	0.14
20	0.20	0.5	0.30	an + mull + liq	7.7×10^{-5}	0.61	0.85
21	0.55	0.4	0.05	C2S + liq	2.7×10^{-2}	4.1×10^{-2}	2.5×10^{-2}
22	0.50	0.4	0.10	liq	1.8×10^{-2}	6.0×10^{-2}	4.2×10^{-2}
23	0.45	0.4	0.15	liq	1.1×10^{-2}	7.2×10^{-2}	0.15
24	0.40	0.4	0.20	liq	6.7×10^{-3}	7.9×10^{-2}	0.42
25	0.35	0.4	0.25	an + liq	4.3×10^{-3}	8.2×10^{-2}	0.83
26	0.30	0.4	0.30	an + CA6 + liq	4.1×10^{-3}	7.9×10^{-2}	0.94

Phase designations:

Designation	Phase	Formula/Constituents
an	anorthite	CaO.2SiO$_2$.Al$_2$O$_3$
CA6		CaO.6Al$_2$O$_3$
C2S		2CaO.SiO$_2$
liq	liquid	CaO,SiO$_2$,Al$_2$O$_3$
mull	mullite	SiO$_2$,Al$_2$O$_3$
trid	tridymite	SiO$_2$

3

PRACTICAL APPLICATIONS OF THERMODYNAMIC DATA IN IRON POWDER METALLURGY

EVA JACOBSSON AND CAROLINE LINDBERG
Höganäs AB, Sweden

ABSTRACT

The market for parts produced by powder metallurgy (PM) is expansive. There is an increasing requirement of sintered components.This is particularly evident within the automobile industry, where the demand for materials resistant to wear, fatigue and contact fatigue is evident for engine and transmission components.

In powder metallurgy it is possible to work with a broad range of materials, from iron powder alloyed with only small amounts of carbon to stainless steel and high speed steel compositions.

PM provides a unique opportunity to produce materials which cannot be produced by conventional metallurgy, thus emphasizing the need for thermodynamic calculations of systems not used in conventional steel metallurgy.

Phase diagram calculations have contributed to the successful evaluation of alloys for new sintered materials in several projects. Some examples are:

- Prealloyed material for surface hardening, for example case-hardening applications and nitriding,

- New types of material, mainly for wear resistant applications,

- Calculations for better understanding of the microstructures of sintered materials, for example analytical description of the systems Fe-P, Fe-P-C and Fe-P-C-Cu.

- Development of different types of stainless steel materials.

Quantitative equilibrium calculations of multicomponent systems including influence of the amounts of all components and different process parameters (e.g.temperature) on the outcome of a process are used to increase the flexibility and to optimize the process.

Results obtained from some of these projects will be presented and other alloying systems believed to be of specific interest for future development in the iron based PM industry.

4

THERMODYNAMIC CALCULATIONS OF THE AUSTENITE/MARTENSITE CONTENTS OF SILICON-CONTAINING DUAL-PHASE STEELS

R.D.LONGBOTTOM* AND F.H.HAYES

Materials Science Centre, University of Manchester and UMIST, Grosvenor Street, Manchester M1 7HS, UK
*present address: British Steel Technical, Swinden Laboratories, Rotherham S60 3AR, UK

ABSTRACT

A series of thermodynamic calculations carried out using Thermo-Calc are presented for four nine-component dual phase steels whose silicon contents range from 0.35 to 1.91 weight percent. Austenite-ferrite equilibria pertinent to intercritical annealing over the temperature range 1000-1150K have been computed for three different conditions, namely full equilibrium, para-equilibrium and local equilibrium. Comparisons with recently published experimental results show that the amount of martensite present in the quenched microstructure for various annealing times can be calculated with some confidence. The 'silicon plateau' effect is explained.

1. INTRODUCTION

1.1 Dual-Phase Steels

Compared with conventional low-carbon formable steels, dual-phase steels exhibit a low elastic limit, high tensile strength, high work-hardening rates in the early stages of plastic deformation and good ductility during forming relative to their strength in the formed condition. These properties result from a composite microstructure consisting of islands of martensite in a soft ductile matrix of ferrite produced by annealing the steel in the austenite plus ferrite phase field followed by rapid quenching to convert austenite into martensite[1]. Although conversion of austenite to martensite is often somewhat less than 100%, it is of interest to be able to calculate for a given steel both the volume fraction and composition of the austenite present at equilibrium for different intercritical annealing temperatures. In addition, due to the nature of the temperature dependence of the austenite plus ferrite phase field boundaries, dual-phase steels arrive at the intercritical annealing temperature containing less than the equilibrium volume fraction of austenite and with an austenite composition that is different from that at equilibrium. Since equilibrium is not acheived instantaneously upon arrival at the intercritical annealing temperature, of equal importance is an understanding of the changes that occur during the actual approach to equilibrium. These changes involve growth of austenite from ferrite with the simultaneous repartitioning of the alloying elements between the two phases.

1.2 Composition

Dual-phase steels are usually based on iron plus 1.5 wt% manganese and 0.1 to 0.2 wt% carbon with silicon additions up to 2 wt%. Silicon is particularly important since it is known that when this element is present in a dual-phase steel, a temperature range exists extending over some 25 degrees over which the volume fraction of austenite present after 20 minutes annealing is virtually independent of the annealing temperature[2]. This effect, known as the 'silicon plateau', is useful commercially because it reduces the degree of furnace temperature control required to produce the same microstructure during heat treatment. Shen and Priestner[3] have recently carried out an experimental programme in which the effects of both time and temperature on the martensite content of a series of silicon containing dual-phase steels were studied. The aims of the present work were to examine various alternative models proposed for the intercritical annealing process by means of thermodynamic computer calculations carried out on the steel compositions studied by Shen and Priestner[3]. By comparing the calculated results with the experimental data an insight into the dominant mechanisms should become apparent leading eventually to a reliable calculation procedure for the design of annealing schedules.

1.3 Equilibrium, Para-equilibrium and Local Equilibrium

Details of the mechanism whereby the low temperature microstructure containing ferrite and pearlite transforms into ferrite plus austenite when heated above the eutectoid region have been a source of some debate in the literature for several years. Liu and Agren[4] have recently reviewed developments since Hultgren originally introduced the idea of paraequilibrium[5]. They conclude that in the case of the ferrite-austenite transformation when controlled by carbon diffusion, the question as to whether the reaction occurs under local equilibrium conditions or under paraequilibrium conditions has yet to be resolved.

(i) Equilibrium defines the condition in which the steel has reached its most stable configuration with the Gibbs energy having achieved its lowest possible value. In this state the chemical potential of each component is the same in both phases. There is no further driving force for change.

(ii) Paraequilibrium is an intermediate condition first recognised by Hultgren[5] in which the rapidly diffusing interstitial solutes, in this case carbon, have redistributed to the point of having equal chemical potentials in both phases whereas the slower diffusing and esentially immobile substitutional solutes still retain their original partitioning.

(iii) Local Equilibrium is a reaction mechanism considered by Hillert[6] who showed that it is possible to have full equilibrium at the moving interface but with zero partitioning at regions remote from the interface. As a consequence a thin spike for each alloying element exists ahead of the moving interface.

2. COMPUTER CALCULATIONS

Calculations were performed on each of the four dual-phase steel compositions given in Table 1 for (i) complete equilibrium (ii) paraequilibrium and (iii) local equilibrium. All calculations were carried out using the Thermo-Calc system[7] version D together with the Fe-Base Database[8] for which the complete two-phase equilibrium calculation is routine and requires no further explanation.

For the paraequilibrium calculations, the steel was treated essentially as a binary system since in this condition there is only one composition variable namely the carbon content. Firstly the molar Gibbs energies of formation of austenite and ferrite were calculated for a fixed temperature at a series of carbon contents ranging from zero to 0.03 mole fraction in steps of 10^{-5} with the substitutional alloying elements present in the same ratios as those in the bulk steel. Next, using a separate computer program, the carbon content at which the molar Gibbs energies of formation of austenite and ferrite are equal is found and the Gibbs energy of formation of one mole of this phase mixture is calculated. This is now used as the starting point for the minimisation of the Gibbs energy with the constraint that the substitutional alloying element ratios remain fixed in both phases. The mole fractions of carbon in the austenite and in the ferrite are each changed iteratively and the Gibbs energy of the system recalculated after each iteration until no further reduction is acheived by increasing or decreasing the carbon content of either phase by 10^{-11}. This procedure gives the carbon contents of austenite and ferrite corresponding to the lowest Gibbs energy of the steel for the condition of zero partitioning of the substitutional alloying elements. Knowing the total amount of carbon in the steel allows the fraction of austenite to be calculated in this paraequilibrium condition by means of the Lever Rule.

If complete equilibrium occurs locally at the interface between austenite and ferrite then the chemical potentials of all of the elements in the system have their equilibrium values in this interfacial region. In areas remote from the interface however where partitioning has not yet occurred the composition will be the same as that of the bulk steel value the chemical potentials of each element will be different from the equilibrium values. This difference in chemical potentials creates a driving force for diffusion which only carbon is able to eliminate within the normal timescale used in intercritical annealing. Thus, a condition will be acheived in which carbon in regions away from the interface, where the substitutional alloying elements are present in the same ratios as in the bulk steel compositon, has a chemical potential equal to that of the carbon at the interface where complete equilibrium has been established. On this basis the amount of austenite present in the steel when the condition of local equilibrium at the interface is fulfilled was determined by first calculating the chemical potential of carbon at complete equilibrium at the interface i.e. with all of the alloying elements fully partitioned. Then the carbon contents required to reproduce the identical value of this chemical potential in both austenite and ferrite with zero substitutional alloying element partitioning were determined. These two carbon contents together with the bulk steel carbon content then give the phase fractions for this local equilibrium condition via the Lever Rule.

3. RESULTS AND DISCUSSION

The fractions of austenite for the four steels in Table 1 calculated over the temperature range 950-1150K for the conditions of (i) complete equilibrium (ii) paraequilibrium and (iii) local equilibrium, are given in Figs. 1-4 together with the experimental points of Shen and Priestner[3] for annealing times of 20 minutes and 48 hours. The measured data clearly show that over this range of temperature:

(i) the amount of austenite present in these steels increases from zero to between 70 and 90% depending upon the silicon content.
(ii) for a given intercritical annealing temperature below about 1070K more austenite is present after the 20 minutes anneal than after 48 hours.
(iii) above about 1070K these two amounts are approximately equal.
(iv) for each steel the difference between the two amounts is higher the lower the temperature and is more pronounced the higher the silicon content.

The calculated results for all four steels obtained with the two partial equilibrium models exhibit the same general trends with regards to the temperature variation of the austenite fraction relative to the equilibrium calculation. That is the amount of austenite calculated at paraequilibrium is less than at equilibrium at the lower end of the temperature range but increases more rapidly with increasing temperature. The calculated results for local equilibrium follow the same basic trends as those for equilibrium converging at the higher temperatures and diverging at the lower range. Figures 5 and 6 show the effects on the calculated austenite fraction of increasing silicon content at 1033 and 1048 respectively. As the silicon content is increased up to approximately 1wt% the amount of austenite decreases sharply whereas beyond 1wt% silicon has little effect. An additional and in the case of dual phase steels a more important effect is the increase in the difference between the amount of austenite at equilibrium and that at local equilbrium with increasing silicon.

Comparison between the calculated and experimental results shows that there is excellent agreement between the equilibrium calculations and the measured amount of martensite after isothermal annealing for 48 hours. There is also good agreement, at low temperatues, between the local equilibrium calculations and the amount of martensite/austenite measured after 20 min. However with increasing temperature the austenite fraction present moves from the local equilibrium line towards the calculated equilibrium austenite content curve. No correlation exists between the paraequilibrium calculations and the experimental observations.

The net result particularly on the high silicon steel is that the fraction of austenite switches from local equilibrium to full equilibrium at about 1050K. This gives rise in this temperature region to a small change in austenite content over a relatively large change in temperature. It is this switch which creates the 'silicon plateau'.

As the amount of austenite present at local equilibrium is greater than that present at full equilibrium then the amount of austenite present must go through a maximum with respect to time. This was found experimentally by Shen and Priestener[3] who measured the variation in the fraction of austenite with time. Comparisons of these results and the calculated amounts of austenite at local and full equilibrium are given in Figs. 7 to 16.

At the low temperature, 1008K, the peak, after approximately 20 min, is quite pronounced and gives the same amount of austenite as that calculated using the local equilibrium model. As time increases the fraction of austenite decreases to the equilbrium amount. The size of the peak reduces with temperature until at 1113K no peak exists. The kinks at approximately 10 s, which occur in nearly all of the curves, match the experimental observations of the times taken for pearlite to dissolve.

It is evident, from comparisons between the results of the present calculations with the experimental work of Shen and Priestner[3], that there are three steps in the growth of austenite.

Step 1.
The first step is the dissolution of the pearlite; this happens rapidly, within 15 s.

Step 2.
The second stage is the growth of austenite with a local equilibrium at the interface. At the lower end of the temperature range, due to the faster diffusion of carbon and the slower diffusion of the substitutional elements, a situation develops where the amount of austenite is controlled by the equilibrium carbon chemical potential. As the temperature increases and the diffusion rates of the substitutional elements increase sufficiently to cause this situation to break down as the volume of the steel which is no longer in a non-partitoned state is not negligble; stage 3 begins before stage 2 has gone to completion.

Step 3.
The final step is the total equilibration of the steel. As the amount of austenite at full equilibrium is less than that at local equilibrium this must entail dissolution of this phase to a certain extent.

These three stages are in general agreement with those proposed by Speich[1] and Agren[9].

Table 1 Compositions of the dual-phase steels studied by Shen and Priestner[3] and in the present work

Steel	Weight Percent Element							
	Mn	Si	C	Ni	Co	Ti	Cu	Al
DP1	1.46	0.35	0.13	0.013	0.016	–	0.007	0.01
DP2	1.33	0.74	0.12	0.010	0.013	0.006	0.007	–
DP3	1.43	1.28	0.12	0.009	0.013	0.007	0.007	–
DP4	1.48	1.91	0.13	0.010	0.014	0.001	0.008	0.01

4. CONCLUSIONS

(a) The rate of austenite growth during intercritical annealing immediately following the dissolution of pearlite is controlled by local equilibrium at the austenite/ferrite interface with the amount of austenite present being controlled by carbon diffusion.

(b) The amount of austenite present in the microstructure after about 20 min can be calculated by firstly obtaining the carbon activity at full equilibrium, secondly finding the carbon contents of austenite and ferrite with zero partitioning of the substitutional elements, and then applying the lever rule.

(c) The 'silicon plateau' phenomenon is the result of the switch from local equilibrium/zero-partitioning growth to growth with some partitioning of the substitutional elements towards full equilibrium.

(d)The paraequilibrium model is not applicable for calculating the amount of austenite in silicon-containing dual-phase steels.

(e) The equilibrium austenite content decreases with increasing silcon content.

(f) For a fixed silicon content the difference in austenite contents for local equilibrum/zero partitioning and full equilibrium decreases with increasing temperature.

(g) The difference between the austenite contents for local equilibrium/zero partititoning and full equilibrium increases with increasing silcon content.

ACKNOWLEDGEMENTS

We would like to express our thanks to Weir Materials Services Ltd for their continued financial support of our work and to Drs.R.Priestner and X-P.Shen both for kindly making their experimental data available to us in advance of publication and for stimulating discussions relating to this project.

REFERENCES

1. G.R.Speich, 'Fundamentals of Dual-Phase Steels', editors R.A.Kot and S.L.Bramfitt, AIME, (1981), p. 3.
2. G.Thomas and J.Y.Koo, ibid, pp.183-201.
3. X-P.Shen and R.Priestner, SERC Report, "Dual Phase Steels", Materials Science Centre, Grosvenor St., Manchester M1 7HS.
4. Z-K.Lui and J.Agren, Acta Met. 12, (1989), 3157.
5. A.Hultgren, Trans.Amer.Soc. Metals 39, (1947), 915.
6. M.Hillert, Internal Report, Swedish Institute of Metals, (1953) (quoted in Reference 4 above).
7. Bo Sundman, Bo Jansson and Jan-Olof Andersson, The Thermo-Calc databank system, Calphad 9, (1985), pp.150-193.
8. B.Uhrenius, see Thermo-Calc Manual, Phys. Met. Divn., R.I.T., S-100 44, Stockholm, Sweden.
9. J.Agren, Acta Met., 30, (1982), 841.

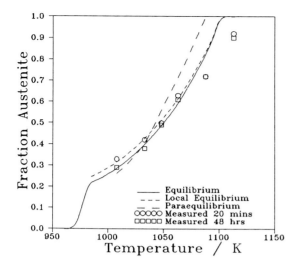

Fig. 1. Measured and calculated austenite contents in DP1.

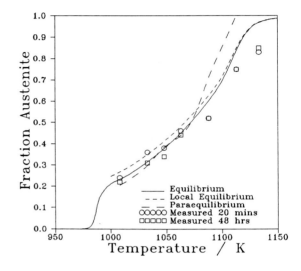

Fig. 2. Measured and calculated austenite contents in DP2.

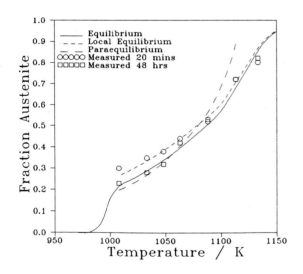

Fig. 3. Measured and calculated austenite contents in DP3.

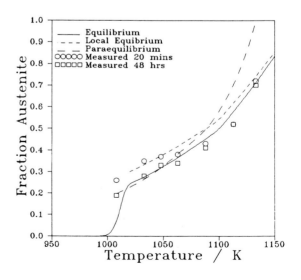

Fig. 4. Measured and calculated austenite contents in DP4.

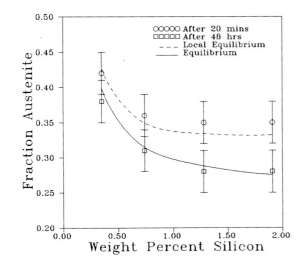

Fig. 5. Austenite fraction versus wt% Si at 1033K.

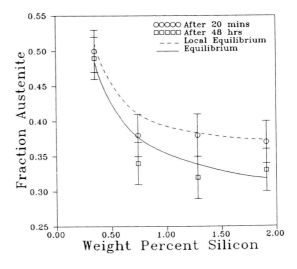

Fig.6. Austenite fraction versus wt% Si at 1048K.

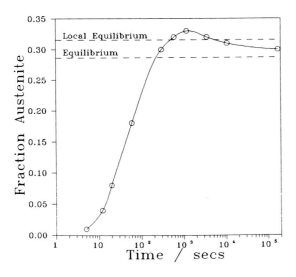

Fig.7. Austenite fraction versus time-DP1 at 1008K.

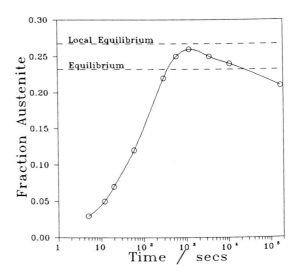

Fig.8. Austenite fraction versus time-DP2 at 1008K.

38

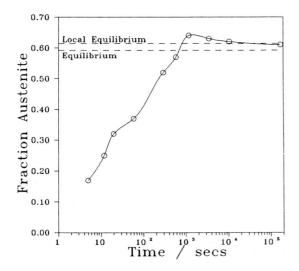

Fig.9. Austenite fraction versus time-DP1 at 1063K.

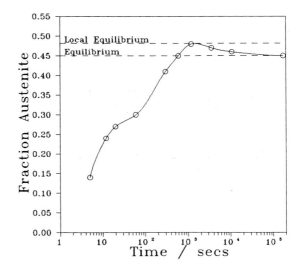

Fig.10 Austenite fraction versus time-DP2 at 1063K.

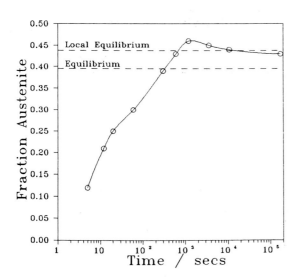

Fig.11. Austenite fraction versus time-DP3 at 1063K.

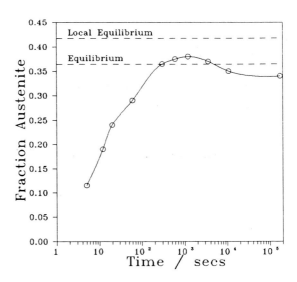

Fig.12. Austenite fraction versus time-DP4 at 1063K.

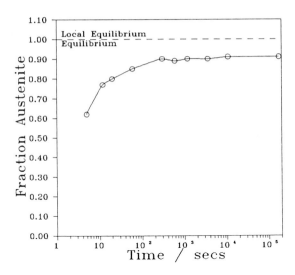

Fig.13. Austenite fraction versus time-DP1 at 1113K.

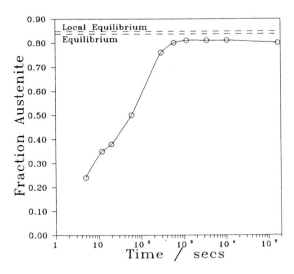

Fig.14. Austenite fraction versus time-DP2 at 1113K.

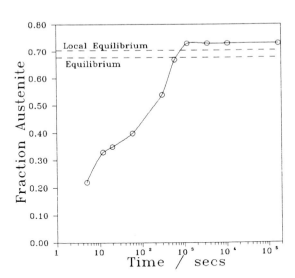

Fig.15. Austenite fraction versus time-DP3 at 1113K.

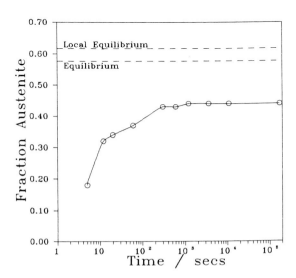

Fig.16. Austenite fraction versus time-DP4 at 1113K.

5

PHASE DIAGRAMS: TOOLS FOR THE DEVELOPMENT OF ADVANCED CERAMICS

G. Petzow and K. G. Nickel

Max-Planck-Institut für Metallforschung, Institut für Werkstoffwissenschaften, Pulvermetallurgisches Laboratorium,
Heisenbergstr. 5, 7000 Stuttgart 80, Germany

ABSTRACT

A range of different types of phase diagrams including the classical isothermal sections in compositional space, phase amount diagrams, free energy plots, P- T- diagrams, volatility diagrams and maximum partial pressure surfaces are shown from case studies of practical use from all steps of advanced ceramic development: System definition, raw material disengagement, processing and property optimisation.

In particular the application of computer assisted thermochemical calculations allow the construction of the type of phase diagram appropriate for a specific problem.

INTRODUCTION

While most basic components of the substances used or developed nowadays as advanced ceramic materials have been discovered accidentally, often a long time ago (e.g. Si_3N_4 in 1857)[1], their technological potential is usually recognized much later (Si_3N_4: 1959)[2].

From the point of this recognition to the point of industrial production and marketing there are four major steps in the development of advanced ceramics:

(a) the substance has to be defined within the context of an appropriate chemical system, (b) the raw material (powder, solution, etc.) has to be disengaged, (c) methods of processing (forming, shaping, sintering, etc.) must be developed and (d) the properties of the product have to be evaluated and optimised.

In this overview we will outline the role of phase diagram work in this development process. It is not our aim to give a complete description of all types of phase diagrams. Rather, we want to highlight the practical use of some important types in the form of case studies and to demonstrate in this way that phase diagrams are much more than a dull compendium of compositional sections.

The term "phase diagram" is used in senso lato and may be equated with the modelling and picturing of phase relations in general, emphasizing the need for thermodynamic calculations and their linkage to the kinetics of the processes.

Substance/System Definition

A substance is often referred to as "known" if we have the knowledge of its chemical composition, crystallographic build-up and some physical data. A substance used as a material is only known if one is able to define its boundaries of use: you have to show under which physical and/or chemical conditions it will still work.

In the simple description of some of these conditions, e.g. melting or decomposition point we have already performed phase diagram work by defining the substance as a part of a thermodynamic system, which can be described by thermophysical data and models.

The details of the thermodynamic models, which are all based on the relation between Gibbs Free Energy (G), Enthalpy (H), Entropy (S), their differences and the equilibrium constant K

$$\Delta G = \Delta H - T \Delta S = - RT \ln K$$

have been reviewed elsewhere[3].

The importance of the ability to describe phase relations not only in graphical form becomes clear if we move to more complex situations.

Figure 1 shows an isothermal section through the Si- Al- O- N- tetrahedron in the Si_3N_4- AlN- Al_2O_3-SiO_2 plane, revealing a number of solid solutions, phases and a liquid. From Fig. 1 it is plain that the desired phase assemblage and material composition can only be obtained with an exact knowledge of the phase diagram.

If we rely on experiments alone a huge number of experiments would be needed to define the system (which becomes even larger if one moves only slightly outside of the plane shown). In addition, experimental studies in this system were debated for a long time[4].

One of the advantages of the calculation methods is their ability to check for consistency of experimental results with other investigations in simpler systems and simultaneously to optimize the thermophysical parameters[5]. In this way the most likely phase diagrams are obtained and one is able to extrapolate to regions where experimental evidence is lacking.

Another advantage is the possibility to show metastable equilibria and thus envisage the results of kinetically hampered reactions.

Material Disengagement

For most advanced ceramics it is necessary to use well-defined, pure raw materials. Again phase diagrams are needed to make sure that a particular phase and not an unknown phase assemblage is yielded.

One method for high-purity material production is chemical vapour deposition (CVD). Thermodynamic calculations are used to find conditions, where the desired substance is the only condensed phase.

In investigating these processes one is often confronted with multivariant equilibria. The calculation results using e.g. Eriksson's SOLGASMIX[6], where a particular bulk composition has to be specified may be reassembled to yield a second type of phase diagram: phase amount diagrams. Figure 2 shows an example from a study on AlN produced by CVD techniques[7]. This type of diagram consists of two parts, one side showing the overall phase amounts, where gas is defined as a single phase vs. temperature, the second shows the composition and amount of the main species of this gas phase vs. temperature.

From diagrams of this type it is immediately visible, at which condition the desired phase is the only stable condensed phase. Using the second part of the phase amount diagram and thus comparing the amount of the elements remaining in the gas phase and those residing in the equilibrium yield one may calculate a maximum possible yield (Fig. 3). Thus phase amount diagrams can be used to discuss whether a process is possible at all, whether it has advantages over others or whether it may require a recycling operation in order to be economic.

Processing

Probably most efforts in advanced ceramic development are directed to processing. Processing is the point where a microstructure may be tailored to yield a certain property level. In advanced ceramics it is also of importance, because many materials of interest are covalently bonded and cannot be sintered without sintering aids.

Thermodynamic calculations help to screen the most effective sintering additives without performing a single experiment. Negita[8] demonstrated this for Si_3N_4 using a third kind of phase diagram: a free energy diagram (Fig. 4).

Negita based his analysis on only two critical conditions. Firstly Si_3N_4 must not be oxidised or decomposed more readily than the sintering additive. Thus defining the free energies of the reactions as

$$\Delta G_1^\circ \; (1/_3 \, Si_3N_4 + O_2 \rightleftharpoons SiO_2 + 2/_3 \, N_2)$$

$$\Delta G_2^\circ \; (2/_3 \, Si_3N_4 + O_2 \rightleftharpoons 2SiO + 4/_3 \, N_2)$$

$$\Delta G_3^\circ \; (Metal + O_2 \rightleftharpoons Oxide)$$

we have the boundary condition 1:

$$\Delta G_3^\circ \leq \Delta G_1^\circ, \Delta G_2^\circ.$$

Condition 2 is that Si_3N_4 must not react with the sintering additive oxide. We get this condition by defining

$$\Delta G_4^\circ \; (Nitride + O_2 \rightleftharpoons Oxide + N_2)$$

and subtracting this from ΔG_1°:

$$\Delta G_1^\circ - \Delta G_4^\circ = Si_3N_4 + MO \rightleftharpoons SiO_2 + MN.$$

Thus the second boundary condition is

$$\Delta G_4^\circ \leqslant \Delta G_1^\circ, \Delta G_2^\circ.$$

The oxides which satisfy both conditions are shown in the shaded area of Fig. 4. Experimental evidence confirms this analysis completely.

It is not only the identification of substances and systems, where phase diagramming is a valuable tool for the ceramic processor. A further example with the utilisation of a fourth kind of phase diagram, a P-T-diagram, is given by Fig. 5. It shows the calculated lines of reactions of Si_3N_4 and SiC in the Si- C- N system[9]. The critical reactions for a desired Si_3N_4- SiC composite are SiC- degradation by N_2 and Si_3N_4 decomposition (which is why one cannot use a N_2- free atmosphere to process the composite properly). The P- T- diagram shows the window of processability.

Evaluation and Optimisation

Phase diagrams can give only a limited information about mechanical properties of materials, because these properties are governed by many factors.

However, there is a complex interaction between chemical reactions and mechanical properties which is currently investigated in the field of advanced ceramics[10]. From such studies it becomes clear that the chemistry and kinetics of these processes need to be understood before a particular high-temperature application can be assessed by life-time prediction methods.

Even if mechanical properties are neglected the interaction of a material and its chemical environment is a property, which sets limits to the use of advanced ceramics in applications at high or very high temperatures (combustion chambers, heat exchangers, filter systems, etc.).

This interaction is accessible to thermodynamic reasoning. For the case of gas corrosion the simple establishment of an equilibrium partial pressure in a defined gas volume may directly be described in terms of a corrosion depth of a material:

$$x = \frac{M}{u \rho R T} \frac{V}{a} P$$

(x: corrosion depth, M: molecular weight of material, u: stoichiometry factor, ρ: density of material, R: gas constant, T: Temperature [K], V: volume of gas chamber, a: exposed surface, P: equilibrium partial pressure of corroding species).

At least for moderate flow rates of an atmosphere, the build-up of a partial pressure is rapid enough to allow the flow of the atmosphere to be approximated as the repeated exchange of the atmosphere of defined volume[11],

$$\dot{V} = \nu \cdot V$$

($\dot{V}$: flow rate, ν : exchange frequency) and hence corrosion kinetics involving gaseous decomposition be modelled by

$$\dot{x} = \frac{M}{u \rho R T} \frac{\dot{V}}{a} P$$

($\dot{x}$: corrosion rate [depths/time]).

If gas flow is too fast to allow the build-up of an near-equilibrium pressure thermodynamic reasoning, neglecting all kinetic hindrances, can give a maximum corrosion mass flux based on gas kinetics and equation of state[11]

$$J = \frac{P}{\sqrt{2 \pi R T M}} \qquad \text{(J: mass flux).}$$

The equations above require a knowledge of the chemical reaction governing the corrosion process to define the equilibrium partial pressure and the stoichiometry factor.

The determination of these reactions is done by means of phase diagrams, in particular by Ellingham-, volatility- and P- T- diagrams.

The classical Ellingham diagrams show the contours of the partial pressure of a species in a T-ΔG-plot. Volatility diagrams are an alternative display method showing the partial pressures of two gas species in equilibrium with several condensed species of a system. They have been reviewed by Lou et al.[12], an example is given in Fig. 6, from which the equilibrium partial pressures and stability fields in the system Al- O are clearly visible.

Technical processes are often between complex gas mixtures with (comparatively) simple materials. This makes most equilibria multivariant. To deduce the main reaction and equilibrium partial pressure in these cases another version of the P- T- diagram is useful: the 'maximum partial pressure surface' plot[13] (Fig. 7). Figure 7 is constructed by calculating the partial pressures of a large number of gas species for a bulk composition of $SiC:H_2O = 1:1$ but plotting only the main species carrying the material elements and superimposing the stability of the condensed phases.

From this type of plot main reactions are easily deduced. In the temperature region of $\approx$1200-1400°C in Fig. 7 CO is dominating over SiO, the condensed phases are SiC, SiO_2 and C. Thus the main reaction is

$$C + H_2O \rightleftharpoons CO + H_2$$

while between $\approx$1700-1850°C SiO and CO are equal over SiC as the single condensed phase. Hence here

$$SiC + 2 H_2O \rightleftharpoons SiO + CO + 2 H_2$$

is the main reaction.

Once the main reaction and the according equilibrium partial pressure is known we can enter the equations above and make predictions about possible corrosion rates in specified physico-chemical environments (Fig. 8).

Investigations and parametric studies can then be used to screen alternative materials or material combinations. Thus the optimisation of properties is again done with the aid of phase diagrams.

SUMMARY

Specific problems of the development of advanced ceramics require different types of phase diagrams. Some important types have been shown.

The use of phase diagrams, in particular those involving thermodynamic calculations make the development of advanced ceramics in many cases easier, if not possible at all.

The application of phase diagrams may not only be viewed as one of practical common sense but as an important point telling materials science from alchemy.

ACKNOWLEDGEMENTS

We gratefully acknowledge support of Dr. Nickel by BMFT- funded project No. 03 M 2012 ('Development of thermally and mechanically highly stressed advanced ceramics').

REFERENCES

1. F. Wohler (1857), Ann. Physik Chem. 102: 317-318.
2. N.L. Parr, G.F. Martin & E. R.W. May (1959), Admiralty Mater. Lab. Report No. A/75.
3. G. Petzow, E.- Th. Henig, U. Kattner & H.L. Lukas (1984), Z.Metallkunde, 75: 3-10.
4. W.Y. Sun & D.S. Yan (1988), Reviews of Solid State Sciences 1: 493-504.
5. H.L. Lukas, E.- Th. Henig & B. Zimmermann (1977), CALPHAD 1: 225-236.
6. G. Erikksson (1975), Chem.Scr. 8: 100-103.
7. K.G. Nickel, R. Riedel & G. Petzow (1989), J.Am. Ceram.Soc. 72: 1804-1810.
8. K. Negita (1985), J.Mat.Sci.Letters 4: 755-758.
9. K.G. Nickel, M.J. Hoffmann, P. Greil & G. Petzow (1988), Advanced Ceramic Materials 3: 557-562.
10. G.A. Gogotsi, Y.G. Gogotsi, V.P. Zavada & V.V. Traskovsky (1989), Ceramics International 15: 305-310.
11. K.G. Nickel, R. Danzer, G. Schneider & G. Petzow (1989), Powder Metallurgy International 3: 29-34.
12. V.L.K. Lou, T.E. Mitchell & A. Heuer (1985), J.Am.Ceram.Soc. 68: 49-58.
13. K.G. Nickel, H.L. Lukas & G. Petzow (submitted), submitted to: K. Hack (Ed): Computer assisted thermochemistry—an introduction and application to practical problems. Springer Verlag, Berlin.

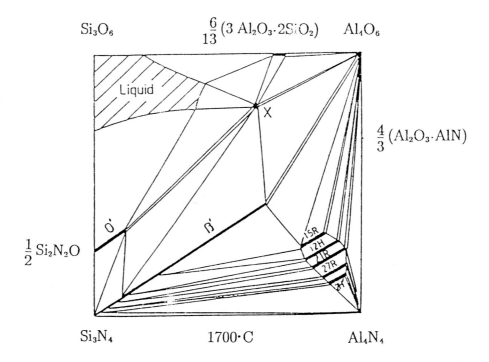

Si_3O_6 $\frac{6}{13}(3\,Al_2O_3 \cdot 2SiO_2)$ Al_4O_6

$\frac{4}{3}(Al_2O_3 \cdot AlN)$

$\frac{1}{2}Si_2N_2O$

Si_3N_4 1700·C Al_4N_4

Fig. 1. Section at 1700°C through the Si- Al- O- N- tertrahedron in the Si_3N_4- AlN- Al_2O_3-SiO_2 plane[4].

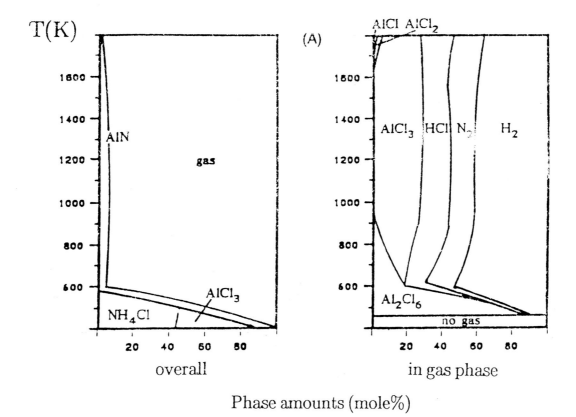

Fig. 2. Phase amount diagram for a bulk composition corresponding to 1 mole $AlCl_3$ + 1.5 mole NH_3 at 0.1 MPa[7].

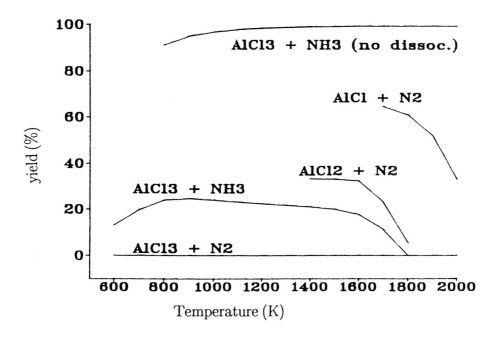

Fig. 3. Maximum possible yields of AlN in weight percent vs. temperature for different bulk compositions and CVD methods as calculated from phase amount diagrams[7].

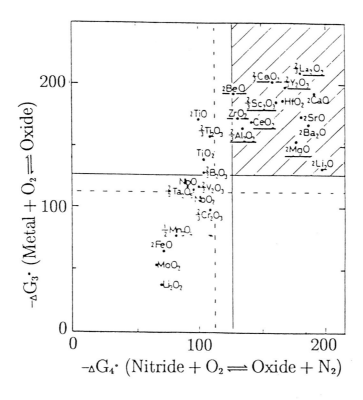

Fig. 4. Free Energies of metal-oxidation (ΔG_3) vs. nitride-oxidation (ΔG_4) for a number of oxides compared to the critical conditions for Si_3N_4 (broken and solid lines) after Negita[8]. The shaded area indicates the useful aids, underlined oxides have been used successfully.

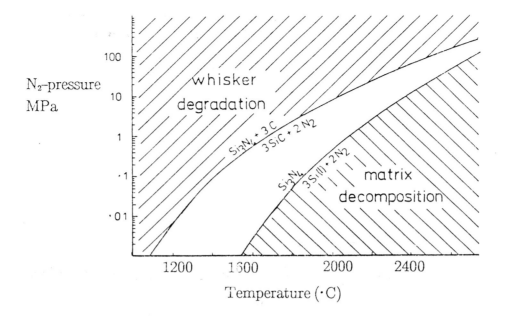

Fig. 5. Pressure-Temperature diagram in the Si- C- N system showing the window of safe Si_3N_4- SiC composite processing[9].

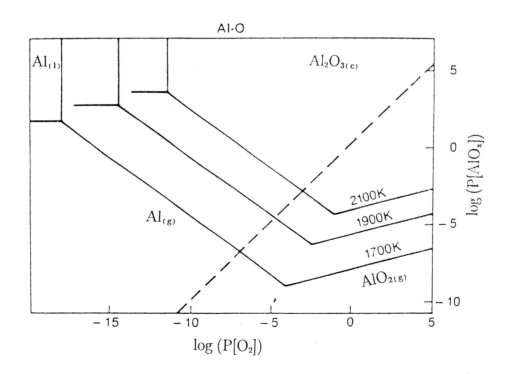

Fig. 6. Volatility diagram of the Al- O system from 1700 to 2100 K[12].

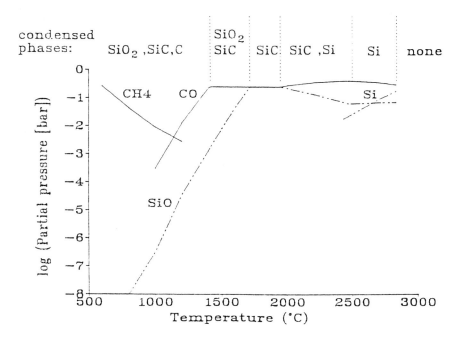

Fig. 7. Maximum partial pressure surface for a bulk composition SiC:H$_2$O = 1:1[13].

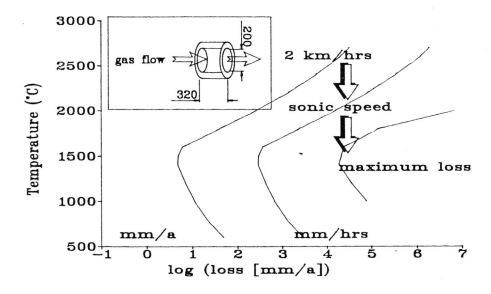

Fig. 8. Corrosion rate prediction for differing gas speeds using the models for moderate (2 km/h and sonic speed) and extremely fast gas flow[13].

6

PHASE EQUILIBRIA IN THE SILICON CARBIDE– ALUMINIUM SYSTEM

E. HEIKINHEIMO, P. SAVOLAINEN AND J. KIVILAHTI

Helsinki University of Technology, Department of Materials Science, SF-02150 Espoo, Finland

INTRODUCTION

The SiC-Al system has interesting applications e. g. in electronics and in coating technology; it is also increasingly employed as a metal-matrix composite. When SiC and Al are in contact at elevated temperatures, several solution phases and compounds may appear depending on kinetic constraints. The interfacial phases which control the properties and behaviour of the SiC-Al components can be predicted by combining experimental and calculated phase diagrams of the system.

At temperatures relevant to the manufacture of compound structures (500-1000°C), the reaction equilibrium in the Al-Si-C ternary system is determined by the following sum reaction:

$$3SiC + 4Al \rightarrow Al_4C_3 + 3\,Si \qquad (1)$$

The free energy change for this reaction is the following:

$$\Delta G_R = \Delta°G_R + RTln\,(a_{Si}{}^3/a_{Al}{}^4) \qquad (2)$$

If Al and Si remain in their pure elemental states (a_{Al}, $a_{Si} = 1$), reaction (1) cannot take place, i. e. $\Delta G_R > 0$. But as soon as Si dissolves in Al, the activity of Si decreases and the reaction proceeds, at a constant temperature and pressure, until the equilibrium condition, determined by the activities, is met. Additionally, dissolved carbon can affect the equilibrium. In order to find out the activity functions of the elements which control the formation of the carbides, we need the interaction parameters which can be obtained through a consistent thermodynamic description of the system.

THERMODYNAMIC CONSIDERATIONS

The preliminary assessment of the Al-Si-C ternary phase diagram was carried out assuming substitutional solution behaviour of the liquid and solid solutions and using Kaufman's data (Al-C[1]; SiC, Al-Si[2]). This, however, produced too stable graphite, and the carbon solubility in the liquid phase at high temperatures was much too low as compared with the experimental results provided[e.g.3]. It was therefore necessary to re-evaluate the respective binaries so that we could produce a ternary description which corresponds better to published phase diagram data[4].

The experimental phase diagram of the Al-Si system is well-known and a comprehensive assessment exists from Murray and McAlister[5] based on Gibbs energies optimised with respect to thermochemical and phase boundary data. We have relied on this with one minor modification related to the liquid phase.

The Al-C binary (Fig. 1a) was calculated with the Thermo-calc program[6] which was modified to comply with the phase diagram published by Yokokawa et al.[4], i.e. the carbon-aluminium interaction was adjusted to increase the stability of the liquid phase. Solubility data for carbon in aluminium at T = 1600-2150°C is taken from Oden and McCune[3] and from Dorward[7].

Our calculation of the Si-C phase diagram (Fig. 1b) is based on the experimental phase diagram given by Olesinski and Abbaschian[8]. Oden and McCune[3] have also determined the solubility of C in Si at 1700-2150°C. The silicon-carbon interaction was also adjusted to the increase the stability of the binary liquid phase.

The isothermal sections of the Al-Si-C system at 1000 and 2000°C (Figs. 2a and 2b, respectively) were re-calculated without any ternary interaction parameters but including the carbides Al_4SiC_4 and Al_8SiC_7 with the aid of the "common tangent plane-method": the Gibbs energies of the ternary compounds were determined by adjusting their stabilities to be marginally below the common tangent plane. The ternary carbides were included in the thermodynamic description because Oden & McCune[3] observed these carbides when studying the system experimentally at 2000 and 2150°C. According to their results, the ternary carbides $2Al_4C_3·SiC$ and $Al_4C_3·SiC$

are stable down to their peritectic decomposition temperatures of 2085 and 2080°C, respectively.

The thermodynamic data used in our calculations are given in Appendix.

EXPERIMENTAL PROCEDURE

Reaction bonded (RB) SiC was supplied by Goodfellow Ltd; commercially pure aluminium, of composition: 99.82 wt% Al, 0.08 wt% Fe, 0.071 wt% Si, 0.0027 wt% Cu, 0.017 wt% Zn, 0.001 wt% MgO, was received from Nokia Co. Preliminary investigation revealed that the RB SiC that we used contains elemental Si.

The samples were annealed in a vacuum furnace (10-5 Pa) or in a vacuum ampoule under conditions given in Table 1.

After annealing the samples were cut and polished flat for microanalysis. Prolonged exposure to ambient atmosphere caused some of the samples to deteriorate: the Al_4C_3 reacted with moisture, as also reported in literature[9, 10]. Repolishing is necessary shortly before introducing the samples into the EPMA specimen chamber. The EPMA instrument used was a JEOL JXA-840A + Tracor automation. The measurements, as point analyses, were done under the following conditions: HV = 12 kV, beam current ≈ 50 nA, O-Kα and C-Kα were analysed with STE crystal using hematite and glassy carbon as standards, respectively, and Al-Kα and Si-Kα with PET crystal using pure elements as standards. The correction program employed was Tracor's PRZ, a version of the Φ(Rho-Z) procedure.

No special measures were taken to avoid carbon contamination within the SEM/EPMA specimen chamber; this is caused by residual hydrocarbons cracking on the sample surface, on the spot of impact of the beam. Accuracy was, however, sufficient to positively identify the phases present and to see significant differences in concentration levels.

RESULTS AND DISCUSSION

SEM observations of polished samples show that in all cases SiC and Al were bonded together and only limited interface areas had developed cracks observable on SEM. The formation of aluminium carbide Al_4C_3 took place in the vicinity of the interface as separate grains within the SiC matrix, the grain size corresponding approximately to that of the SiC grains, see Fig. 3. Polishing the samples which contained Al_4C_3 revealed its inferior hardness as compared with SiC.

The EPMA results show that elemental Si is present in samples annealed at 500-600°C but it was not detected in the 1000°C samples. This is understandable considering the large difference between the solubilities of silicon in solid and liquid aluminium. The aluminium carbide, formed in the samples annealed at 1000°C, contained some silicon, and it also showed a tendency to pick up oxygen from the atmosphere, depending on the time between polishing and measurement. Therefore the possibility of aluminium oxy-carbide formation cannot be totally excluded as reported by Yokokawa et al.[4]. Only very low levels of Al were measured in the SiC phase, in agreement with the phase diagram[4]. Different annealing times at 1000°C (1 vs. 16 h) and cooling rates (3 vs. ≈ 100°C/min) did not lead to marked differences in compositions and microstructures except for the eutectic structure and for precipitated Si visible in the Al matrix in the latter sample (16 h/fast cooling).

Isaikin et al.[11] have previously studied the effect of silicon activity in the formation of aluminium carbide assuming that the activity of Al (or activity coefficient) is equal to one and that carbon can be ignored. According to their calculations, the equilibrium silicon content of the liquid solution at 1200K is 17.2 at.%. Moreover, they have assumed that aluminium activity is equal to one. In the experiments of Iseki et al.[10] the reaction equilibrium is obtained when X_{Si} ≈ 0.125 at 1273 K which agrees qualitatively with the calculations of Isaikin et al., particularly when the activity of Al is taken into account. In the present work the actvities of Al, Si and C are concentration dependent and temperature dependent (except C). But in our case the experimental situation is complicated by the elemental silicon which also dissolves in liquid aluminium, in principle, up to the saturation value of 44.6 at.% Si. Thus, in our case Si dissolves in liquid Al from the two different sources but elemental silicon present affects only the relative amounts of phases formed, not the position of point P (Fig. 2a). When Iseki et al. added silicon (10 wt%) into aluminium prior to the heat treatment, the silicon content detected in the aluminium after the equilibration experiment was X_{Si} ≈ 0.092. In view of what is mentioned above, this may be accounted for by experimental inaccuracy.

The corresponding phase equilibria in the Al-Si-C system can be seen in the isothermal section in Fig. 2a. If the silicon content of the liquid phase is less than that indicated by point P, aluminium carbide formation takes place. Instead, when the silicon content is higher than that indicated by point P, the stable system is composed of silicon carbide and the liquid phase, in which case the aluminium carbide is metastable. As can be seen from Fig. 2a, the effect of carbon on the equilibrium is negligible at 1000°C. In the case, when the ternary carbides are considered stable, the phase equilibria correspond to situation presented in Fig. 2b, where equilibrium silicon contents of the liquid phase are considerably lower than in the case of only binary carbides.

CONCLUSIONS

A thermodynamic description of the Al-Si-C has been produced, and it has been used for the interpretation of experimental results, in particular, the effect of the activity of aluminium, silicon and carbon on the phase equilibria. The phases found experimentally were SiC, Al_4C_3, Si, and Al-Si solution. In principle, the effect of other alloying elements on the activity of silicon and aluminium can now be calculated in a similar manner with the aid of computer programs such as Thermo-Calc.

REFERENCES

1. Kaufman, L., Nesor, H. Coupled phase diagrams and thermochemical data for transition metal binary systems—IV. Calphad 2 (1978) 4, 295-318.

2. Kaufman, L. Coupled phase diagrams and thermochemical data for transition metal binary systems. Calphad 3 (1979) 1, 45-76.

3. Oden, L. L., McCune, R. A. Phase equilibria in the Al-Si-C system. Met. Trans. 18A (1987), 2005-2014.

4. Yokokawa, H., Fujishige, M., Ujiie, S., Dokiya, M. Phase relations associated with the aluminium blast furnace: aluminium oxycarbide melts and Al-C-X (X=Fe, Si) liquid alloys. Met. Trans. 18B (1987) 2, 433-444.

5. Murray, J. L., McAlister, A. J. The Al-Si (aluminium-silicon) system. Bull. Alloy Phase Diagrams 5 (1984) 1, 74-89.

6. Sundman, B., Jansson, B., Andersson, J.-O. The Thermo-Calc databank system, Calphad 9 (1985), 150-193.

7. Dorward, R. C. Discussions of "Comments on the solubility of carbon in molten aluminium", Met. Trans. 21A (1990), 255-257.

8. Olesinski, R. W., Abbaschian, G. J. The C-Si (carbon-silicon) system. Bull. Alloy Phase Diagrams 5 (1984) 5, 486-498, 527.

9. Chernyshova, T. A., Rebrov, A. V. Interaction kinetics of boron carbide and silicon carbide with liquid aluminium. J. Less-Common Metals 117 (1986), 203-207.

10. Iseki, T., Kameda, T., Maruyama, T. Interfacial reactions between SiC and aluminium during joining. J. Mat. Sc. 19 (1984), 1692-1698.

11. Isaikin, A. S., Chubarov, V. M., Trefilov, B. F., Silayev, V. M., Gorelov, Yu. A. Compatibility of carbide coated carbon fibers with aluminium matrix. Metallovedenie i termicheskaya obrabotka metallov (1980) 11, 32-33 (in Russian).

APPENDIX

Listing of thermodynamic data used in the Thermo-Calc calculations:

OUTPUT FROM GIBBS ENERGY SYSTEM ON VAX/VMS HUT   DATE 90-05-15

All data in SI units
Functions valid for 298.15<T<6000.00 K unless other limits
stated

AL4C3
     2 sublattices, sites .571 : .429
     constituents: AL : C

     G(AL4C3,AL:C;0)-0.571G(C,AL;0)-0.429G(GRAPHITE,C;0) =
     -29617+6.9417*T

AL4SIC4
     3 sublattices, sites .444 : .111 : .445
     constituents: AL : SI : C

     G(AL4SIC4,AL:SI:C;0)-0.444G(C,AL;0)-0.445G(GRAPHITE,C;0)
     -0.111G(DIAMOND,SI;0) = -34284+6.284*T

AL8SIC7
     3 sublattices, sites 0.5 : 0.0625 : 0.4375
     constituents: AL : SI : C

     G(AL8SIC7,AL:SI:C;0)-0.5G(C,AL;0)-0.4375G(GRAPHITE,C;0)
     -0.0625G(DIAMOND,SI;0) = -31500+5.89*T

DIA
     constituents: SI

     G(DIA,SI;0)-G(DIAMOND,SI;0) = 0.0

FCC
Excess model is Redlich-Kister-Muggianu
     constituents: AL,C,SI

     G(FCC,AL;0)-G(FCC,AL;0) = 0.0
     G(FCC,C;0)-G(GRAPHITE,C;0) = +138490-14.644*T
     G(FCC,SI;0)-G(DIAMOND,SI;0) = +50600-17.88*T
     L(FCC,AL,SI;0) = -200-7.594*T
     L(FCC,AL,C;0) = -101494
     L(FCC,AL,C;1) = -22835

GRAPHITE
     constituents: C

     G(GRAPHITE,C;0)-G(GRAPHITE,C;0) = 0.0

52

LIQ
Excess model is Redlich-Kister-Muggianu
      constituents: AL,C,SI

      $G(LIQ,AL;0)-G(C,AL;0) = +10792-11.56*T$
      $G(LIQ,C;0)-G(GRAPHITE,C;0) = +114223-27.196*T$
      $G(LIQ,SI;0)-G(DIAMOND,SI;0) = +50600-30*T$
      $L(LIQ,AL,SI;0) = -10695.4-1.823*T$
      $L(LIQ,AL,SI;1) = -4274.5+3.044*T$
      $L(LIQ,AL,C;0) = -21500$
      $L(LIQ,C,SI;0) = -25069+4.452*T$

SIC
      2 sublattices, sites .5 : .5
      constituents: SI : C

      $G((SIC,SI:C;0)-0.5G(GRAPHITE,C;0)-0.5G(DIAMOND,SI;0) =$
      $-33473+4.133*T$

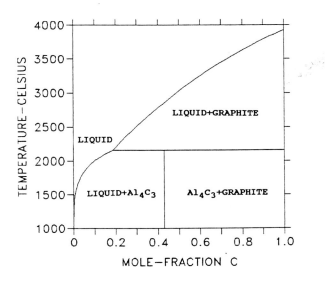

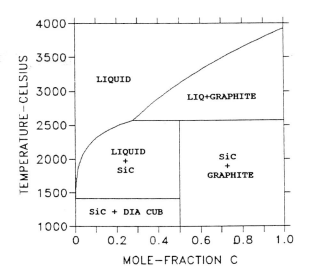

Fig. 1(a). The binary Al-C phase diagram.

Fig. 1(b). The binary Si-C phase diagram.

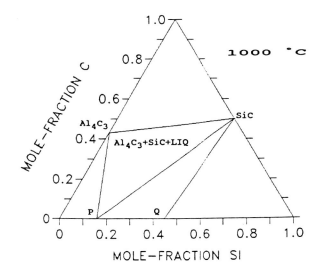

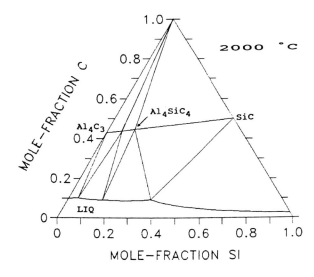

Fig. 2(a). Isothermal section at 1000°C of the Al-Si-C ternary phase diagram based on the corresponding binaries.

Fig. 2(b). Isothermal section at 2000°C of the Al-Si-C ternary phase diagram and including the ternary phases Al4SiC4 and Al8SiC8.

Fig. 3. BSE image of SiC-Al sample #3 annealed for 1 h at 1000°C, 4°C/min cooling, vacuum furnace. Left: SiC (light gray) and Al_4C_3 (dark gray), right: Al.

Table 1 Experimental conditions for the SiC-Al samples

#	Temp./°C	Time	Load/kg	Cool. rate/°/min
3	1000	60 min	–	4
4	500	100 h	20	3
5	600	100 h	20	3
6	1000	16 h	–	≈ 100

7

THERMODYNAMICAL CALCULATION OF PHASE EQUILIBRIA IN A QUINARY OXIDE SYSTEM OF FIRST INTEREST IN NUCLEAR ENERGY FIELD UO_2-ZrO_2-SiO_2-CaO-Al_2O_3

O. RELAVE, P.Y. CHEVALIER, B. CHEYNET AND G. CENERINO*

THERMODATA, B. P. 66, 38402 Saint Martin d'Heres Cedex, France
*CEA-CEN FAR, IPSN/DAS/SASC, B. P. 6, 92265 Pontenay aux Roses Cedex, France

ABSTRACT

As part of a study on nuclear reactor safety, on request from the Institute for Health and Nuclear Safety (IPSN) of the Commissariat a l'Energie Atomique, THERMODATA is presenting a model of the thermodynamical properties of an oxide mixture called "corium" obtained in the unlikely event of a severe accident in a nuclear power plant, if the reactor core melts through the vessel and slumps into the concrete reactor cavity.

The aim of this thermodynamic work is to allow the calculation of the equilibrium state of the corium, i.e. to determine the nature and the composition of the coexisting phases at thermodynamic equilibrium, as well as the constituent activities for given compositions and temperatures or the liquidus and solidus temperatures of the corium.

The present paper deals with the optimization results of the thermochemical properties of ten limiting quasi-binary systems; this is a necessary step to calculate the thermodynamical equilibrium of the five component oxide component (UO_2-ZrO_2-SiO_2-CaO-Al_2O_3).

These five oxides are the main components of the oxide phase of the corium (Fe oxide will be added later).

INTRODUCTION

In the unlikely event of a severe accident in a nuclear power plant, the core may melt through the vessel and slump into the concrete reactor cavity. Then, the hot mixture of core material interacts thermally with the concrete basemat. This mixture, called "corium", is therefore composed of molten fuel, control rods, stainless steel structures, part of the bottom of the vessel and molten concrete.

Up to now, many attempts have been made to model the thermochemical behaviour during the corium-concrete interaction (i.e. to predict the equilibrium state : nature and composition of the equilibrium phases, activity and vapour pressure of the multicomponent system as a function of temperature and melt composition) mainly in order to predict the fission products release from the melt during the interaction. However, in most cases, some simplified hypothesis had to be chosen, such as the initial choice of the reaction process for the calculation steps and the neglecting of some of the interaction phenomena in condensed phases (activity coefficients set to 1).

These methods need to be updated by means of available experiments. They are sometimes enough but can lead on other occasions to some unexpected results if some reactions are not taken into account.

They also do not allow the determination of the transition temperatures between the condensed phases, i.e. the liquidus and solidus temperatures.

As a consequence, THERMODATA was asked by the Institute for Health and Nuclear Safety (IPSN) of the Commissariat a l'Energie Atomique to calculate a more reliable thermochemical equilibrium state of the corium (in order to predict in a better way not only for obtaining the volatile forms of fission products compounds that could exist in the corium, but also the liquidus and solidus temperature of this melt).

This information is useful for IPSN to calculate the progression of core concrete interaction and the fission product release from the corium during that phase.

The first step of that work is the thermochemical equilibrium calculation of a corium composed of only five oxides : UO_2, ZrO_2, SiO_2, CaO, Al_2O_3.

In this paper, we present the optimization of the thermodynamical properties of the ten limiting quasi-binary systems obtained with these five oxides.

THERMODYNAMICAL DATA OF THE BASIC STOICHIOMETRIC COMPOUNDS

As a first step, a very important effort has been made on the "critical assessment" of thermodynamical data of the basic oxides. This work was necessary, because though many classical compilations are giving data sets for these oxides, these do not take into account the most recent experimental information and thus are not updated. This work of up-dating is more of prime importance as these data will at a later stage constitute the reference state to which all the calculations in the binary or higher-order systems will be made, thus giving a data base that is entirely self-consistent.

In other respects, if these new data are agreed by the international community, it will then be possible to exchange data for the higher-order systems, without any obligation to recommence the work, which would be tedious and time consuming.

Description of the analytical expressions used for the data storage
We have taken great care to ensure perfect consistency of the data formats stored on different files.

Calculation codes allow the storage format used in the THERMOCOMP database [1] to be changed automatically to the one used for calculations that include condensed multicomponent phases.

In the THERMOCOMP data base format, the fundamental thermodynamic quantities stored are: the enthalpy of formation at 298.15K — $\Delta H^\circ_{298.15K}$, the entropy at 298.15K — $S^\circ_{298.15K}$, the heat capacity at constant pressure — $Cp(T)$, expressed with an analytical relation (1) as a function of temperature. The transition enthalpies — L_{tr}, and entropies — S_{tr}, are also stored when the compound shows structural transformations in the temperature range to be studied.

The agreed heat capacity relation is, according to the SGTE [2], the following:

$$\text{for } T_k < T < T_{k+1}$$
$$Cp(\Phi) = C_k + D_k T + F_k T^2 + E_k T^{-2} + I_k T^3 + J_k T^6 + K_k T^{-10} \tag{1}$$

In the format used for phase diagrams calculations, the stored quantity is the Gibbs free energy of the stoichiometric compound Φ referred to a given reference state, called SER - $(G_T-H_{SER})(\Phi)$. SER means "Stable Element Reference" and is defined by the use of $H_{(298.15K)}$, and $S_{(0K)}$ for the stable state of the pure elements at 298.15K and 1 bar, from Hillert [3]. This quantity can be written as:

$$(G_T - H_{SER})(\Phi) = G_T(\Phi) - \sum_i X_i * H_{298.15K}(i)$$
$$= G_T(\Phi) - H_{298.15K}(\Phi) + \Delta H_{298.15K}(\Phi) \tag{2}$$
$$= G_T(\Phi) - G_{298.15K}(\Phi) + \Delta H_{298.15K}(\Phi) - T S^\circ_{298.15K}(\Phi)$$

i = pure elements index, and x = molar fraction.

When the heat capacity is described by the relation (1), the previous quantity can be written according to the following relation:

$$(G_T - H_{SER})(\Phi) = A_k - B_k T + C_k T(1 - \ln T) - D_k T^2/2 - E_k/2T - F_k T^3/6$$
$$- I_k T^4/12 - J_k T^7/42 - K_k T^{-9}/90 \tag{3}$$

in which the coefficient $C_k, D_k, E_k, F_k, I_k, J_k, K_k$ are directly those of the heat capacity from equation (1), while A_k and B_k are two integration constants depending on all the fundamental thermodynamic quantities in the THERMOCOMP database format.

OPTIMIZATION OF THE THERMODYNAMICAL PROPERTIES OF THE TEN LIMITING QUASI-BINARY SYSTEMS

In a second step, the ten quasi-binary oxides systems were analyzed. The Gibbs free energies of the different solid and liquid phases were modelled, and the model parameters were optimized using the program of Lukas et al.[4]. This allows one to take into account all of the available experimental information simultaneously, whether it comes from calorimetric, e.m.f. or vapour pressure measurements, or whether they are phase diagram experimental points. It also allows the introduction of experimental values uncertainties. The method of the program has been previously detailed by the authors. The thermodynamical modelling used will be briefly described, and optimization results are presented graphically.

Short description of the thermodynamical modelling used
The descriptions used do not necessarily correspond with the most physically realistic models. As a matter of fact, this topic still remains in the field of research, and for example, the respective advantages of the ionic or associate models are not decisive. More, the most important part of this work consists in the very exhaustive compilation of all of the existing experimental information, their storage as files on computers, and particularly studying their self-consistency by means of optimization procedures from rigorously selected reference states. Once this stage is completed, taking into account new models does not create any major problem, except to repeat the optimization program.

The Gibbs free energy of all the intermediate stoichiometric compounds has been represented in a similar manner to the basic oxides by the relation (3).

The Gibbs free energy of the solid or liquid solution phases has been represented with a classical substitutional model, the Excess Gibbs free energy being itself described by a Redlich-Kister type polynomial expression (4).

$$G^{Ex} = X_i X_j \sum_v L^{(v)}_{i,j} (X_i - X_j)^v \qquad (4)$$

The interaction parameters $L^{(v)}_{i,j}$ depend on temperature and are described with a similar analytical expression as (3).

Graphical Presentation of the Optimized Phase Diagrams
Some phase diagrams calculated with the optimized interaction parameters, consistent with both the selected basic stoichiometric oxides and with all the available experimental information taken from literature, are presented on Figs. 1 to 5.

The agreement between the calculated and experimental phase diagrams is very satisfactory.

For the two systems CaO-SiO_2 and CaO-ZrO_2, the specific shape of the Gibbs free energy function needs to use more complex models in order to describe the liquid phase (ionic or associate type) and the solid solutions (sub-lattice type).

Only the quasi-binary systems have been modelled; the non-stoichiometry range of the two oxides UO_2 and ZrO_2 have not been considered.

TERNARY AND HIGHER-ORDER SYSTEMS

The parameters obtained after the thermodynamical modelling of the limiting quasi-binary systems are used to model the multicomponent phases, and to make calculations in higher-order systems. However, the solid phases which do not exist in the sub-systems will have to be complementarily evaluated. In some cases, ternary sub-systems are experimentally rather well known (for example the Al_2O_3-CaO-SiO_2 system), and it happens that the directly calculated phase diagram, without any ternary interaction parameter, does not agree with the experimental results, whatever the model used. For such cases, it is necessary to proceed to a ternary interaction parameter optimization, similar to that made for the binary systems.

FUTURE WORK

The obtained data base will be extended in the future to other oxides (FeO, fission products) and used for calculations including a complex gas phase and/or a multicomponent metallic phase.

ACKNOWLEDGEMENTS

The present work was financially supported by the Institute for Health and Nuclear Safety (IPSN) of The Commissariat à l'Energie Atomique (C.E.A.), France. Contract No. BC4983.

REFERENCES

1. B. Cheynet, Reprinted from Computerized Metallurgical Databases, Proceedings of a symposium sponsored by the ASM-MSD Alloy Phase Diagram Data Committee. Fall Meeting of the Metallurgical Society in Cincinnati, Ohio, USA, (October 12-13, 1987). A publication of the Metallurgical Society, Inc., J.R. Cuthill (1988).
2. S.G.T.E : The Scientific Group Thermodata Europe was founded in 1974 by the professor E. Bonnier and reassembles seven research laboratories working in Europe on thermochemical data bases.

3. M. Hillert, Reprinted from Computer Modelling of Phase Diagrams, Proceedings of a symposium sponsored by the Alloy Phase Diagram Data Committee of the Materials Science Division of the A.S.M. - Fall Meeting of the Metallurgical Society in Toronto, Canada, October 13-17, 1985. Edited by L.H. Bennett.

4. H.L. Lukas, E.Th. Henig, B. Zimmermann, Calphad, 1,(3), 225-235, (1977).

5. O. Redlich and A.D. Kister, Ing. Eng. Chem., 40, 345, (1948).

6. E.S. Shepherd, G.A. Rankin, F.E. Wright, Amer. J. Sci., [4th series], 28, (166), 293-333,(1909).

7. G.A. Rankin and F.E. Wright, Amer. J. Sci.,[4th series], 39 (229), 1-79,(1915).

8. F.C. Langenberg and J. Chipman, Journal of the American Ceramic Society, 39, 432-433,(1956).

9. R. W. Nurse, J.H. Welch, A.J. Majumdar, Trans. Brit. Ceram. Soc., 64, 323-332, (1965) and 64, 409-418,(1965).

10. M. Rolin and Phamh Huu Thanh, Rev. Hautes Temper. et Refract., 2, (2), 175-185,(1965).

11. N. Nityanand and H. Alan Fine, Met. Trans. B, Vol.14B, 685-692, (December 1983).

12. P.P. Budkinov, S.G. Tresvyatsky and V.I. Kushakovsky, Proceedings of the Second Conference on Peaceful Uses of Atomic Energy, Geneve, 6, 124-131,(1958). (Paper P/2193 USSR).

13. W.A. Lambertson and M.H. Mueller, Journal of the American Ceramic Society, 36, (10), 329-331, (1953).

14. A.M. Alper, R.C. Doman, R.N. McNally and H.C.Yeh. The use of Phase diagrams in fusion-cast refractory materials research. Vol.2, chap.IV, 127-130, (1970) in "Phase Diagrams", edited by A.M.Alper (Academic Press, New York and London, 1970).

15. G.R. Fischer, L.J. Manfredo, R.N. McNally and R.C.Doman, Journal of Materials Science, 16, 3447-3251, (1981).

16a. S. Lungu, Acad. R.P.R. 13, 739, (1962) ; Rev. Phys. de l'Acad. R.P.R., 7, 419, (1962).

16b. S. Lungu, Rev. Roum. Phys., 9, 895-910, (1964).

16c. S. Lungu, Journal of Nuclear Materials, 19, 155-159, (1966).

17. I. Cohen and B.E. Schaner, Journal of the Nuclear Materials, 9, (1), 18-52, (1963).

18. N.H. Voronov, E.R. Voitekhova and A.S. Danilin, Proc. 2nd Conf. Peaceful Uses of Atomic Energy, Geneva, 6, 221-225, (1958). (paper P/2490 USSR).

19. K.A. Romberger, C.F. Beas Jr. and H.H. Stones, Journal of Inorganic Nuclear Chemistry, 29, 1619-1630, (1967).

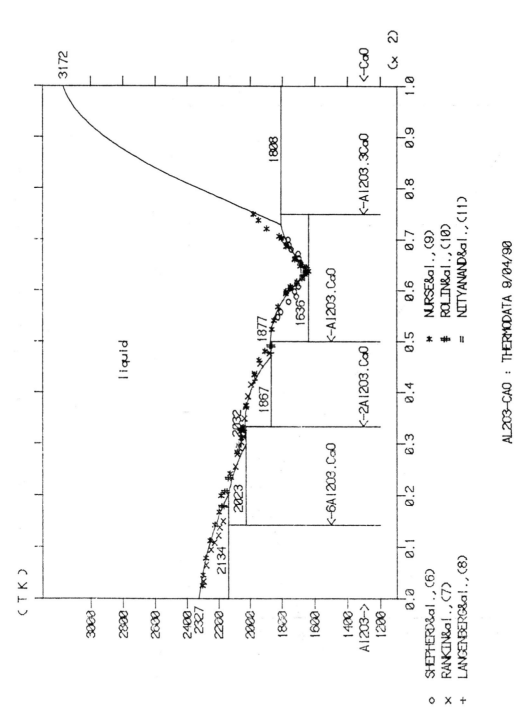

Fig. 1. Al₂O₃–CaO phase diagram, compared with the selected experimental values.

60

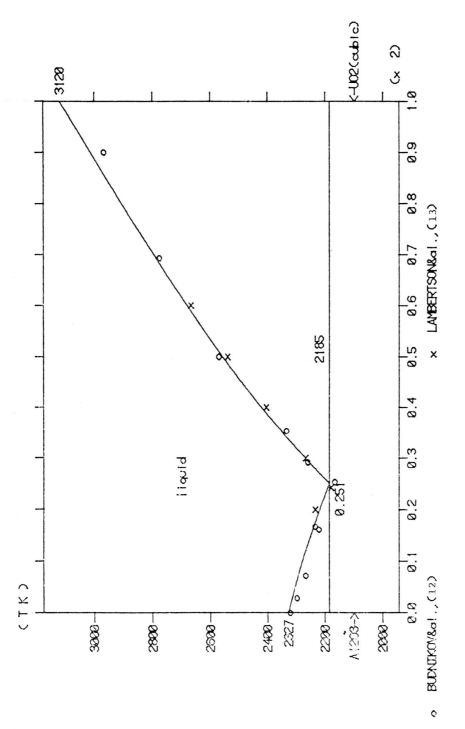

Fig. 2. Optimized Al$_2$O$_3$-UO$_2$ phase diagram, compared with the selected experimental values.

61

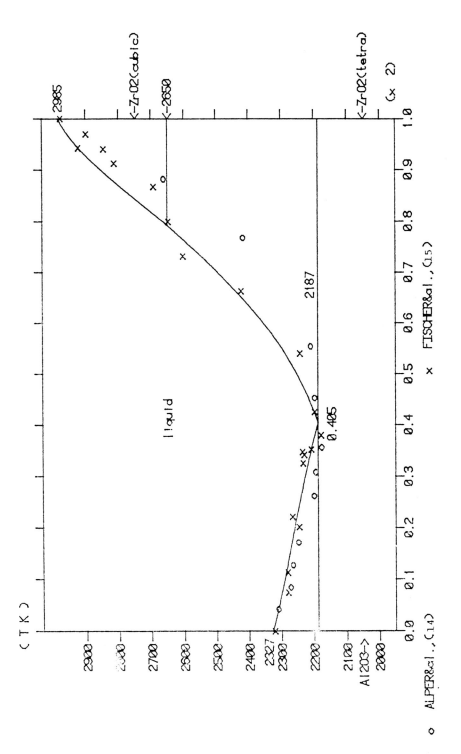

Fig. 3. Optimized Al₂O₃-ZrO₂ phase diagram, compared with the selected experimental values.

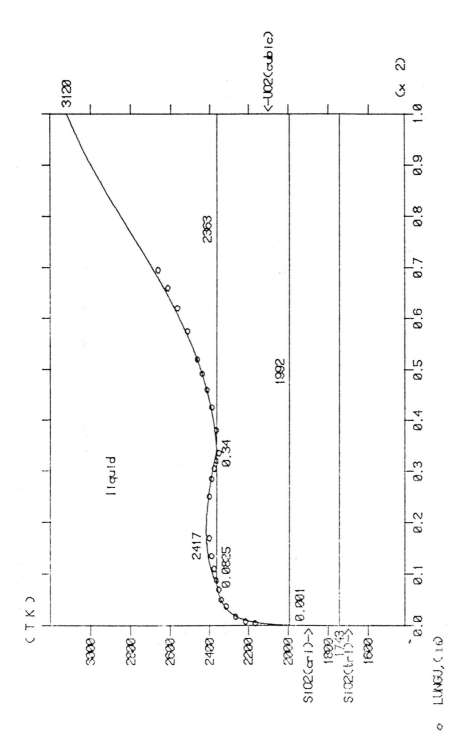

Fig. 4. Optimized SiO₂-UO₂ phase diagram, compared with the selected experimental values.

63

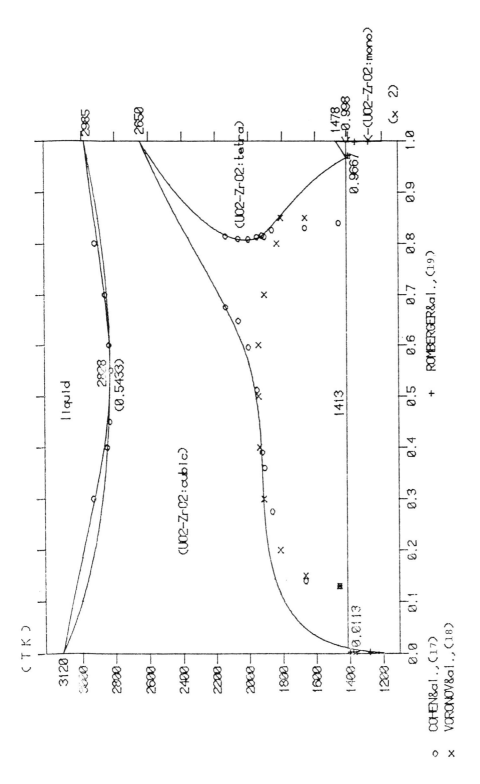

Fig. 5. Optimized UO₂-ZrO₂ phase diagram, compared with the selected experimental values.

8

MICROSTRUCTURE AND PROPERTIES OF A RAPIDLY SOLIDIFIED Fe-Cr–Mo–B ALLOY

N. Saunders, L.-M. Pan, K. Clay*, C. Small* and A. P. Miodownik

Department of Materials Science and Engineering, University of Surrey, Guildford, Surrey GU2 5XH, UK
*Rolls-Royce plc, PO Box 31, Derby DE2 8BJ, UK

ABSTRACT

A $Fe_{70}Cr_{18}Mo_2B_{10}$ alloy has been produced by rapid solidification and subsequent hot extrusion into rod form. The initial amorphous structure of the melt spun ribbons breaks down during the high temperature extrusion to give a fine dispersion of boride phases in a ductile Fe-based matrix. Initial electron microscopy showed that the borides consisted of two populations (i) a Cr-rich orthorhombic boride based in the Fe–Cr–B system and (ii) a Mo-rich boride of unknown origin. The Fe-based matrix was found to be martensitic, which was very surprising because the amount of Cr and Mo in the alloy was expected to stabilise the α-Fe structure.

As part of this work phase diagram calculations were performed to try and understand the microstructural evolution in the alloy. The calculations correctly predicted the appearance of the Cr-rich boride and also predicted that the Mo-rich boride should be an M_3B_2 type originating in the Fe–Mo–B system. The latter was subsequently confirmed by electron and x-ray diffraction. The calculations also showed that, at the extrusion temperature, the Fe-based matrix is actually austenitic and on cooling would transform to ferrite hence explaining its martensitic structure.

INTRODUCTION

The mechanical properties at elevated temperatures of Fe-based alloys with addition of up to 2.5wt%B produced via a rapid solidification (RS) route holds substantial promise for use in new generation gas turbine engines [82Ray,83Ray,85Hah,87Hah,88Raw]. The alloys are first produced by an RS process such as melt spinning and subsequently consolidated at high temperatures. During the consolidation process the initial amorphous structure breaks down and forms a fine dispersion of stable borides in a ductile Fe-based matrix (Fig.1).

Of this class of alloys the $Fe_{70}Cr_{18}Mo_2B_{10}$ alloy was chosen for evaluation by Rolls Royce as a candidate for shaft materials where a combination of high strength and modulus at elevated temperatures is required. The alloy was first produced by chill block melt spinning and continuous ribbons of 30–80 μm in thickness were produced. These were subsequently pulverised to <150μm. The resulting powder was consolidated by hot extrusion at a ratio of 10:1 following a 2hr pre-soak at 950°C. During the extrusion process adiabatic heating caused the temperature to rise to ~1050°C.

Earlier work on a similar alloy [85Xu] indicated that in the final product the matrix was α-ferrite and the boride which provides the strength and modulus increase was an orthorhombic M_2B type phase originating in the Cr–B binary which extends considerably into the Fe–Cr–B ternary system. As part of a detailed evaluation process microstructural characterisation by transmission electron microscopy (TEM)[†] and x-ray diffraction (XRD) were performed and to further understand how the development of microstructure occurred phase diagram predictions via the CALPHAD route were performed.

The TEM and XRD results confirmed the earlier work which stated that an orthorhombic boride M_2B was present and its composition was Cr-rich. However, they also showed that a proportion of the borides (~10%) were Mo-rich and that the Fe-based matrix was martensitic. The latter result was particularly surprising because of the high level (20at.%) of α-ferrite stabilisers Cr and Mo. Furthermore initial analysis of diffraction patterns from the TEM work indicated that the structure of the Mo-rich boride was a tetragonal type whose structure had not been reported in previous literature.

The phase diagram calculations performed as part of this work were based on original thermodynamic characterisations for the binary and ternary boride phase diagrams [89Pan] and work by [88And] for the Fe–Mo–Cr system. Following usual CALPHAD procedures [70Kau,79Ans,80Hill] these were combined to give predictions for phase equilibria in the quaternary Fe–Cr–Mo–B system.

[†]W.T.Kim and B.Cantor, Dept of Metallurgy and Science of Materials, Oxford University, Parks Rd, Oxford OX1 3PH, UK

RESULTS AND DISCUSSION

Figures 2–4 show the calculated phase diagrams for the Fe–B, Cr–B and Mo–B systems with the experimentally determined phase boundaries superimposed. These were used as the basis for the ternary characterisations of the boride systems. The ternary phase diagrams of which (Fe,Cr,Mo)–B are subsets are reasonably well determined. They are characterised by substantial solubility of ternary elements in the binary compounds and, in the composition range of interest, the formation of two types of ternary compounds. Of these the M_3B_2 type compounds in Fe–Mo–B and Cr–Mo–B are of the U_3Si_2-type and the $Fe_{13}Mo_2B_5$ phase is isostructural with the Fe_3B phase often observed as a metastable product in devitrified Fe–B glasses. The calculated diagrams are shown in Figs.5–7 and are generally in very good agreement with experimental observation. These were combined to give predictions for phase equilibria in the quaternary Fe–Cr–Mo–B system.

Predicted phase equilibria were checked against experimental data reported by Takagi et al. [86Tak] on the partitioning of elements to the M_3B_2 compound in CERMET alloys based on this quaternary. Figure 8a-d shows the agreement between the predicted and experimentally determined composition of the M_3B_2 phase in equilibrium with austenite for a wide series of alloys. Good agreement is found between the predicted values and those experimentally observed.

The calculations were then extended to the $Fe_{70}Cr_{18}Mo_2B_{10}$ alloy and it was predicted that:-

(1) The M_2B phase exists in the alloy and that it is chromium rich,

(2) There is a substantial proportion (7.3%) of the U_3Si_2-type, M_3B_2 boride in the alloy with high levels of Mo,

(3) The Fe-based matrix is austenitic at 1000°C and transforms below 857°C to ferrite thus giving the reason for the observed martensitic structure.

Based on the prediction for the M_3B_2 boride and its observed lattice parameter variation with Fe, Cr and Mo levels the structure of the Mo-rich phase was re-evaluated and clearly shown to be primitive tetragonal of the U_3Si_2-type [89Kim]. This work [89Kim] also gives results for the Fe:Cr:Mo ratio of the phases which are in excellent agreement with those predicted at 1000°C.

It is clear that the input of the phase diagram predictions has greatly helped in the understanding of the evolution of microstructure in this alloy. They have also shown that the phases present after extrusion are stable and problems initially foreseen with breakdown of metastable borides at service temperatures will not be evident.

A further advantage of the calculation route is that it is now possible to cost effectively design the alloy to give the best microstructure/property combination. For example, the formation of martensite on cooling from the extrusion process gives rise to internal stresses which are problematical for the final product shape. The calculations can now be used to predict the composition range where the alloy is ferritic at the extrusion temperature and the composition further refined to provide the optimum level of Fe, Cr and Mo in terms of density and cost of alloy addition.

It is now also possible to optimise properties of the alloy in terms of the make-up of borides. The size of the M_3B_2 boride is substantially smaller than that of the M_2B boride and its proportion could be maximised to optimise its effect on hardness and yield strength. Also there appears to be a property/composition relationship with respect to the borides themselves. For example, the M_3B_2 boride can adopt a variety of aspect ratios dependent on its composition [86Tak]. Now that phase equilibria and tie-lines can be established with confidence in the quaternary, the properties of the alloy can be designed in a realistic and cost effective fashion.

CONCLUSIONS

The combination of experiment and phase diagram calculation has been shown to be a powerful tool in understanding the microstructural evolution in a rapidly solidified $Fe_{70}Cr_{18}Mo_2B_{10}$ alloy. The phases present and their composition were accurately predicted by the calculations and explained the unexpected appearance of a Mo-rich boride and the martensitic nature of the ferrite matrix. A further distinct advantage of using the calculation route is that the properties of the alloy as a function of its composition and thermal-mechanical treatment can now be optimised far more cost effectively.

ACKNOWLEDGEMENTS

The authors would like to thank Drs W.T.Kim and B.Cantor for useful discussions.

REFERENCES

82Ray R. Ray, "Rapidly Solidified Amorphous and Crystalline Alloys", eds. B.H. Kear, B. C. Giessen and M. Cohen, (North-Holland, NY, 1982), 435.

83Ray R. Ray, V.Panchanathan and S.Isserow, J. Metals, 36(6), (1983), 30.

85Hah S. Hahn, S.I sserow and R. Ray, J. Mater. Sci. Letters, 4, (1985), 972.

85Xu J. Xu, Y. Lin, J. Zhang, Y.You, X. Zhang, W. Huang, W. Zhang and J. Ke, "Rapidly Solidified Materials", eds P.W. Lee and R. J. Carbonara, (ASM, Ohio, 1985), 283.

86Tak H. Takagi, M. Komai, T. Ide, T. Watanabe and Y. Kondo, in "Horizons of Powder Metallurgy: Vol 3", eds W. A. Kaysser and W. J. Huppmann, (Pull.Verlag Schmid, 1986), 1077.

87Hah S. Hahn, S. Isserow and R. Ray, J. Mater. Sci., 22, (1987), 3395.

88And J-O. Andersson, Met.Trans. A, 19A, (1988).

88Raw M. J. Rawson et al., "Rapidly Solidified Materials — Properties and Processing", eds. P.W. Lee and J. H. Mole, (Metals Park, Ohio, 1988) 37.

89Pan L.M. Pan and N. Saunders, "A Review and Phase Diagram Calculations for the Fe-Cr-Mo-B system", report to Rolls-Royce plc, October 1989.

90Kim W.T. Kim, B. Cantor, K. Clay and C. Small, in "Fundamental Relationships Between Microstructure and Mechanical Properties of Metal Matrix Composites", eds P.K. Liaw and M.N. Gungor, (Minerals, Metals and Materials Society, 1990), 89.

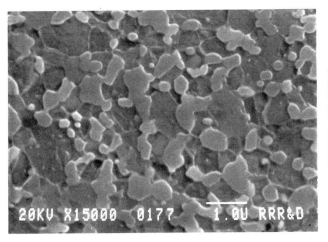

Fig. 1 Microstructure after consolidation showing a fine dispersion of stable boride in a ductile Fe-based matrix.

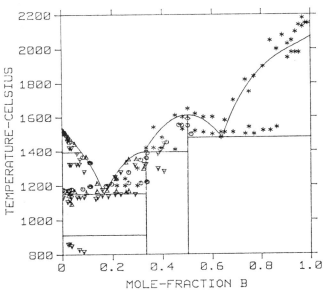

Fig. 2 Calculated phase diagram of the Fe-B system with experimentally determined phase boundaries superimposed.

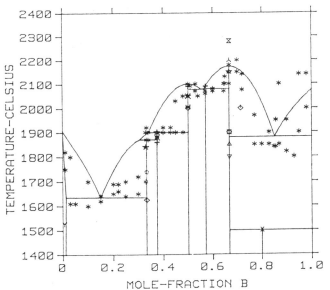

Fig. 3 Calculated phase diagram of the Cr-B system with experimentally determined phase boundaries superimposed.

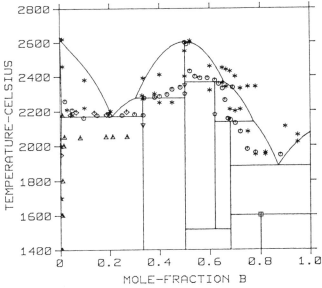

Fig. 4 Calculated phase diagram of the Mo-B system with experimentally determined phase boundary data superimposed.

68

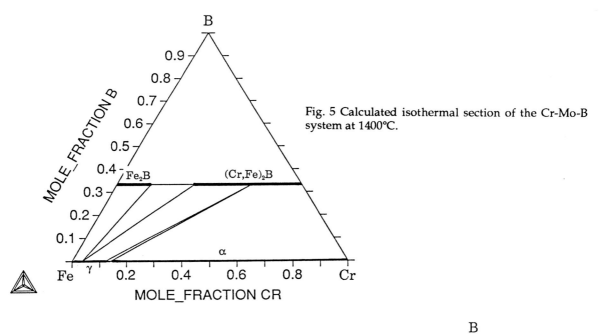

Fig. 5 Calculated isothermal section of the Cr-Mo-B system at 1400°C.

Fig.6 Calculated isothermal section of the Fe-Mo-B system at 1050°C.

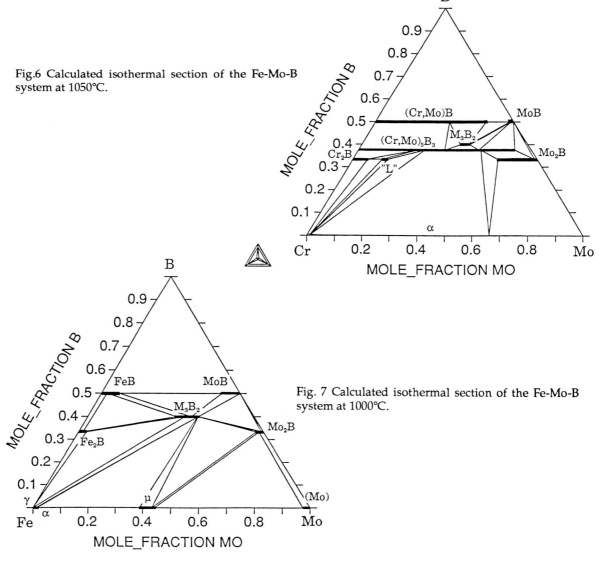

Fig. 7 Calculated isothermal section of the Fe-Mo-B system at 1000°C.

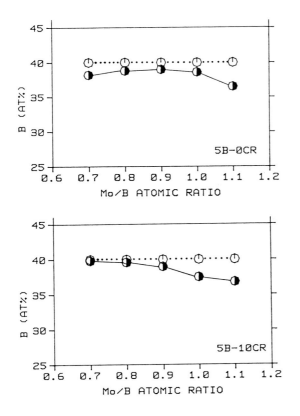

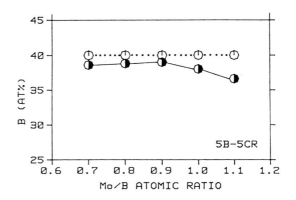

Fig. 8(a) B contents in M_3B_2 as a function of Mo/B atomic ratio.

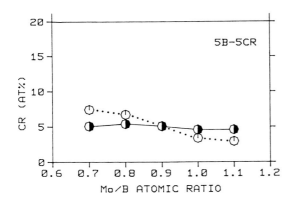

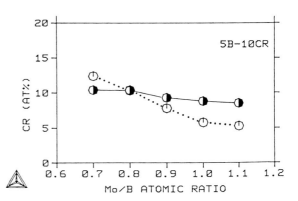

Fig. 8(b) Cr contents in M_3B_2 as a function of Mo/B atomic ratio.

Fig. 8 Comparison of the alloy partitioning in the phase M_3B_2 between the calculated results (open circles with dotted lines) from this work and the experimental data from Takagi et al. [86Tak] (half filled circles with solid lines).

70

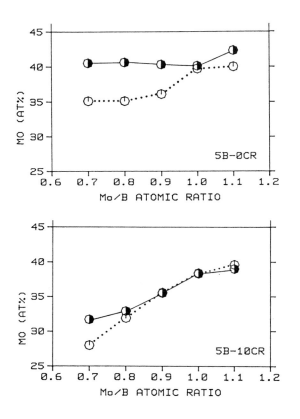

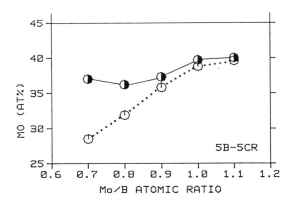

Fig.8(c) Mo contents in M_3B_2 as a function of Mo/B atomic ratio.

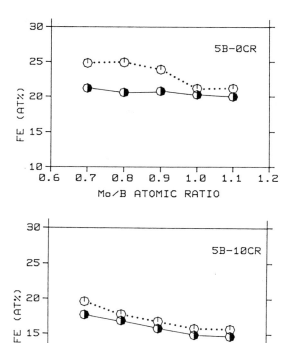

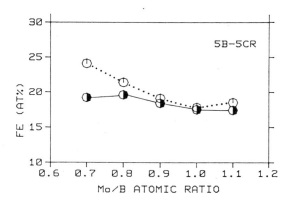

Fig.8(d) Fe contents in M_3B_2 as a function of Mo/B atomic ratio.

9

PROGRESS IN THE DETERMINATION AND INDUSTRIAL APPLICATIONS OF EQUILIBRIUM PHASE DIAGRAMS FOR LIGHT ALLOYS

B. Dubost

PECHINEY, 10 Place des Vosges, Cedex 68, 92048 Paris La Défense 5, France

ABSTRACT

The current and future uses of equilibrium phase diagrams in the metallurgy of multicomponent light metals are reviewed, with emphasis on industrial applications for aluminium and titanium alloys. In addition to traditional uses of phase diagrams in the qualitative study of alloy constitution, microstructural analysis and heat treating, quantitative approaches can also be applied to process metallurgy (solidification, casting and welding), using simple analytical criteria derived from equilibrium phase diagrams as a first approximation. Progress in the methodology of determination of feasible multicomponent phase diagrams results from the combination of experimental and thermodynamic approaches, which allows advanced alloy design. Advanced process metallurgy needs further coupling with kinetic modelling, fluid and solid mechanics and heat transfer.

INTRODUCTION

All light alloys of industrial interest belong to multicomponent systems which imply at least a matrix element and two or more elements (as primary, secondary alloying elements, impurity or trace elements).

Typically five components are added as primary alloying elements (Mg, Mn, Cu, Si, Zn), secondary elements (Mn, Cr, Ti, Zr) or left as impurities (Fe) in current aluminium alloys to provide the property/cost combinations that have led to their wide use in packaging, transportation, building, domestic and electrical applications, either as cast or wrought products in strain hardened or precipitation hardened tempers (Table 1).

At a much lower tonnage level, cast or extruded magnesium alloys can be strenghtened by several solute eutectiferous elements (Al, Zn, Li, Ca) whereas secondary or impurity elements may affect corrosion resistance (Table 1).

Titanium alloys may have up to 10 alloying elements (e.g. in advanced aerospace alloys), which govern their static and dynamic mechanical property combination at both low and high temperatures through phase distribution after casting, thermomechanical processing and heat treating across the α/β transus surface. Alpha or near-alpha alloys (as stabilized by Al) exhibit high temperature strength (due to the high stability of the α (hcp) needle-like phase), flow stress and weldability. Near-beta alloys exhibit better room temperature strength, (due to finer and more homogeneous microstructures after heat treatment) and forming ability when the β phase (bcc) is stabilized, e.g. by vanadium and molybdnenum (Table 2).

Ever increasing quality requirements and cost effectiveness of semi-products and finished parts can no longer be met without a close intelligent control of all critical processing parameters in order to achieve consistent property levels. On the other hand, alloy design can no longer be performed using a purely empirical approach to develop innovative products, as it has been the case for nearly one century.

As a first step a good knowledge of equilibrium phase diagrams within wide composition and temperature ranges is traditionally needed to provide the basic parameters needed by metallurgists: phase status, critical temperatures and phase compositions for phase transformations during solidification, thermomechanical processing and heat treating.

A qualitative understanding of the constitution of industrial alloys, based merely on the implementation of few ternary or quaternary isothermal sections or projections of equilibrium phase diagrams, is no longer a sufficient tool due to the narrow temperature ranges investigated, the absence of tie-lines, and their inability to provide usable graphical representations for higher order alloys. The need for a quantitative approach has opened the way to computer calculations and new uses of phase diagrams.

The goal of this paper is to give first an overview of current applications, actual limitations and future uses of equilibrium phase diagrams by research metallurgists as a tool for metallurgical understanding, processing and design of light multicomponent alloys and engineered materials of industrial interest, with emphasis on aluminium and titanium alloys. Finally, the progress in the methodology of determination of multicomponent equilibrium phase diagrams since the 1970s will be illustrated by recent developments on advanced light alloys for aircraft. Prospects for further applications in the field of metallurgical modelling will be outlined.

CURRENT METALLURGICAL USES AND LIMITATIONS OF PHASE DIAGRAMS FOR LIGHT ALLOYS

Alloy Composition and Process Temperature Ranges
Knowledge of liquidus temperatures is used to rule overheating practice in foundries and cast shops. Eutectic compositions and temperatures of widely used ternary and quaternary alloy phase diagrams, e.g. Al–Cu–Mg–(Si), Al–Mg–Si–(Mn), Al–Zn–Mg–(Cu), are well known. Hence, thermal analysis can be performed and interpreted to account for solidification paths in cast parts.

The composition fields for primary crystallisation are also determined (e.g. Si in Al–Si–Mg–Cu(Fe) hypereutectic cast alloys). In wrought aluminium alloys (2xxx, 6xxx, 7xxx series), alloying with transition elements (Cr, Mn, Zr) must be carefully mastered to prevent crystallisation of coarse primary phases $Al_7(Cr,Fe,Mn)$ and $Al_3(Zr,Ti)$ during solidification. However, a better knowledge of higher order phase diagrams is necessary to take into account chemical interactions between soluble primary alloying elements (Cu,Mg,Si,Zn), less soluble secondary alloying elements (Cr,Mn,Ti,V,Zr) and impurity elements (Fe, Si) in aluminium alloys.

With knowledge of eutectic melting temperatures (mostly temperatures of invariant equilibria) and solvus temperatures, internal composition ranges are fixed within the broad specified composition ranges of commercial heat treatable alloys, to allow feasible homogenization and solution heat treating temperature ranges. Simple or multistep heating or homogenisation practices are derived to achieve microsegregation resorption in cast slabs and billets without risk of incipient melting before hot working. Solution heat treating temperature ranges of newly developed alloys are also defined to ensure high quality by avoiding incipient melting, quench induced cracking, high temperature oxidation or inadequate mechanical property levels. For these applications, phase diagrams are the best tool for the interpretation of differential scanning calorimetry measurements before pilot scale or in-plant validations.

Constitution–Property Relationships
Phase diagrams can be used efficiently in industrial practice to optimise the composition and distribution of stable intermetallic compounds which exert a first order effect on product properties:

In non-heat treatable aluminium alloys (3xxx and 5xxx series alloys) for packaging applications, formability in strain-hardened or recovered tempers is essential. The solid solution content (Cu, Mg, Si) governs strain hardening, strain rate sensitivity and necking (through control of dynamic recovery). The distribution of equilibrium precipitates (e.g. AlFeMnSi dispersoids) governs recrystallisation and shear or ductile fracture in addition to dislocation densities and crystallographic structures[1]. In the case of rigid container sheets for the packaging industry (by far the largest application for aluminium alloys), a critical parameter for canmaking ability of sheets of alloy 3004 (Al–Mn–Mg–Fe–Si) in as strain hardened temper (H19), is the distribution of coarse constituent particles of $Al_6(Fe,Mn)$ and $Al_{12}(Fe, Mn)_3Si$ phases which result from the breakup and transformation of the eutectic structure formed during casting. These particles (1–10 μm size) govern the resistance to galling during ironing through prevention of Al transfer to the ironing dies[2].

In the case of aluminium alloys for building, architectural and domestic applications (1xxx, 3xxx, 5xxx, 6xxx series alloys), the nature and distribution of intermetallic constituent particles, stable precipitates and secondary dispersoids strongly affect the anodic oxidation behaviour. Heterogeneities in the response of cast and wrought products to anodic oxydation may arise from in-depth or lateral matrix concentration and intermetallic constituent particles gradients (e.g. after semi-continuous DC casting). A comprehensive study of these phenomena must be based on equilibrium phase diagrams implementation in solidification and heat treating but it should also involve metastable phase knowledge for Al_6Fe and AlFe compounds in Al–Fe–Si alloys, when the stable Al_3Fe compound does not nucleate[3].

In the case of heat treatable high strength aluminium alloys for aerospace, transportation and building applications, most mechanical properties are governed by metastable precipitation. However equilibrium phase diagrams can be used for metallurgical understanding when equilibrium phases govern some product properties. Stable, ternary dispersoids ($Al_{18}Cr_2Mg_3$, $Al_{12}Mn_3Si$, $Al_{20}Cu_2Mn_3$) inhibit recrystallisation and govern anisotropy. Together with grain boundaries, they also act as sites for heterogeneous precipitation of equilibrium phases ($MgZn_2$, Mg_2Si, Al_2CuMg, Al_2Cu) during slow quenching, which may affect fracture toughness and corrosion performance. Coarse intermetallic constituent particles (Mg_2Si, Al_7Cu_2Fe) formed during the last stages of eutectic solidification have a deleterious effect on fracture toughness[4, 5] and fatigue crack initiation resistance[6] of aerospace aluminium alloys. This drawback can be overcome by using high purity alloys (e.g. 7475) or special processing at high temperatures to decrease the volume fraction of coarse constituents through phase transformation of Al_7Cu_2Fe into $Al_{23}CuFe_4$ and optimised dissolution of Mg_2Si[7].

Another application of phase diagrams is the study of chemical reactivity of light metals with refractories and gaseous environments in the liquid state (especially for titanium alloys) and in the solid state due (to contamination problems by hydrogen or oxygen). These thermochemical aspects will not be discussed in this paper.

PHASE DIAGRAMS AS A TOOL FOR ADVANCED METALLURGICAL MODELLING IN PROCESSING

Prediction of Solidification Microstructures
All casting processes in foundries and cast shops involve near-rapid or semi-rapid solidification, with typical cooling rates between 0.1 and 1000°C/s. The latter cannot be considered as equilibrium conditions from a thermodynamical viewpoint. Assuming no diffusion in the solid phase and constant partition ratio, the classical Scheil-Gulliver approach[8], based on phase diagram data (Table 3), may nevertheless be considered a fair analytical approximation to account for microsegregation in industrial alloys when rapid cooling conditions are involved. This critical cooling rate equal to 100°C/sec. in Al-Cu alloys[9] is probably lower when lower diffusivity elements (e.g. transition elements) are considered.

At the other extreme, assuming complete solid state diffusion with constant diffusivity within the solidification range and little variation of concentration gradient at the interface, the model given by Brody and Flemings[10] provides a quantitative understanding for slow solidification paths (e.g. during gravity casting in insulating ceramic molds) and yields and extremum value for microsegregation profiles since it overestimates diffusion (Table 3). The analytical model derived by Clyne and Kurz[11] and numerical approches recently reviewed by Flemings[12] would account for most industrial solidification conditions, especially with rapidly diffusing species and longer solidification times (e.g. in permanent mold casting, continuous casting, DC casting).

A practical result for users is the prediction of microsegregations, terminal solidification temperature, evolution of solid fraction versus temperature, and mass fraction of metastable eutectic (Table 3), which rule casting ability and homogenisation parameters. These analytical models have been extensively applied to the eutectic solidification of binary alloys (Al–Cu, Al–Si) and, more recently (Fig. 1), to quaternary Al–Cu–Mg–Si alloys[13].

However it has to be remembered that these models can be only implemented on multicomponent alloy systems with well-known equilibrium phase diagrams, assuming constant binary partition ratios (unless thermodynamic calculations are considered). They remain valid approximations except when metastable intermetallic phases are nucleated during rapid or semi-rapid solidification (e.g. in subsurface areas of DC cast slabs and powders), as it is often the case with transition alloying in aluminium alloys elements (Fe, Mn, Ni). Metastable monovariant lines can be graphically extrapolated (Fig. 2) by omitting the unnucleated compound and assuming solid solubility extension in the other compounds[14].

This thermodynamic approach is also of limited use when nucleation and growth kinetics are of primary importance during solidification: the modification of eutectic acicular morphologies of Al–Si cast alloys through limited alloying with Na and Sr (to get globular or fibrous eutectic) or Sb(to get lamellar eutectic) results from increased undercooling and smaller recalescence effects, with minor changes in phase diagrams[15, 16].

In the same way, much work has yet to be completed in order to apply phase diagrams to grain refinement mastership during casting through dense heterogeneous nucleation of aluminium crystals at primary $Al_3(Ti,Zr)$ and insoluble TiB_2 or $(Ti,Al)B_2$ particles in industrial alloys[17,18]: complex peritectic reactions occur in the ternary Al–Ti–B system[19] and strong interactions between Al, Si and Ti promote additional peritectic reactions involving the ternary $Al_3Ti_2Si_2$ compound[20]. Thermal analysis[21] implementing phase diagram knowledge is a powerful tool to monitor solidification in castings.

Hot Tearing Sensitivity and Porosity of Castings
Hot tearing (i.e. cracking during terminal solidification) results from hot shortness of cast alloys near solidus (or eutectic) temperatures, due to nil ductility in the mushy zone. Beside grain size, dendrite morphology and shrinkage sensitivity of the solid phase, the solidification range is empirically known to govern hot tearing sensitivity: alloys with compositions slightly below solid solubility limit and concentrated or proeutectic solid solution alloys, such as Al–5% Cu cast alloys[22] exhibit the worst hot tearing behaviour (Fig. 3). The evolution of solid fraction (f_s) versus temperature (T) under actual (non-equilibrium) solidification conditions exerts a major

effect on hot tearing sensitivity, which is a maximum with the lowest df_s/dT value[23].

The feeding ability of residual liquid through the dendrite network is a key parameter for the problem of hot tearing, which arises when the volume fraction of the liquid phase is lower than 30%. The final solidification path and the morphology of the dendrites, ruled by alloy constitution and heat transfer through the mold and solid metal shell, exert therefore a major effect on crack generation in castings. Using alloy phase diagram data, Feurer[24, 25] has derived a hot tearing tendency parameter (W), taken inversely proportional to the critical cooling rate for which the maximum volumetric flow rate of liquid per unit volume through an interdendritic network is equal to the velocity of volume shrinkage during primary dendrite formation (Table 3).

This approach has been applied rather successfully to binary Al–Cu and Al–Si alloys and ternary Al–Cu–Mg alloys[24-26], thereby allowing a qualitative classification of hot tearing behaviour of aluminium alloys based merely on their constitution (Fig. 4) in accordance with prior experimental results of Pumphrey[27]. Of course, a complete theory of hot tearing should next require an extensive numerical approach also taking into account heat transfer, fluid mechanics (in the liquid and the mushy zone) and solid mechanics (rheology of mushy zones, fracture mechanics in interdendritic areas), especially in the case of DC casting of large slabs and billets.

Interdendritic microporosity occurring during the terminal solidification stage of cast aluminium products results from the additional deleterious effects of poor interdendritic feeding and hydrogen segregation above solid solubility limit (about 0.1 cm³/100 g). Hydrogen solubilities are well-known in binary aluminium alloys: the liquid solubility is increased by Mg, Li, Ti and Zr and decreased by Si, Cu, Mn and Ni, respectively. Using binary partition ratios for hydrogen taken from the Al–H diagram, Poirier et al.[28] have calculated the dependence of the volume fraction of porosity on hydrogen concentration and volume fraction of eutectic liquid in an Al–Cu alloy (Table 3).

A better knowledge of thermochemical parameters for ternary (Al–X–H) and quaternary (Al–X–Y–H) alloys, including phase diagrams, is needed. This would allow a better understanding of the effects of hydrogen and specific primary or secondary elements (eg. Sr, Zr) and inclusions on pore formation in commercial cast alloys. However, the major kinetic effects of thermal gradient and solidification rate on microporosity must also be considered, to account for the favourable effect of narrow mushy zones on soundness (with high cooling rates), in addition to the effects of liquid metal cleanliness.

Other prospects might also concern the thermodynamic prediction of phase equilibrium in such Al–X–Y–H systems under superimposed pressure to allow porosity prevention through pressure and enhanced heat transfer effects (e.g. in low pressure and squeeze casting processes).

Semi-solid Processing of Thixotropic Alloys

Phase diagrams might also be of use to master semi-solid processing of alloys with thixotropic behaviour, which should lead to lower energy costs, higher productivity and better part quality (improved soundness, lower porosity, more homogeneous grain structure) than gravity casting[29]: the rheological behaviour of semi-solid slurries is critically dependent on the volume fraction of the solid phase surrounded by the network of liquid phase, together with the shear rate and the microstructure (grain morphology, particle agglomeration).

PHASE DIAGRAMS AS A TOOL FOR ADVANCED ALLOY AND ENGINEERED MATERIAL DESIGN

Alloy design can be performed by using phase diagrams in a number of industrial cases, provided that the effects of stable phases on product properties and/or behaviour are established.

Cast Aluminum and Magnesium Alloys

Hot tearing and porosity sensitivities are current limiting properties for most high strength light alloy castings. The design of new aluminium and magnesium cast alloys for automotive and aircraft applications will take advantage of recent progress in solidification modelling. It is necessary to take into account the aforementioned effects of parameters taken from phase diagrams (solidification range, eutectic volume fraction and concentration, non equilibrium terminal solidification path) and the effect of kinetics of microstructure generation on interdendritic feeding ability.

On-going progress in numerical approaches of heat transfer, fluid flow and solid mechanics should bring an additional contribution to this problem in the next decade. Quaternary or higher order phase diagrams of Al and Mg alloys with major elements (Si, Cu, Zn, Ag, Li), secondary elements (Sr, Sb, Ti, Zr, rare earths) and impurity elements must therefore be determined but trace element effects on microstructure and casting ability must also be mastered.

Wrought High Strength Aluminium Alloys

In the case of heat-treatable alloys, only high angle grain boundary and other heterogeneously nucleated precipitates rely strictly on equilibrium phase diagram analysis. Structural hardening is known to be governed by the square root of the volume fraction (f_v) of metastable matrix precipitates. Since the metastable solvus surface is shifted from the stable solvus surface towards the right of phase diagrams, f_v is overvalued by the volume fraction of equilibrium precipitates (calculated by the lever rule) if the compound density and the phase diagram are known and if the compositions of both metastable and equilibrium precipitates are assumed to be identical. This the case for Al_2CuMg, Mg_2Si and $MgZn_2$ phases in Al–Cu–Mg, Al-Mg-Si and Al-Mg-Zn-(Cu) systems, (2xxx, 6xxx or 7xxx series alloys), respectively.

In multicomponent alloys with favourable solution hardening response (e.g. Al-Cu-Mg) and/or efficient precipitation or coprecipitation hardening systems, maximum strength is therefore achieved with maximum solute supersaturation. This occurs when the concentrations of alloying elements are equal to those of the saturation limit curve, where solvus and solidus surfaces intersect (Fig. 5). This is also necessary to avoid the occurrence of coarse potentially soluble constituent particles that are deleterious for short transverse elongation and fracture toughness in artificially aged aerospace wrought alloys [5]. All high strength aluminium alloys empirically designed since the 1930s are actually saturated (e.g. alloy 2024) or very slighty undersaturated at ternary or quaternary eutectic melting temperature(e.g. alloys 2214: Fig. 1; 7049: Fig. 6; 7150), in order to allow full solution heat treatment within a narrow but industrially acceptable temperature range. This criterion is actually fulfilled by the new light alloys that have been recently been designed for maximum strength/density performance, such as the commercial alloy 2090, the alloy CP276[30] and the weldable alloy Weldalite™ 049[31], which belong to the Al–Cu–Li–Zr, Al–Cu–Li–Mg–Zr and Al–Cu–Li–Mg–Ag–Zr systems, respectively.

Using a similar alloy design approach dedicated to spray-deposition (a semi-rapid solidification process), an outstanding ultimate strength/elongation combination has also been achieved recently on extruded bars of an optimised Al–Zn–Mg–Cu–(Cr, Mn, Zr) alloy system[32], thereby confirming the interest of the pioneering work of Bower et al.[33] on phase diagram implementation.

Weldable Wrought Aluminum Alloys

Brazing and welding metallurgy are relevant to phase diagram use for joining. Only the latter will be discussed here. Although it is little documented on aluminium alloys, welding metallurgy is of great interest for the development and use of light, energy efficient parts for transportation and building requiring high specific strength and good welding ability (i.e. no cracking and high joint efficiency) . Since weld cracking is governed by mainly hot tearing, the effect of alloy constitution on weldability is similar to that already mentioned in the study of casting defect generation, with interdendritic feeding ability and solidification shrinkage having a major effects.

Therefore it is not surprising to find qualitatively identical classifications in hot tearing behaviour of cast parts (Fig. 2) and weld cracking behaviour (Fig. 7) under similar cooling conditions and thermomechanical environment[23]: binary eutectiferous alloys (Al–Cu, Al–Mg, Al–Li) exhibit increased cracking sensitivity with increasing solute content, up to slightly undersaturated compositions, which results from near maximum solidification range and insufficient interdendritic feeding ability[34].

Sharp improvements in weldability occur in the case of saturated or slightly oversaturated alloys (such as the Al–Cu–Zr alloy 2219), when the flow of interdendritic liquid (mostly the metastable eutectic) at the end of solidification is able to provide sufficient interdendritic feeding and hot tear healing effects. However, phase diagrams fail to account for morphological effects: crack healing is reportedly favoured when finely dispersed eutectic pools are formed and solidify throughout the weld as in the case of alloys 2219 and Weldalite™ 049[35]. A continuous network of interdendritic eutectic promotes cracking in the case of wrought alloys designed for high strength, not for weldability (e.g. 2014, 2090).

Hence, achieving both high strengh ductility/toughness combinations and good weldability remains a challenge because of the apparent incompatibility of the constitutional criteria discussed previously in the case of wrought aluminium alloys. This metallurgical approach, should also be completed by taking into account thermomechanical effects induced by actual welding technologies and part configurations (e.g. clamping).

Metal Matrix Composites

Stiffness, high temperature strength, fatigue resistance and low thermal expansion are the main property improvement goals for light metal matrix composites. A major problem in the research and development of these advanced engineered materials is the control of surface interaction between the metal matrix and the reinforcing particles (e.g. during rheocasting) or fibres (e.g. during pressure infiltration). This may harm the quality of the final parts, yielding poor soundness, low elongation and unacceptable deviations from rule of mixtures predictions of strength and Young's modulus. Equilibrium phase diagrams are therefore a useful tool in understanding the thermodynamics of interfacial reactions. This work has been recently been undertaken on aluminum alloys reinforced with either SiC or TiC[36, 37] in order to determine temperature and composition ranges of invariant reactions favouring the binary (Al_4C_3) or ternary (Al_4SiC_4) carbides or aluminides (Al_3Ti). A wide research field remains open for titanium matrix composites, due to the high reactivity of titanium alloys in the liquid state.

PROGRESS IN THE DETERMINATION OF EQUILIBRIUM PHASE DIAGRAMS

The Traditional Metallographic Approach
Our knowledge of equilibrium phase diagrams of light alloys has traditionally relied on experimental determination, using a metallographic approach performed on numerous cast and heat treated samples (Fig. 8). In the case of aluminium alloys, all binary diagrams, many projections of isothermal sections of ternary diagrams of industrial use and a number of isopleth or isothermal sections of important quaternary diagrams have been reviewed by Phillips[39] and by Mondolfo[40]. Drits et al. have published an extensive review of ternary phase diagrams for aluminium and magnesium alloys[41]. In the case of titanium alloys, most binary systems are known and the major ternary Ti–Al–V diagram[42] has been revisited. Considerable research work is currently devoted to the design of titanium aluminides for light engine parts, which has led to re-examination of the binary Ti–Al phase diagram near the ordered γ-TiAl and α_2(Ti$_3$Al) compounds.

This work, mainly experimental, is pursued on higher order Ti–Al–(V, Nb,...) intermetallic alloys with improved high temperature strength–ductility–creep properties.

The Thermodynamic Approach
Due to cost efficiency and manpower availability, the purely experimental approach can no longer be affordable to solve higher order or unknown phase diagrams of industrial use. Thermodynamic calculation of phase diagrams has been an alternative route since the 1970s due to enhanced thermodynamic modelling and computation capabilities, using for instance the CALPHAD method and the THERMOCALC program[43]. A thermodynamic approach based on critical assessment of binary and phase diagrams has been performed under ASM supervision in the 1980s which has resulted in dedicated monographs series on binary alloys[44]. However limited thermodynamic properties were available at this time for Al, Ti and Mg alloys and this approach was, by nature, not oriented experimentally. This approach is being currently pursued on ternary aluminium alloys at the Max Planck Institute in Stuttgart, which results in a comprehensive compendium of evaluated constitutional data and phase diagrams[45].

Very little thermodynamic information has been published on magnesium alloys, except in the case of the Al–Mg–Si system[46] and Mg- and Li-rich corners [47] of the Mg–Li–Al system (Fig. 9).

Relatively few thermodynamic data are also available on titanium alloys except those resulting from the pioneering work of Kaufman and co-workers[48,49] on binary systems Ti–(Al,Co,Sr,Cu,Hf, Ir, Mo, Nb, Ni, Os, Pb, Pd, Re, Rh, Ru, V, W, Zr). This might result from the difficulty of direct measurement of thermodynamic parameters (due to high reactivity of titanium in the liquid state). Usual methods may also prove inadequate even in the solid state, due to slow evaporation (during partial vapour pressure measurements), multivalence of Ti (with electrolytical methods) and contamination problems (by the surrounding atmosphere). *Ab initio* phase diagram calculations are expected to bring useful answers to this serious problem in the future.

Thermodynamic calculations of high-order phase diagrams must obviously be used very carefully when limited experimental data are available. In the case of ternary systems, interactions leading to ternary compound formation cannot be predicted by a thermodynamic calculation based merely on the binary constitutive system. This can be seen by comparing both purely experimental and thermodynamic approaches in the case of the Al–Li–Mg system (Figs. 8 and 9, respectively): due to the lack of data, thermodynamic calculations of phase diagrams can yield metastable phase fields near unpredicted ternary compound phase fields (e.g. Al_2LiMg), which would limit their practical value to composition fields near the edges and perhaps the corners of the ternary diagram.

Depending on the validity of the thermodynamic data and the availability of satisfactory descriptions for metastable phases, thermodynamic calculations can also provide useful information for metastable phase diagram and T_0 curve predictions, yielding upper limits for solid solution extension in rapid solidification.

In any case, thermodynamic calculation will nevertheless remain a modern and efficient teaching tool to educate future users of phase diagrams.

The Coupled Thermodynamic and Experimental Approach
To prevent these drawbacks and achieve comprehensive equilibrium phase diagrams determination and implementation within a reasonable time, it is advantageous to combine the previous approaches into an interactive process (Fig. 10):

- metallogaphy based on literature survey and limited experimental alloy study (e.g. near composition of equilibrium compounds, monovariant lines, invariant eutectic or peritectic-type equilibria).

- experimental measurement of thermodynamic parameters of binary and higher order phases (e.g. solution enthalpy of intermetallic compound samples) to derive their enthalpies of formation, if assessed thermodynamic data are not available in current databases (e.g. THERMODATA, SGTE)

- computer calculation of equilibrium phase diagrams (e.g. with the THERMO-CALC program[43], using optimisation procedures to comply with both metallographic results and thermodynamic data.

It must be emphasized that the experimental work can remain a time consuming step when numerous stable intermetallic compounds are identified in multicomponent systems, especially when they are not stoichiometric and exhibit large composition ranges (e.g. in aluminium alloys). In such a case, thermodynamic description through sublattice modelling may be arduous.

Provided reasonable agreement is found (which depends on the quality of data and the limitations of semi-empirical models for non stoichiometric compounds), such a modern approach is a very powerful way to achieve feasible phase diagrams within wide composition and temperature ranges. It allows an efficient use of isothermal sections (with tielines) and isopleth sections for process modelling and alloy design.

Recent achievements of industrial interest are the constitution and phase diagrams calculation in the Al-rich corners of the systems Al–Cu–Mg–Si[50], Al–Li–Mg[51] and Al–Cu–Li[52], the calculation of phase equilibria and α/β transus surface in the multicomponent Ti–(Al, Sn, Zr, V, Mo, Cr, Fe) alloys[53] and the constitution of ternary and quaternary Ti–Al–(Mo,Zr,Cr), Ti–Mo,(Zr,Cr), Ti–Al–Mo–Zr, Ti–Al–Mo–Zr–Cr in the Ti-rich corner[54]. These diagrams (Figs. 11 and 12) have been implemented at PECHINEY RHENALU to design new Al–Cu–Li–Mg–Zr alloys[30] at CEZUS and to optimise the new high strengh titanium alloy BetaCez (Ti–Al–Mo–Zr–Sn–Cr–Fe) for aircraft engines[55].

This approach is also of great interest for metal matrix composites and liquid alloy reactivity studies in cast shop technology and foundry applications, provided assessed thermodynamic data on both metals and refractories are available.

Finally, it should be mentioned that phase diagrams determined according to this method can also be used to derive metastable phase fields or monovariant lines (e.g. eutectic alloys) for metastable equilibria involving the liquid phase and intermetallic compounds, which provides a more powerful tool than graphical extrapolation (Fig. 2).

CONCLUSIONS

Equilibrum phase diagrams will definitively remain a major tool for metallurgical process understanding (solidification, heat treating, welding) as well as for modern design of advanced cast and wrought aluminium, magnesium and titanium alloys. The recycling issue is also likely to speed up the rate of phase diagram determination and implementation.

Due to the complexity of multicomponent systems of industrial interest, the determination of feasible equilibrium phase diagrams is a huge task. It should therefore be continued, preferably on a continental or world scale, by using a feasible and powerful approach combining metallography of selected alloy samples, direct measurement and assessment of thermodynamic parameters and interactive computer calculation of phase equilibria. In Europe, such an approach is currently being conducted on binary and ternary alloys in the frame of the COST 507 program in order to provide a thermochemical database for light alloys.

However, advanced metallurgical process modelling requires further implementation of phase diagrams with kinetic modelling, fluid flow, heat transfer and solid mechanics studies, especially in the case of semi-rapid solidification and solid state phase transformations. This would also require further progress in the determination and applications of metastable phase diagrams for light alloys.

ACKNOWLEDGEMENTS

The author thanks Dr I. Ansara for helpful comments.

REFERENCES

1. F. R. Boutin and Ph. Lequeu, 6th Yugoslav Int. Symposium on Aluminium, Titovo, Uzice, 1990, in press.
2. R.E. Sanders, D. J. Lege and T.L.Hartman, Aluminium, 65, 1989, p 941.
3. Y. Langsrud, this conference.
4. M.V. Hyatt, Aluminium Alloys in the Aircraft Industries, Proc. Symp. Turin, 1976, Technicopy Ltd, 1978, p 31.
5. J.T. Staley, Properties Related to Fracture Toughness, ASTM STP 605, 1976, p 71.
6. T. H. Sanders, Jr. and J.T. Staley Jr, Fatigue and Microstructure, ASM, 1979, p 467.
7. B. Dubost, R. Macé and J. Bouvaist, Aluminium Technology '86, The Institute of Metals, 1986, p 459.
8. E.Z. Scheil, Z. Metallkde 34, 1942, p 70.
9. A.B. Michael and M.B. Bever, Trans. AIME 200, 1954, p 47.
10. H.D. Brody and H.C. Flemings, Trans. AIME 236, 1966, p 615.
11. T.W. Clyne and W. Kurz, Met. Trans. 12A 1961, p 965.
12. H.C. Flemings, F. Weinberg Conference on Solidification Processing, Canadian Inst. of Metals, Ontario, Canada, August, 1990, in the press.
13. J. Lacaze and G. Lesoult, State of the Art of Computer Simulation of Casting and Solidification Process, Proc. European Materials Society Conference, Strasbourg, 1986, Les Editions de Physique,1987, p 119.
14. J.H. Perepezko, Mat. Res. Soc. Symp. Vol. 19, Elsevier,1983, p 223.
15. M.D. Hanna and A. Hellawell, Mat. Res. Soc. Symp., Vol. 19, Elsevier, 1983, p 411.
16. O.N. Semenova, I.N. Ganiev and A.V. Vabhokov, Russian Metallurgy, 1987, p 99.
17. 1. Maxwell and A. Hellawell, Acta Met., 23, 1975, p 895.
18. F.H. Hayes, H.L. Lukas, G. Effenberg and G. Petzow, Z. Metallkde 80, 1989, p 361.
19. A. Abdel- Hamid and F. Durand, Z. Metallkde 76, 1985, p 739 and p 744.
20. A.M. Zakharov, I.T. Gul'din, A.A. Arnold and Y.A. Mat'senko, Russian Metallurgy, 1988, p 185.
21. J. Charbonnier, J. Morice and R. Portalier, AFS International Cast Metals Journal, Sept. 1979, p 39.
22. R.A. Rosenberg, M.C. Flemings and H.F. Taylor, Modern Castings, Sept. 1960, p 112.
23. C. Bracale, Aluminio y Nuova Metallurgia, 38, (8), 1969, p 391.
24. U. Feurer, Proc. Intern. Symp. on Quality Control of Engineered Alloys and the Role of Metal Science, Delft (NL), 1977, p 131.
25. U. Feurer, Giesserei Forschung, 2,1976, p 75.
26. J.C. Ramseyer, U. Feurer, J.B. Gabathuler and R. Wunderlin, Proc. 7th ILM Congress, Leoben, Vienna, 1981, p 122.
27. W.l. Pumphrey, Aluminium Dev. Assoc. Report N° 27, London, 1955.
28. D.R. Poirier, K. Yeum and A.L. Maples, Met. Trans. 18A, 1987, p 1979.
29. M.C. Flemings, Edward Campbell Memorial Lecture, ASM International Fall Meeting, Detroit, October 1990, Preprint.
30. P. Meyer and B. Dubost, Aluminum-Lithium Alloys III, ed. by C. Baker, P.J. Gregson, S.J. Harris, and C.J. Peel, The Institute of Metals, 1986, p 37.
31. J.R. Pickens, F.H. Neubaum, T.J. Langan and L.S. Kramer, Aluminum-Lithium Alloys Vol III, ed. by T.H. Sanders, Jr. and E.A. Starke, Jr., MCEP Ltd, Birmingham, 1989, p 1397.
32. J.F. Faure and B. Dubost, Proc. ASM Conference on Advanced Aluminium and Magnesium Alloys, Amsterdam, June 1990, in the press.
33. T. F. Bower, S.N. Singh and M.C. Flemings, Met. Tans. 1,1970, p 191.
34. J.C. Lippold, Aluminium-Lithium Alloys Vol III, op.cit., p 1365.
35. L.S. Kramer, F.H. Neubaum and J.R. Pickens, Aluminium-Lithium Alloys Vol III, op. cit., 1986, p 1415.
36. J.C. Viala, C. Vincent, H. Vincent and J. Bouix, Mat. Res. Bull, 25, 4, 1990.'
37. J.C. Viala, P. Fortier and J. Bouix, J. Mat. Sc., 25, 1990, p 1842.
38. D.W. Levinson and D.J. Mc Pherson, Trans. ASM, 48, 1956, p 689.
39. H.W.L. Phillips, Annotated Equilibrium Phase Diagrams of some Aluminium Alloy Systems, The Institute of Metals, London, 1959.
40. L.F. Mondolfo, Aluminium Alloys: Structure and Properties, Butterworths, London, 1976.
41. M.E. Drits, N.R. Bochvar and E.S. Kadaner, Phase Diagrams for Aluminium and Magnesium-Based Alloys, ed. Nauka, Moscow, 1977.
42. K. Hashimoto, H. Doi and T. Tsujimoto, J. Jpn Inst. Met. 49 (6), 1985, p 410.
43. B. Sundman, B. Jansson and J.O. Andersson, Calphad, 9, 2, 1985, (2), p 153. See also B. Jonsson and B. Sundman, this conference.

44. T.B. Massalski, J.L. Murray, L.H. Bennett, H. Baker, Binary Alloy Phase Diagrams, ASM, 1986. J.L. Murray, Phase diagrams of Binary Titanium Alloys, ASM, (1987).

45. G. Petzow and G. Effenberg, Ternary Alloys: A Comprehensive Compendium of Evaluated Constititutional Data and Phase Diagrams Vol III- VCH, Germany, (1990).

46. D. Ludecke, Z. Metallkde 77, (5), 1986, p 278.

47. M.L. Saboungi and C.C. Hsu, Calphad 1(3), 1977, p 237.

48. L. Kaufman and H. Bernstein, Computer Calculation of Phase Diagrams with Special Reference to Refractory Metals, Academic Press, New York, 1970.

49. L. Kaufman and H. Nesor, Computer Analysis of Alloy Systems, Technical Report AFML - TR - 73 - 56, March 1973.

50. J. Lacaze, G. Lesoult, O. Relave, T. Ansara and J.P. Riquet, Z. Metallkde 78 (2),1987, p 141.

51. B. Dubost, P. Bompard and I. Ansara, Proc. 4th. International Aluminium-Lithium Conference, ed. by G. Champier, B. Dubost, D. Miannay, and L. Sabetay, Journal de Physique Coll C3, 48, 1987, p 473.

52. B. Dubost, C. Colinet and I. Ansara, Aluminium-Lithium Alloys vol II, ed. by T.H. Sanders, Jr. and E.A. Starke, Jr., MCEP Ltd, U.K, 1989, p 623.

Quasi crystalline Materials, ed. by Ch. Janot and J.M. Dubois, World Scientific, Co. Ltd, Singapore, 1988, p 39.

53. H. Onodera, K. Ohno, T. Yamagata and M. Yamazaki, Trans. Iron. Steel Inst. Jpn, 28 (10), 1988, p 802.

54. J.P. Gros, 1. Ansara and M. Allibert, Proc. 6th World Conference on Titanium, Vol III, ed. by P. Lacombe, R. Tricot and G. Beranger, Les Editions de Physique, 1989, pp 1553 and 1559.

55. B. Prandi, E. Alhéritière, F. Schwartz and M. Thomas, Proc. 6th World Conference on Titanium, Vol II, op.cit, p 811.

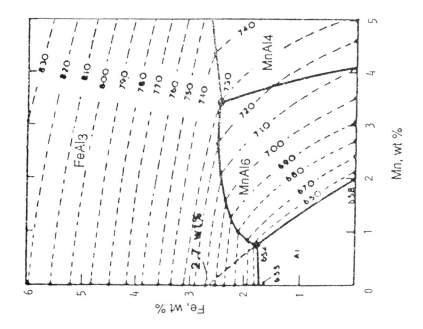

Fig. 2. Graphical extrapolation of stable monovariant lines on liquidus to account for metastable phase formation in the Al-Fe-Mn system: if the phase Al₃Fe does not nucleate and grow, the eutectic valley L-α + Al₆(Fe, Mn) extends to the Al-Fe binary alloy (dashed line). From Perepezko (14).

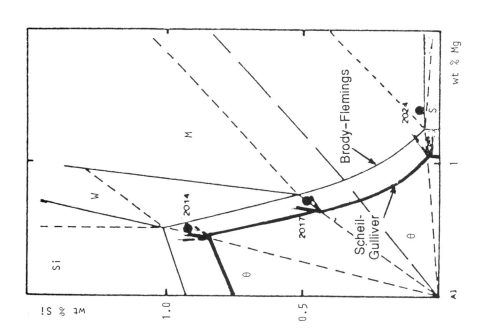

Fig. 1. Predicted domains of alloy composition related to a specific sequence of equilibrium phase formation for quaternary Al-Cu-Mg-Si alloys containing 4.2 % Cu for a cooling rate equal to 100 K/mn by implementing phase diagram with Scheil-Gulliver and Brody-Flemings models. From Lacaze and Lesoult (13).

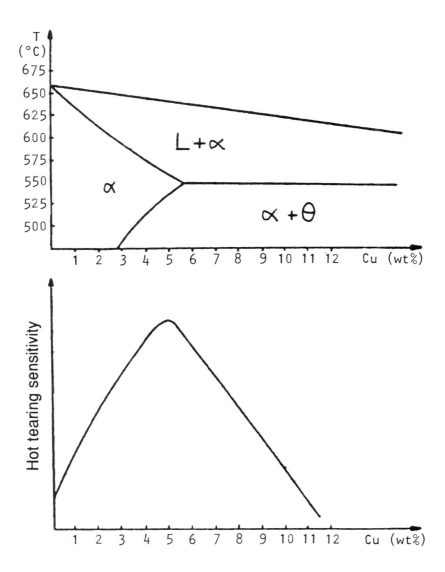

Fig. 3. Effect of solidification internal on hot tearing sensitivity of cast binary Al-Cu alloys. From Rosenberg, Flemings and Taylor (22).

82

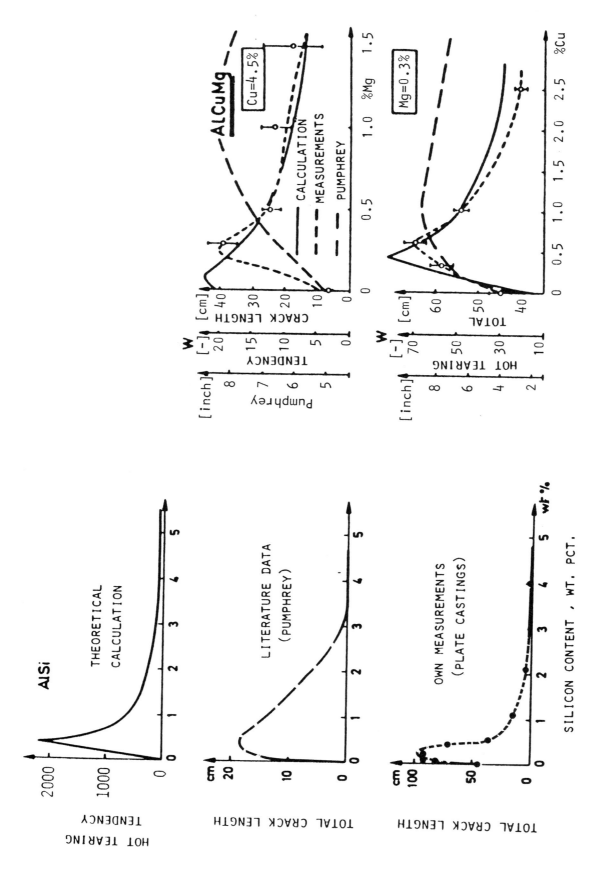

Fig. 4. Experimental measurements and theoretical prediction (according to Feurer's model) of hot tearing tendency in binary Al-Si and ternary AlCu-Mg alloy castings (24-26).

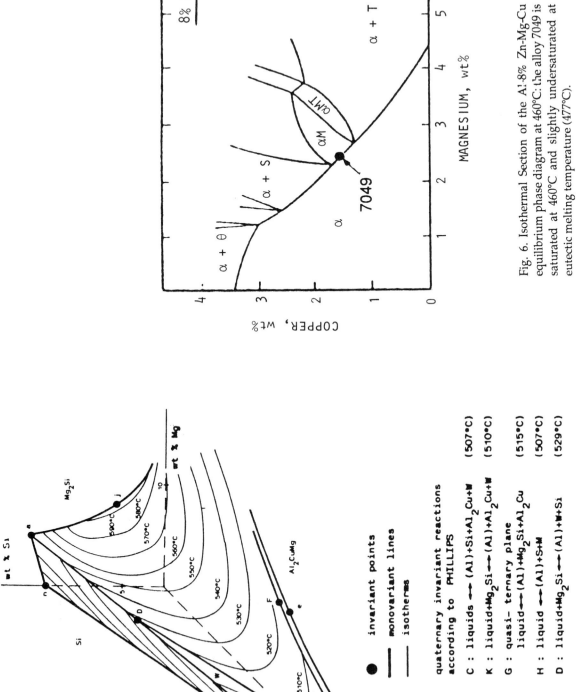

8% ZINC

7049

α + θ

α + S

α

αM

αMT

α + T

COPPER, wt%

MAGNESIUM, wt%

Fig. 6. Isothermal Section of the Al-8% Zn-Mg-Cu equilibrium phase diagram at 460°C: the alloy 7049 is saturated at 460°C and slightly undersaturated at eutectic melting temperature (477°C).

wt % Mg

Mg₂Si

wt % Si

Si

590°C

580°C

570°C

560°C

550°C

540°C

530°C

520°C

510°C

Al₂CuMg

Al₂Cu

wt % Cu

● invariant points
— monovariant lines
— isotherms

quaternary invariant reactions
according to PHILLIPS

C : liquids → (Al)+Si+Al₂Cu+W   (507°C)

K : liquid+Mg₂Si → (Al)+Al₂Cu+W   (510°C)

G : quasi- ternary plane
    liquid → (Al)+Mg₂Si+Al₂Cu   (515°C)

H : liquid → (Al)+S+M   (507°C)

D : liquid+Mg₂Si → (Al)+W+Si   (529°C)

Fig.5. The saturation surface at the Al-rich corner of the Al-Cu-Mg-Si system. From Lacaze and Lesoult (13).

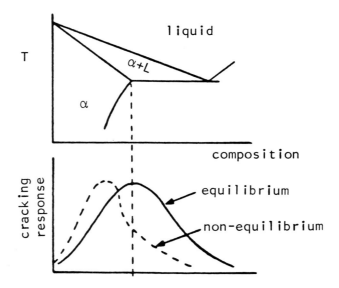

Fig.7. Schematic effect of alloy content on weld sensitivity of binary eutectiferous aluminum alloys under equilibrium and non equilibrium solidification conditions. From Lippold (34).

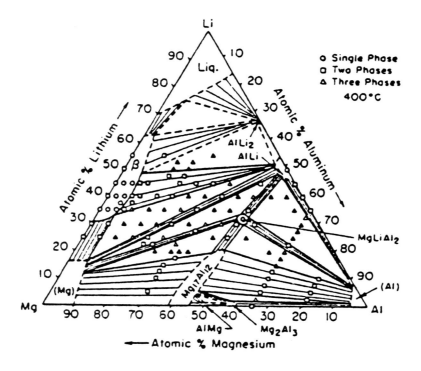

Fig. 8. The traditional metallographic approach for phase diagram determination in the case of the Al-Li-Mg system: experimental isothermal section at 400°C. From Levinson and Mc Pherson (38)

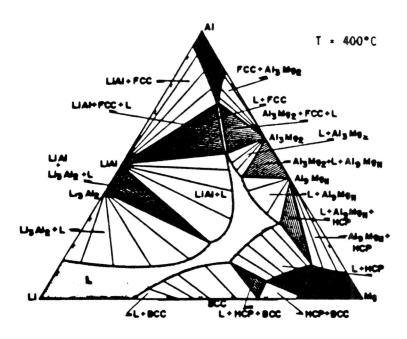

Fig. 9. The thermodynamic approach for phase diagram calculation of the ternary system based on binary constitutive diagrams computed isothermal section at 400°C. From Saboungi and Hsu (47)

EXPERIMENTAL ANALYSIS OF SYSTEM CONSTITUTION
(METALLOGRAPHY, MICROANALYSIS, THERMAL, ANALYSIS) :
PHASE STATUS, INVARIANT AND MONOVARIANT EQUILIBRIA, ISOTHERMAL SECTIONS

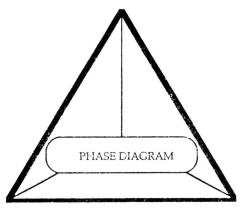

THERMODYNAMIC MEASUREMENTS,
E.M.F., VAPOUR PRESSURE :
DERIVATION OF ENTHALPY OF FORMATION AND
GIBBS ENERGIES OF PHASES.

THERMODYNAMIC CALCULATION
(GIBBS ENERGY MINIMIZATION) :
ASSESSMENT, OPTIMIZATION OF PARAMETERS

Fig. 10. Progressive phase diagram determination methodology for multicomponent systems.

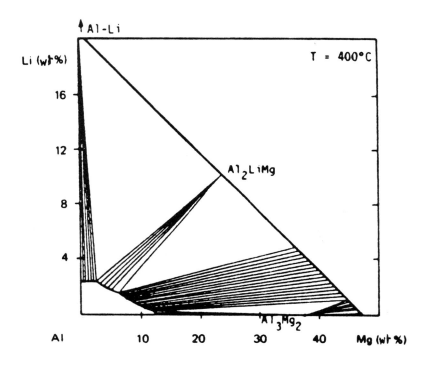

Fig. 11.Computed isothermal sections of the Al-Li-Mg equilibrium phase diagram at 400°C in the Alrich corner, using experimental data taken from 12 alloy samples and the THERMOCALC program. From Dubost, Bompard and Ansara (51).

Table 1 Phase diagram classification of current alloy-ing elements for aluminium and magnesium alloys versus solid solubility at high temperature

k = partition ratio	k ≤ 1	k ≥1
<u>Aluminium alloys</u>		
- high solubility (x ≥ 2wt%)	Li, Mg, Cu, Zn, Ag	
- medium to low solubility (0,1 ≤ x<2wt%)	Mn, Co, Ni, Si	Cr, V, Zr
- very low solubility (x <0,1wt%)	Fe	
<u>Magnesium alloys</u>		
- high solubility (x > 1 wt%)	Al, Zn, Li	
- medium solubility (0,2 ≤ x≤1wt%)	Ca,Cu	Mn,Zr
- low solubility (x<0,2wt%)	Si,Co,Ni,Fe	Ti

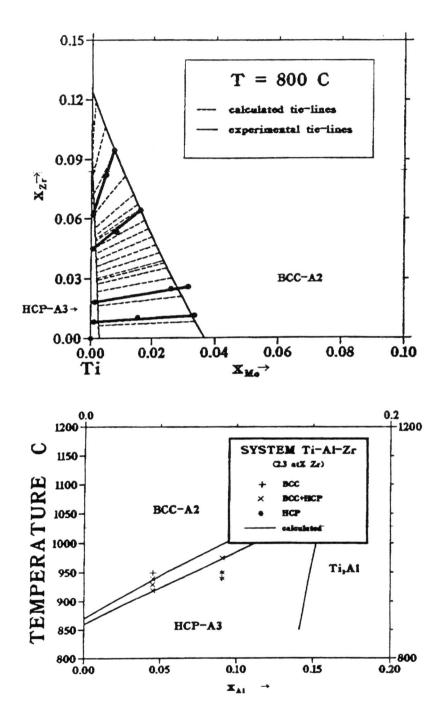

Fig. 12. Thermodynamic calculation of ternary titanium alloy phase diagrams in the Ti-rich corner with the program THERMO-CALC as compared to experimental data:
• Ti-Mo-Zr system: isothermal section at 800°C   with computed and experimental tielines (top)
• Ti-Al-Zr system: vertical (isopleth) section. From Gros, Ansara and Allibert (54).

88

Table 2 Phase diagram classification of alloying elements for titanium alloys versus allotropic α/β transformation

		Equilibrium Phase Diagram	Insertion Solid sol.	Substitut. Solid Sol.
α stabilizers			O_2, N_2, B, C	Al
β stabilizers	Isomorphous			Mo V Nb, Ta
	Eutectoïd		H_2	Mn, Fe, Cr, Co, W, Ni, Cu, Au, Ag, Si
Neutral				Sn, Zr,

Table 3 Synopsis of analytical equations derived from equilibrium phase diagram analysis to account for non-equilibrium solidification, microsegregation, hot tearing and microporosity generation

---

MICROSEGREGATION IN SOLIDIFICATION

SCHEIL-GULLIVER (8) no diffusion in the solid

$$C_s^* = k\,C_0\,(1-f_s)^{k-1}$$

$$C_l = C_0 f_l^{(k-1)}$$

$$g_e = \left(\frac{C_e}{C_0}\right)^{-\frac{1}{1-k}}$$

C_s^* = Composition of solid at liquid-solid-interface
k = equilibrium partition ratio
C_0 = initial composition
C_l = liquid composition
f_s = solid fraction
C_e = eutectic composition
f_l = liquid fraction
g_e = liquid eutectic volume fraction

BRODY-FLEMINGS (10) : diffusion in the solid with solute redistribution

$$C_s^* = kC_0\left(1-\frac{f_s}{1+\alpha k}\right)^{k-1}$$

$$\alpha = \frac{4D_s\,t_f}{d^2}$$

$$g_e^D = g_e - (1-g_e)\,\alpha k$$

D_s = solute diffusion coefficient in the solid
t_f = solidification time
d = dendrite arm spacing

g_e^D = eutectic volume fraction with solid diffusion

HOT TEARING IN SOLIDIFICATION
FEURER (24,25) : Hot tearing tendency parameter related to feeding ability and shrinkage velocity

$$W = \max(W_1, W_2)$$

$$W_1 = \left(\frac{(\rho_0-\rho_s + ak\,C_0)^3 (1-ge)^6}{(1-k)(\ln(Ce/Co))^2 (m_lCo)^3\,g_e^{3(1+k)}}\right)^{1/5}$$

$$W_2 = -m_l\,C_0\,(1-k)$$

ρ_0 = density of liquid aluminium at melting point
a = composition coefficient of liquid density
ρ_l = density of the liquid ($\rho_l\,\rho_0 + aC_\rho$)
ρ_s = density of the solid phase
m_l = slope of liquidus line
W = relative value of hot tearing tendency, units neglected (to be multiplied by 1000 to obtain numbers above 1)

MICRO POROSITY DUE TO SOLIDIFICATION SHRINKAGE AND GAS POROSITY
POIRIER (28) : volume fraction of porosity, effects of process parameters

$$\varepsilon = \left| 1 + \{ (1-f_e)/\rho'_s + f_e/\rho_e \}\{MP_g^e/RT_e\phi\}\right|^{-1}$$

$$f_e = g_e / \left(g_e + (\rho'_s + \rho_e)(1-g_e)\right)$$

ε = volume fraction of porosity
f_e = mass fraction of eutectic liquid

T_e = eutectic temperature (absolute scale)
ρ_e = masse density of liquid eutectic at temperature T_e
ρ'_s = mass density of primary solid at eutectic temperature T_e
P_g^e = pressure inside the gas phase when $f_e = f_e$
ϕ : mass of the gas phase per 1g of alloy
g_e = volume fraction of eutectic liquid
M = molecular weight of the gas (hydrogen)
R = gas constant

10
THE USE OF PHASE DIAGRAMS FOR CALCULATING SOLIDIFICATION PATHS

Y. LANGSRUD

Centre for Industrial Research, Postbox 124 Blindern, 0314 Oslo 3, Norway

ABSTRACT

The general problem of using phase diagrams and solidification path calculations in order to predict phase distributions is discussed. An example from the ternary Aluminum-Iron-Silicon system is given, comparing observed and calculated phase distributions. It is shown that metastable phase diagrams must be required. These phase diagrams are constructed through a process in which calculated phase distributions are compared continuously with the observed ones.

1. IDENTIFYING THE PROBLEM OF CALCULATING A MICROSTRUCTURE

AFTER SOLIDIFICATION

Microstructure control is essential in every alloy treatment process, casting being no exception. A solidified microstructure can be characterized by several parameters, which we may classify into two groups. The first group, the "structural fineness" group, comprises parameters such as grain size, particle sizes, cell-size or secondary dendrite arm spacing, lamellae spacings in eutectic colonies, etc. The other group, the "chemical" group, includes parameters describing the element distribution, i.e., matrix solid solution profiles and the amount and type of different phases. We shall restrict our problem identification to the latter group, asking the following question: For a given metal melt of uniform and known composition, solidifying under well-known temperature conditions, what do we need to know to calculate the phase distribution in the cast sample? We will make the important assumption that no reactions take place in our sample below the solidus temperature T_s.

2. A GENERAL SOLUTION TO THE PROBLEM OF CALCULATING THE PHASE

DISTRIBUTION AFTER SOLIDIFICATION

On the assumption that our microstructure remains unchanged when cooling from T_s, our problem can be solved knowing the following:

(i) A correct solidification equation from which a solidification path can be calculated.

(ii) A correct phase diagram, taking into account that equilibrium conditions are not attained during solidification.

A solidification equation describes how the composition on the solidification front alters during solidification. If cooling is not too fast, the solution by Brody and Flemings[1], and later slightly modified by Clyne and Kurz[2] is quite realistic, providing correct solidification coefficients (k-values): '

(1) $C_1 = C_0 \{ 1 - (1-2\alpha k) f_s \}^{(k-1)(1-2\alpha k)}$

(2) $C_s = K C_0 \{ 1 - (1-2\alpha k) f_s \}^{(k-1)(1-2\alpha k)}$

$C_1 = $ Liquid concentration on the solidification front

$C_s = $ Solid concentration on the solidification front

$\alpha = $ Back diffusion parameter (eq. 3 and 4)

K = Partition coefficients

f_s = Fraction solid

Clyne and Kurz registered that eqs. (1) and (2) would not be valid for high α values, and suggested that α' be substituted for α with

(3) $\alpha' = \alpha [1 - \exp(-1/\alpha)] - 0.5 \exp(-1/2\alpha)$,

whereas α was given by

(4) $\qquad \alpha = \dfrac{D_s\, t_f}{L^2}$

D_s = Solid state diffusion coefficient

t_f = Local solidification time

L = Size of the solidifying system (see appendix)

The two well known equations, equilibrium or lever rule solidification (complete solid state diffusion) and Scheil (no solid state diffusion) become special cases for $\alpha' = 0$ ($\alpha = 0$) and $\alpha' = 0.5$ ($\alpha = \infty$), respectively. By using equations (3) and (4) a realistic back diffusion can be calculated. Figure 1 shows solidification paths for 4 different binary aluminium alloys, according to (1), for the equilibrium case, the Scheil case, and for the calculated back diffusion case. Figure 1 clearly demonstrates that for the typical industrial solidification (local solidification time of 100 s) the equilibrium path diverges substantially from the path calculated for "realistic" back diffusion for elements with high solid state diffusion coefficients (Mg and Si). The calculations do not, however, deviate greatly at low solid fractions. This is further illustrated in Fig. 2, showing that for the highly alloyed material the choice of solidification equation in the α-precipitating phase field is unimportant, while for the low alloyed material the "Scheil" and "Lever rule" equation give quite different microstructures.

Let us assume that we know the correct solidification path (for a ternary system), in the two phase region Liquid (L) + α, i.e. along the line ab in Fig. 3. Further solidification is now restricted to take place along the α - β phase boundary, in direction of decreasing temperature. Solidification along the line bo must also be calculated, as equilibrium conditions cannot be assumed to be fulfilled. The crystallization path is now fixed (along line bo), and the interesting parameter to calculate is the rate of crystallization along the phase boundary, i.e. how the fraction of solid f_s alters as solidification takes place along bo. At this stage it is very δ important to include back diffusion into the solid in the calculations. It can be shown[3] that

(5) $\qquad \dfrac{\text{Amount of back diffusion}}{\text{Amount rejected into liquid}} = \dfrac{k}{(1 - f_s)} \cdot \alpha'$

k = partition coefficient, f_s = fraction of solid, α' = back diffusion parameter (eq. 3)

When $f_s \to 1$ the amount that is back diffusing dominates, with the effect that complete solidification takes place earlier than if back diffusion had been neglected. This will be illustrated in the next section.

Let us assume that a "realistic" calculation (taking back diffusion into account) allow some liquid (L) to reach point o in Fig. 3, and that o is peritectic in nature. Then, if equilibrium conditions are fulfilled, the reaction

(6) $L + \beta \to \lambda$ (+ rest L)

takes place in o (and eventually along op).

Again, one has to look for the event that is most likely to take place during solidification at a given rate. Probably, the peritectic reaction indicated in eq. (6) will not have sufficient time to be completed (or even started), and along op the dominating reaction is

(7) $L \to \lambda + \alpha$ (+ rest L),

and the β-phase already precipitated along bo remains in the structure.

Calculating mass balance in eq. (7), back diffusion should again be included through eq. (5). Final solidification may take place between o and p, or at p (assuming p is a eutectic point).

So far, non-equilibrium conditions have been included in the solidification path calculation, but the phase diagram itself has been regarded as an equilibrium diagram. A non-equilibrium phase diagram, taking into account that phase formation may be suppressed, has to be considered as well, in order to provide a realistic phase distribution calculation. The suppression of phases following a fast cooling always expands the α-field (the primary solid phase field), and additionally new metastable phases may occur. It is difficult to calculate a non-equilibrium phase diagram, as the necessary thermodynamic details will normally not be available, especially not for the metastable phases. One way to overcome this problem is to experimentally determine different phase distributions from samples solidified at known conditions, and then look for the phase diagrams needed in order to calculate the observed phase distributions.

To summarize, calculation of a phase distribution in a solidified microstructure involves the following steps:

1. The calculation of a solidification path in the two phase region L (liquid) + α (primary solid phase). If a major part of the solidification is precipition of α from L, a non-equilibrium calculation taking solid back diffusion into account should be used (eq. 1 and 3).

2. A metastable phase diagram valid for the cooling rate in question will probably be needed. Such a phase diagram will not normally be available, and must be constructed from experiments and "intelligent" guessing, or alternatively, from thermodynamical calculations.

3. The solidification path in the three phase region L (Liquid) + α (primary solid phase) and β (secondary solid phase) is restricted to follow a fixed line in the ternary (metastable) phase-diagram. In order to determine the position on the liquidus surface corresponding to complete solidification, the mass balance calculation have to take into consideration:

(i) Back diffusion into solid α (and solid β, if β is a solution phase).

(ii) That equilibrium reactions fail to take place; normally, that late solidifying phases are precipitated directly from liquid instead of being formed peritectically.

The next section provides an example of calculating ternary phase distributions after solidification at different cooling rates, including all the steps mentioned above.

3. CALCULATING THE PHASE DISTRIBUTION AFTER DC-CASTING
OF AlFeSi-ALLOYS

When large aluminium slabs of commercial purity (i.e. AlFeSi-alloys) are D.C. cast, solidification is faster near the surface of the slab than in the centre, causing different phases to be formed in the different parts of the slab. In Table 1, examples are given for experimentally determined phase distributions (based on TEM-investigations using energy dispersive spectrometry and selected area diffraction on intermetallic particles extracted from the matrix[4]).

To calculate the phase distributions given in Table 1, we first need a solidification path calculation in the two phase region L + α-Al. Such a calculation in shown in Fig.4 for the alloy 0.24wt%Fe, 0.24wt%Si. Input for the calculations are given in the appendix. In Fig.4 the phase boundaries, as suggested by Phillips[5], are also indicated. It is observed that regardless of the solidification model being used the phase boundary between α-Al and the Al₃Fe phase is reached, but at slightly different positions for the three cases the "Lever rule", Scheil and Clyne and Kurz. From Table 1 it evident that α-AlFeSi is dominating in the structure, at least for high cooling rates. This observation could be explained in two ways, of which only one is believed to be correct:

1. The phase boundaries, as suggested by Phillips, are displaced.

2. The phase diagram is valid, and the melt follows the phase boundary between α-Al and Al₃Fe until the ternary point o is reached. At this point α-AlFeSi is formed peritectically from Al₃Fe + Liquid.

Regarding (1) it should be noticed that although Phillips is the phase diagram most frequently used, other experiments have failed to reproduce Phillips' diagram[6]. Additionally, and probably more importantly, the high cooling rate suppresses the formation of the intermetallic phases and thereby causes displacement of the phase boundary.

Regarding (2), it is evident that a complete peritectic reaction must take place if the phase diagram in Fig. 4 is valid and the only intermetallic phase observed in the structure is the α-AlFeSi. It is, however, very unlikely that a peritectic reaction should go to completion in o, since the time available for this reaction in D.C. casting would be very short, probably less than 1 s and definitely less than 10 s (10 s equals the total solidification time at positions near the slab surface). The fact that the incomplete peritectic reaction, i.e a coating of α-AlFeSi and a core of Al₃Fe, is never observed in the structure, supports the view that peritectic reactions do not take place. It can be questioned[7] whether peritectic reactions happen at all. When the α-AlFeSi phase is observed, it is then concluded that the phase is being precipitated directly from the liquid.

If only α-AlFeSi particles are observed, as is the case for the 0.24Fe, 0.24Si-alloy at the highest cooling rate, the reasonable explanation is that the solidification path in Fig. 4 reaches the boundary between α-Al and α-AlFeSi, and not the boundary between α-Al and Al₃Fe. In other words, the phase boundaries, as suggested by Phillips, must be displaced, and the α-AlFeSi phase field must be shifted to the left.

In Fig. 4, it is seen that the solidification path is not significantly different for the three different solidification equations. Continuing the calculation along the phase boundaries shows, however, that final solidification takes place at quite different positions assuming different levels of back diffusion, Fig. 5. A local solidification time of 0.1 s corresponds nearly to Scheil, and the final solidification will be eutectic. A local solidification time of 100 s corresponds to a typical industrial case, and the result is that solidification ends at a position on the phase boundary between α-Al and β-AlFeSi. A local solidification time of 10 000s approaches the Lever rule, and solidification terminates earlier than for the two other cases.

When calculating the curves in Fig.5, it is assumed that no peritectic reaction takes place, i.e an intermetallic phase is always precipitating directly from the liquid. Further, growth of an intermetallic phase beyond its peritectic point is allowed. This is illustrated in Fig.6. In the upper diagram, α-AlFeSi particles, that have been formed along ao, continue to grow in the β-AlFeSi phase field. In the lower diagram, however, α-AlFeSi is not formed at all, and only β-AlFeSi precipitates in point b.

Calculations have been carried out for all the alloys in Table 1, as described above. It is quite obvious that the observed phase distributions listed in Table 1 cannot be calculated assuming that Phillips' phase diagram is valid, and, at the same time, no peritectic reactions take place. It is necessary to allow phase field displacements based on an expansion of the α-Al. region, and to introduce metastable phase fields. By repeating calculations for different metastable phase diagrams, always comparing with the observed phase distributions in Table 1, a set of phase diagrams have been suggested, as a continuous function of the cooling rate. The resulting diagrams for cooling rates 0.5°C/s, 3°C/s and 10°C/s are shown in Fig.7. Comparing with Phillips (Fig.4), it is observed that the triple points are shifted towards higher Fe-values and lower Si-values. Additionally, the Al₆Fe phase field appears at 3°C/s. At 10°C/s both the Al₆Fe and the Al_mFe phase fields are observed, while the Al₃Fe phase has vanished.

Calculated phase distributions are shown in Table 2. When comparing with the observed distributions in Table 1, the calculated results for a cooling rate of 10°C/s and local solidification time of 5 s should be compared with the position 20 mm from slab surface, as cooling rate is observed to be highest here (due to secondary water cooling). For the last two alloys 5°C/s cooling rate and 10 s solidification time is used, as these alloys were cast at lower casting speed. The 1°C/s cooling rate and 100 s local solidification time is representative for the centre of the slab (300 mm from slab surface). Complete agreement is not achieved, but one should bear in mind that the cooling rates (and local solidification times) are not exactly known for the experimental observations. The criterion for growth of a phase beyond its peritectic point (Fig.6) is at the moment only based on the cooling rate, but should probably also be a function of the fraction of the phase, as the possibility for continued growth when crossing the triple point is likely to be dependent of the amount of the phase.

4. CONCLUSIONS

Observed phase distributions in DC-cast AlFeSi-alloys have been used to determine metastable phase diagrams. The process used involves the following steps:

1. A phase diagram is suggested.

2. A solidification path in the two phase region α-Al + Liquid in the suggested diagram is calculated. The calculation is based upon the model of Clyne and Kurz, allowing back diffusion into the solid at the solidification front.

3. Solidification along the phase boundaries in the suggested phase diagram is calculated. The calculation is still based upon back diffusion into solid aluminium, but aditionally the mass balance equations include the precipitation of a stoichiometric intermetallic phase.

94

4. When a peritectic point is reached, the peritectic reaction is assumed to be completely suppressed. Precipitation of the new phase (directly from liquid) is, however, not believed to happen exactly at the peritectic point. Instead, the old phase is believed to grow a certain distance beyond the peritectic point. A value for this "additional" undercooling is suggested.

5. The calculated phase distribution is compared with experimental observations. If the calculated and experimental distributions differ significantly, a new phase diagram and/or a new "additional" undercooling (described in 4)) is suggested.

6. Steps 1 - 5 are repeated until reasonable agreement between measured and calculated phase distributions are achieved for all compositions and cooling rates available.

ACKNOWLEDGEMENTS

Based upon the work a computer model is developed. This program calculates the phase distributions, solid solution profiles and solidification intervals in DC-cast 1000-series alloys. Hydro Aluminum, Elkem Aluminum and the Royal Norwegian Council for Scientific and Industrial Research have financed the work, and the model is being used in the two aluminium companies.

REFERENCES

1. H.D. Brody and M.C. Flemings: Trans. of the Metallurgical Society of AIME, vol. 236 (May 1966) p. 615.
2. T.W. Clyne and W. Kurz: Met.Trans., vol. 12A (1981) p. 965.
3. W. Kurz and D.J. Fischer: "Fundamentals of solidification" . Trans.Tech.Publ. (1986).
4. C.J. Simensen et al.: Fresenius Z. Anal. Chem., 319 (1984) p. 286.
5. The Aluminium Development Association by H.W.L. Phillips: "Equilibrium Phase Diagrams of Aluminium Alloy Systems" (1961).
6. V.G. Rivlin and G.V. Raynor: International Metals Reviews (1981) no. 3 p. 133.
7. D.H. St John: Acta Metall Mater. vol 38, no. 4 (1990) p. 631.

APPENDIX

Input data for the calculations presented in Table 2, and in Figs. 1, 4 and 5 are given in the appendix

(I) Partition coefficients (k-values):

$k_{Fe} = 0.027$
$k_{Si} = 0.12$
$k_{Mg} = 0.28$-0.32 (function of liquid composition)
$k_{Mn} = 0.7$-0.9 (function of liquid composition)

(II) Diffusion coefficients:

$D_{Fe} = 5.3 \times 10^{-3} \exp(-183.4 \text{ kJ Mol}^{-1}/RT)$ [m²/s]
$D_{Si} = 2.0 \times 10^{-4} \exp(-133.5 \text{ kJ Mol}^{-1}/RT)$ [m²/s]
$D_{Mg} = 3.7 \times 10^{-5} \exp(-123.5 \text{ kJ Mol}^{-1}/RT)$ [m²/s]
$D_{Mn} = 3.5 \times 10^{-2} \exp(-220 \text{ kJ Mol}^{-1}/RT)$ [m²/s]

(III) Size of the solidifying system (eq. 4):

$L = DAS/2$, DAS = Secondary dendrite arm spacing
$DAS = 15 t_f^{0.33}$, t_f = Local solidification time.

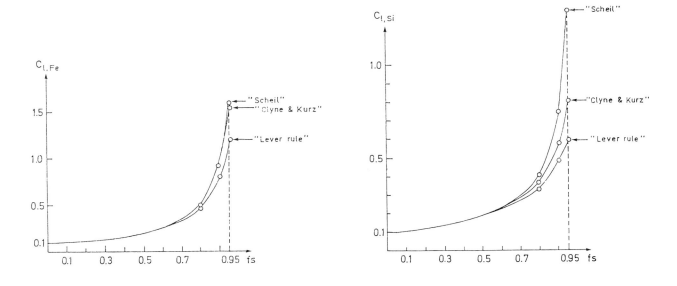

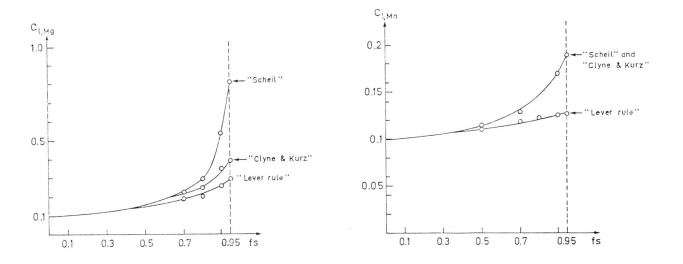

Fig. 1. Concentration of Fe, Si, Mg and Mn in liquid aluminium as a function of fraction solidified aluminium, calculated on basis of the Scheil, Clyne & Kurz and Lever rule equations. Input data to the calculations are given in the appendix.

96

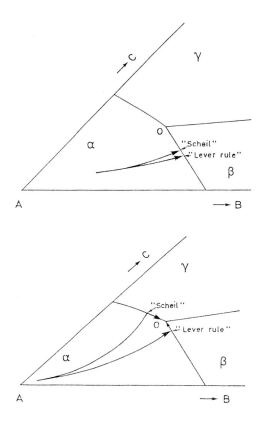

Fig. 2. Hypothetic ternary phase diagram illustrating that the choice of solidification equation is not critical for the high concentration alloy (upper diagram), while for a low concentration alloy (lower diagram) the Scheil and Lever rule calculation provide quite different microstructures.

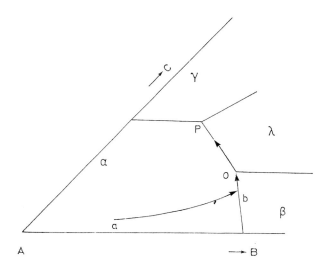

Fig. 3. Hypothetic phase diagram with a peritectic point o, and a eutectic point, p. End of solidification can be anywhere along b-o-p, and must be calculated.

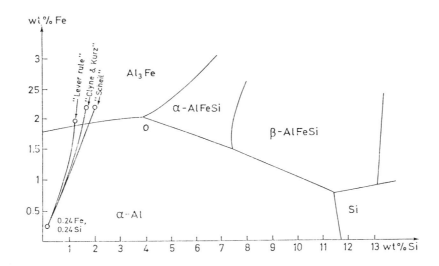

Fig. 4. Solidification path calculations for an 0.24wt% Fe, 0.24wt% Si alloy. The phase diagram is the one suggested by Phillips /6/.

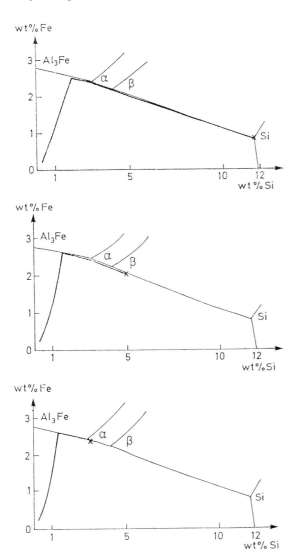

Fig. 5. Calculation of the point of final solidification, as a function of solidification time, using a fixed phase diagram. Solidification time is 0.1 sec in the upper diagram, 100 sec in the middle diagram and 10000 sec in the lower diagram.

98

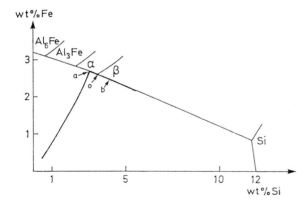

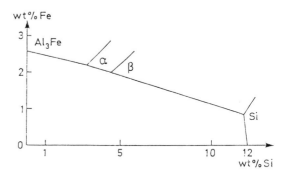

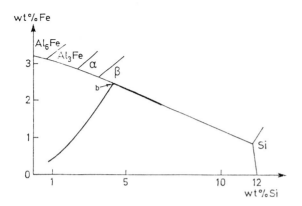

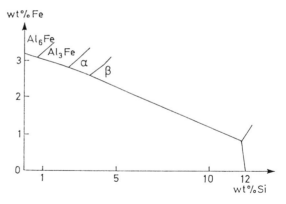

Fig. 6. Illustration of growth of a phase beyond its triple point. In the upper diagram α-AlFeSi may still grow for a liquid concentration corresponding to point b, while in the lower diagram only β-AlFeSi will be formed in point b.

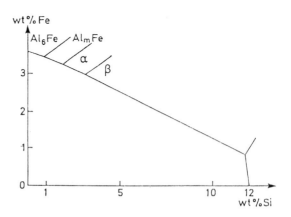

Fig. 7. Metastable phase diagrams suggested for cooling rates 0.5°C/sec (upper diagram), 3°C/sec (middle diagram) and 10°C/sec (lower diagram).

Table 1 Observed phase distributions in D. C. cast AlFeSi-alloys

Alloy	Distance from slab surf. (mm)	wt% Al₆Fe/Al_xFe	wt% Al_mFe	wt% Al₃Fe	wt% α-types	wt% β	wt% Si	Assumed Fe in solid solution	Calculated Si in solid solution
0.32 Fe, 0.05 Si	20	0.91	0.02	0.13	-	-	-	0.03	0.049
	100	0.23	0.04	0.59	-	-	-	0.03	0.048
	260	0.63	0.02	0.32	-	-	-	0.03	0.049
0.27 Fe, 0.09 Si	20	0.58	0.18	0.10	Trace	-	-	0.03	0.083
	100	-	0.28	0.40	-	-	-	0.03	0.077
0.25 Fe, 0.13 Si	20	-	0.25	-	0.45	-	-	0.03	0.102
	100	0.12	0.19	0.33	0.02	-	-	0.03	0.113
0.24 Fe, 0.24 Si	20	-	-	-	0.69	-	Trace	0.03	0.19
	100	-	-	0.22	0.43	-	Trace	0.03	0.20
0.24 Fe, 0.48 Si	20	-	-	-	-	0.77	0.01	0.03	0.35
	100	-	-	-	-	0.77	0.01	0.03	0.35
0.54 Fe, 0.15 Si	20	1.38	-	0.39	-	-	-	0.04	0.123
	100	0.20	-	1.21	-	-	-	0.04	0.124
	300	0.37	-	1.11	-	-	-	0.04	0.124
0.74 Fe, 0.76 Si	20	-	-	-	0.37	2.13	0.05	0.04	0.41
	100	-	-	-	1.86	0.44	0.05	0.04	0.51
	300	-	-	0.72	0.73	0.73	0.05	0.04	0.54

Table 2 Calculated phase distributions in solidified AlFeSi-alloys

Alloy	Cooling rate (°C/s) Local sol. time (s) in paranthesis	wt% Al$_6$Fe/Al$_x$Fe	wt% Al$_m$Fe	wt% Al$_3$Fe	wt% α-types	wt% β	wt% Si	Calculated Fe in solid solution	Calculated Si in solid solution
0.32 Fe,	10 (5)	1.01	0.08	-	-	-	-	0.03	0.05
0.05 Si	1 (100)	-	-	0.78	-	-	-	0.03	0.05
0.27 Fe,	10 (5)	0.54	0.22	-	0.09	-	-	0.02	0.08
0.09 Si	1 (100)	-	-	0.66	-	-	-	0.02	0.09
0.25 Fe,	10 (5)	-	0.46	-	0.23	-	-	0.02	0.10
0.13 Si	1 (100)	-	-	0.63	0.02	-	-	0.02	0.12
0.24 Fe,	10 (5)	-	-	-	0.58	0.14	-	0.02	0.18
0.24 Si	1 (100)	-	-	0.45	0.14	0.03		0.02	0.22
0.24 Fe,	10 (5)	-	-	-	-	0.82	0.06	0.02	0.31
0.48 Si	1 (100)	-	-	0.04	0.35	0.35	-	0.02	0.40
0.54 Fe,	5 (10)	1.05	0.09	0.31	0.25	-	-	0.04	0.12
0.15 Si	1 (100)	-	-	1.33	0.02	-	-	0.04	0.14
0.74 Fe,	5 (10)	-	-	-	1.28	1.12	0.14	0.04	0.38
0.76 Si	1 (100)	-	-	0.78	0.66	0.76	0.13	0.04	0.47

11

PHASE DIAGRAMS FOR THE SOLIDIFICATION OF METASTABLE PHASES IN THE Al-Fe-Si SYSTEM

J.E. TIBBALLS

Senter for Industriforskning, P.B. 124, Blindern, N-0314 Oslo 3, Norway

ABSTRACT

Solidification during the continuous or semi-continuous casting of aluminium occurs at cooling rates which cause the precipitation of phases and phase variants that do not appear on the equilibrium phase diagram. The provision of phase diagram information that can be used in these circumstances has long been required. Techniques exist for estimating the extent of enhanced solid solubility of the aluminium matrix using Gibbs energies estimated from thermochemical data and equilibrium phase diagrams. It is the purpose of this presentation to show how Gibbs energies of metastable phases and metastable variants of stable phases can be estimated and employed in solidification calculations.

In order to specify the phases which precipitate from the solidifying aluminium one needs to know their Gibbs energies for the composition and at the temperature at which they form. A model for solute redistribution in the solidifying aluminium can then be invoked together with the assumption of local (metastable) equilibrium to determine the precipitating phases.

Several techniques exist for estimating the Gibbs energy of a metastable phase. If the phase is a crystallographic variant of a stable phase, knowledge of the crystallographic differences can be employed to estimate the relative Gibbs energies. In the Al-Fe-Si system, the cubic α-AlFeSi and Al_6Fe phases are the metastable end-members of the α-Al(Fe,Mn)Si and Al_6(Fe,Mn) solid solutions, respectively. In addition to the Fe-Mn and Al-Si substitutions, vacancies occur on the transition metal sites. The variant alpha phases in the Al-Fe-Si system represent ordered superstructures of the cubic phase. Their Gibbs energies have been estimated from equilibria between a series of alpha Al(Fe,Mn)Si compositions and aluminium melts on the basis of changes in configurational entropy and enthalpy.

In this case of Al_6Fe, knowledge of the metastable eutectic temperature, confirmed by studies of the eutectic composition provide a sound basis for estimating the Gibbs energy at this temperature. By separating metastable Al_6Fe from the aluminium solid solution it is possible to study the thermal decomposition of the phase and thereby estimate its Gibbs energy at the temperature of decomposition.

Systematic studies of the phases precipitating from solidifying solutions provide a further means of estimating the Gibbs energies of the metastable phases. Metastable liquidus boundaries represent local equilibria between the respective phases and the liquid, the temperature of which can be estimated be extrapolation from the equilibrium liquidus using CALPHAD techniques.

With the Gibbs energies of the possible participating phases, the problem of describing the nucleation of solid phases reduces to estimating a single parameter representing the interface energy for the nucleating phase.

12

THE CONSTITUTION OF Ti-RICH ALLOYS OF THE Ti-V-Fe-Al SYSTEM

A. Nwobu, T. Maeda, H. M. Flower and D. R. F. West

Department of Materials, Imperial College of Science, Technology and Medicine, London SW7 2BP, UK

ABSTRACT

An investigation has been made of the solid state constitution of titanium-rich Ti-V-Al and Ti-Fe-Al alloys. Data are presented in the form of isothermal sections at 800 and 700°C for the titanium rich regions of the phase diagrams, together with data on transus temperatures. Partitioning of V, Fe and Al between the α and β phases in a Ti-10V-2Fe-3Al alloy has also been investigated for correlation with the ternary alloy constitution. The relevance of the ternary alloy data to alloy development and to the control of microstructure and properties through processing and thermal treatments is discussed .

INTRODUCTION

Aluminium (as an α stabiliser) and vanadium (as a β stabiliser) are two of the most important alloy additions in industrial titanium alloys, while iron is also an important stabiliser. In combination, aluminium and vanadium form the basis of a range of medium strength $\alpha + \beta$ alloys, notably Ti-6Al-4V (wt %), which is the most widely used of all the titanium alloys. Aluminium and iron in combination are used in the Ti-5Al-2.5Fe alloy developed for biomedical applications as implants.

In designing high strength titanium alloys of the near-β or β types, vanadium as a β stabiliser is commonly added in substantial amounts; often another β stabilising element such as iron is used to increase the strengthening. Aluminium is always present to reduce the formation of omega phase, and to strengthen the α phase. Ti-10V-2Fe-3Al and Ti-8V-5Fe-1Al are examples of the near-β and β types respectively. Alloy development and design, including the field of titanium based composites, requires accurate data of phase equilibria in the form of phase diagrams. There is an extensive body of literature on the constitution of titanium alloys, and the binary phase diagrams, including those relevant to the work represented here, namely Ti-Al, Ti-V and Ti-Fe, have recently been critically assessed[1,2]. Phase diagram information is also available for a wide range of ternary Ti-rich alloys; this includes a substantial amount of data on the Ti-V-Al system[e.g. 3-5] but less on the Ti-Fe-Al system[6]. There appears to be no previously published phase diagram for the Ti-V-Fe-Al system.

Study of the literature on Ti-alloy systems shows that it is not uncommon to find significant differences between the results of various investigations. A particularly striking example is provided by the Ti-Al system[1], including those regions involving the ordered intermetallic compound Ti$_3$Al; since the publication of the critical assessment[1] further modifications to the liquidus/solidus regions have been proposed[7]. Differences may arise from factors such as oxygen contents of the alloys, failure to attain equilibrium during annealing, particularly at relatively low temperatures, and lack of accurate compositional data on small phase particles. In relatively recent years, the availability of phase compositional analysis in thin foil samples (e.g. by STEM-EDX) has been an important factor in allowing greater accuracy of phase diagram determination. Hitherto, the phase diagrams for Ti-V-Al and Ti-Fe-Al have not been investigated using thin foil analytical techniques. The investigation reported here forms part of a wider programme to determine the solid state constitution of the Ti rich portions of the Ti-Al-X and Ti-Al-X-Y systems, where X and Y include a range of α and β stabilising elements for temperatures ranging from ~1200 to 600°C; in the present paper isothermal sections are reported for temperatures of 800 and 700°C in the ternary Ti-V-Al and Ti-Fe-Al systems, The experimental approach is primarily to establish tie-lines (particularly for α/β mixtures) and tie triangles by phase compositional analysis on equilibrated samples.

Also, in order to correlate the ternary equilibria with the quaternary Ti-V-Fe-Al system, partitioning of V, Fe and Al between the α and β phases has been investigated in a Ti-10-2-3 alloy, equilibrated at temperatures between 800 and 650°C. The relevance of the ternary and quaternary data to alloy processing and development is also discussed.

EXPERIMENTAL PROCEDURE

The alloys were made as arc-melted ingots (~30g in weight). Initially, titanium sponge (~99.5% purity) was melted into ingots whose hardness (95-115 Hv) showed a low level of oxygen contamination by the melting process. This titanium together with aluminium (99.99% purity), vanadium (99.7% purity), and iron (99.9 + % purity) were used for alloy making. Ingots were re-melted several times to improve homogeneity; weight losses after melting were < ~ 0.4%, indicating good agreement between nominal and actual compositions. Ti-V-Al alloys containing 6-23 wt%V and 0-14 wt%Al, and Ti-Fe-Al alloys containing 3-5.3 wt%Fe and 2-6 wt%Al were used in the study. The typical oxygen level in the alloys is about 0.12-0.15 wt%.

Homogenization treatments were given to samples wrapped in molybdenum foil and sealed in partially evacuated silica tubes containing one third of an atmosphere of argon.

The Ti-V-Al alloys containing < 8 wt% Al were hot rolled to 1 mm thick prior to homogenizing for 30 min at 1000°C; Ti-V-Al alloys with the higher aluminium contents were too hard to roll, and were homogenized for 30 mins at ~1180°C. On completion of the homogenisation treatment samples were water quenched. For the Ti-Fe-Al alloys, homogenization was applied to the ingots for 24 h at 1000°C followed by water quenching. The homogenized ingots were then warm rolled (600-700°C) into sheet ~1 mm thick.

Annealing for the determination of isothermal sections was carried out with samples sealed, as for homogenization. The annealing times were for 800°C 4 weeks for Ti-V-Al, 10 days for Ti-Fe-Al alloys, 700°C 5-6 weeks for Ti-V-Al alloys, 30 days for Ti-Fe-Al alloys. An experiment on one of the Ti-Fe-Al alloys showed there was good consistency between the phase compositions after annealing for 21 and 31 days respectively at 800°C, indicating that local equilibrium had been attained. In the investigation of the Ti-V-Al alloys, an alternative approach to equilibrium was used in addition to that involving reheating after quenching from the homogenisation treatment: samples were step cooled from 1180°C to 1100°C and lower temperatures at 100°C intervals, being held for 1 day at each temperature followed by water quenching. Experiments were also made on a commercial Ti-10-2-3 alloy (of analysed composition Ti-9.8V-1.8Fe-3Al) to determine the compositions of α and β phases after equilibration at 800, 750, 700 and 650°C for periods of 4, 5, 6 and 7 weeks respectively.

β transus temperatures were measured by annealing samples in the $\alpha + \beta$ phase region and then reheating to various temperatures close to the predicted transus at temperature intervals of 5°C. Samples were quenched and examined metallographically to detect the disappearance of α phase.

Specimens were examined by light, scanning (SEM) and transmission (TEM) electron microscopy. Volume fractions of α and β phases were determined from SEM observations either by using an image analyser or from a linear intercept method. Quantitative compositional microanalysis of bulk samples and of phase particles was carried out using a JEOL JSM 35CF scanning electron microscope with an attached X ray EDS system. The α and β grains used for analysis were at least 10 μm in size. The ZAF correction procedure of the analytical system was checked by comparison with the wet chemically analysed compositions of "standard" alloys and in some cases minor changes were made to the procedure to achieve the highest possible accuracy. STEM-EDX analysis was also carried out using JEOL TEMSCAN 120 CX or JEM 2000 FX equipment. X ray diffractometry using CuK$_\alpha$ radiation was carried out on bulk samples for phase identification.

RESULTS

(i) Ti-V-Al and Ti-Fe-Al Isothermal Sections

Figures 1 and 2 show the isothermal sections of parts of the Ti-V-Al and Ti-Fe-Al systems at 800°C and 700°C; the relevant boundaries for the binary systems Ti-Al, Ti-Fe, are taken from the recently assessed phase diagrams[1], and for Ti-V the β isomorphous diagram shown in [2], since work by one of the authors[9] indicates that this, rather than the monotectoid form shown in [1] is correct.

The data for the Ti-V-Al system (Fig. 1), includes the $\alpha + \alpha_2 + \beta$ phase region; the limits of this region were accurately determined from the consistency of data between the alloys lying within the region, in terms of phase compositions and volume fractions. In the two phase $\alpha + \beta$ region, the tie-lines show the partitioning of Al to the α phase. There is good consistency with the phase boundaries of the Ti-V system. The main difference between the 800 and 700°C data is the change in composition of the β phase, co-existing with α and α_2, to considerably higher V content and slightly lower Al content; thus the three phase region increased in extent.

Figure 2 shows the isothermal section data for the Ti-Fe-Al system at 800 and 700°C representing the α, β equilibria. The tie-line data for the ternary alloys show reasonable agreement with the $\alpha/\alpha + \beta$ boundary in the Ti-Fe binary system. However, the solubility limit of Fe in α in the ternary alloys (~0.3 wt%) is greater than the very small value of ~ 0.05 wt% reported for the binary system. There is preferential partitioning of Al to α.

(ii) Solute Partitioning in the Ti-10-2-3 Alloy

Figure 3 shows the solute partitioning between α and β in the range 800-600°C. As expected, there is preferential partitioning of the β-stabilisers V and Fe to the β-phase; only V shows a marked temperature dependence of partitioning ratio with decreasing temperature. The partitioning ratio X_α/X_β of V or Fe in the quaternary alloy is higher than that observed in the ternary alloys containing a similar level of Al. For example, at 700°C, it is 3/14 for the V in the Ti-10-2-3 alloy compared to the value of about 3/18 for the metastable β Ti-V-3Al alloys with V between 11-18 wt%. For Fe in Ti-10-2-3 alloy it is about 0.1/3 compared to about 0.1/14 for a Ti-Fe-3Al based alloy with Fe between 3 and 5 wt%.

(iii) β - Transus

Figures 4 and 5 show partial isopleths presenting β transus and α solubility data for selected ternary alloys together with the data for the binary systems. The data for Ti-V-Al alloys of corresponding Al and V contents were obtained by interpolation from isothermal section data at 100°C intervals from 900-600°C[9]. The transus temperatures for the binary Ti-Al compositions of 2 and 4 wt% Al are taken from the assessed phase diagram[1]. Some of the Ti-Fe-Al data were experimentally determined for the alloys containing 2 and 4 wt% of Al respectively with Fe contents of 3-5 wt%. The main feature is the influence of Al in raising the β transus.

(iv) Microstructure and Hardness Values

In the Ti-V-Al alloys, furnace cooling from the solution treatment temperature produces relatively coarse structures which proved to be convenient for phase compositional analysis involving α and α_2. The Ti-6V-10Al alloy, which X-ray diffraction shows contained $\alpha + \alpha_2 + \beta$ after annealing at 800°C provides an illustration of this. Figure 6(a,b) show the coarse structure after furnace cooling to 800°C, consisting of β (transformed on water quenching from 800°C) plus α/α_2. Comparing the furnace cooled condition (Fig.6a,b) with that produced by quenching from 1180°C and then annealing at 800°C (Fig. 6c), the latter shows a retained martensitic morphology resulting from tempering of the martensite (produced by the initial quenching). This occurs rather than reversion of martensite to β followed by precipitation of α/α_2 since the As temperature is > 800°C.

The Ti-Fe-Al alloys which were warm worked prior to annealing showed globular/equiaxed α + β microstructures (Fig. 7a) on annealing at temperatures below the β-transus. The as-received Ti-10-2-3 alloy, which had been hot-worked in the α + β field, produced similar α + β microstructures ie globular/equiaxed α on annealing between 700 and 800°C e.g. Fig. 7(b).

Table 1 shows the structures and hardness values of the β quenched Ti-10-2-3 and some of the ternary Ti-V-Al and Ti-Fe-Al alloys. The β solute contents, 11-23 wt%V and 3-5 wt%Fe, of the ternary alloys are typical of those in titanium alloys in which the embrittling ω-phase is formed; the Al contents of up to 7 wt% are typical in commercial alloys to suppress the ω phase. Table 1 shows that the addition of 2-3 wt% of Al is required to completely suppress athermal ω phase in the ternary Ti-V-Al alloys. On the other hand, a higher amount, about 4 wt%, of Al is needed to achieve similar results in the Ti-Fe-Al alloys. Table 1 also shows that the β quenched Ti-10-2-3 alloy contains athermal ω-phase.

The hardness values of the β quenched alloys indicate a minimum value at about the minimum Al composition required to completely suppress the athermal ω-phase. Higher Al solute additions tended to increase the hardness even though they encouraged the retention of more β phase with associated suppression of α" martensite (Table 1). Generally the β quenched Ti-Fe-Al alloys show higher levels of hardness as compared to the β quenched Ti-V-Fe alloys within this composition range, 11-23 wt%V, 3-5 wt%Fe and 0-7 wt% Al.

DISCUSSION

(i) Comparison of data with previous work

Previous work on the Ti-V-Al system includes investigations by Kornilov and Volkova[3] Tsujimoto[4] and Hashimoto et al.[5]. There is general agreement as to the nature of the phase equilibria between these three investigations and the present work. However, there are significant differences relating to the phase field boundaries, particularly with respect to the vanadium content of β phase; the data of [5] show much higher V values. The compositions of α and α_2 respectively show relatively small differences between the various investigations. Previous investigations used optical microscopy[4], optical microscopy, X-ray diffraction and thermal analysis[3], scanning electron microscopy, including EDX, and X-ray diffraction[5]. A distinguishing feature of the present work is the phase compositional analysis on thin foils.

Considering the 19V 7Al and 23V 7Al alloys to illustrate specific examples of differences between investigators, the present work, using X-ray diffraction, showed $\alpha_2 + \beta$ at 700°C; this contrasts with the location of such compositions in the α + β region according to the work of Kornilov and Volkova[3] which employed shorter annealing times, and which did not use phase compositional analysis. The data from the thin foil analyses in the present work together with the X ray data obtained give strong support to the composition being much lower in V than that reported by Kornilov and Volkova and more consistent with the data of Tsujimoto[4]. The data of

Hashimoto[5], using SEM EDX shows a smaller $\alpha + \alpha_2 + \beta$ field than has been found by the present investigation. This difference can be attributed to the limitations of SEM as compared with TEM when analysing fine scale microstructures, including $\alpha + \alpha_2 + \beta$ mixtures; with SEM the interaction volume will incorporate contributions from more than one phase, leading to an apparently smaller three-phase region.

(ii) Application of Phase Diagram Data to Alloy Development and Processing

Titanium alloy metallurgy provides good examples of the importance of phase diagram data in relation to alloy development involving control of chemical composition and processing, including mechanical and thermal treatments. The $\beta = \alpha$ transformation and the equilibria centred on the α - β relationships are critical features. The isopleth shown in Fig. 8 represents schematically these relationships for multicomponent titanium alloys, covering the range from α to β alloys[10].

Depending on cooling rate and composition β transforms to various microstructures by diffusion and/or shear mechanisms. Solution treatments and mechanical working are commonly carried out in the β field e.g. for near α alloys. However, important benefits in controlling structure and properties are derived in some alloys e.g. Ti-6Al-4V from treatments in the $\alpha + \beta$ field and this treatment is now also being applied to near α alloys. Selection of alloy composition and temperature provides control of phase compositions and proportions, while the phase morphology dominantly evolves from the complete cycle of mechanical and thermal treatments. For example, after $\alpha + \beta$ working to produce a duplex equiaxed structure, rapid cooling leads to martensite formation or retention of metastable β (possibly with some ω phase) depending on whether or not the β composition exceeds that for which M_s is at room temperature; slower cooling allows the β to transform diffusionally to $\alpha + \beta$ mixtures. After rapid cooling alloys such as Ti-6Al-4V or Ti-10V-2Fe-3Al, martensite or retained β can be aged to form a fine scale dispersion of α with consequent strengthening.

The present data for the Ti-V-Al and Ti-Fe-Al equilibrium diagrams, together with knowledge of microstructures produced by phase transformations can be applied to consider possibilities for optimising alloy compositions and treatments. A key factor is the greater β stabilising effect of Fe as compared with V in terms of equilibrium relationships (Table 1). Thus, the total β-solute content of an $\alpha + \beta$ or a metastable β Ti-V-Fe-Al alloy can be reduced by using more of the Fe than V as the β stabiliser. This could be beneficial to the hot processing of the metastable β alloy where the flow stress decreases with the decrease in the amount of solute[7]. Since Fe reduces M_s more effectively than V[10] the β phase is retained at lower solute contents with Fe as compared with V when β phases are quenched. For example, only α'' is present in the β-quenched Ti-15V-2Al alloy compared with the β-quenched Ti-3.4Fe-2Al alloy which contains retained β in addition to some α''-martensite, and from the point of view of mechanical properties of age-hardened $\alpha + \beta$ and metastable β alloys retained β-phase is considered preferable to α''-martensite[13].

Research by the authors to be reported elsewhere[14] shows that in the quaternary system the presence of both V and Fe produces a synergistic effect which increases β stabilisation. Literature data present equivalent values for the β stabilising effects of V and Fe based on Ms depression showing Fe to be more effective than V[13]. The authors' work[14] has shown that the equivalent values do not fully represent the effects of these elements in combination and further research is needed to establish more accurate equilibrium β stabilising equivalents for alloy development.

A further feature relevant to control of structure by $\alpha + \beta$ working is the β approach curve. Precise control of β content is aided when the proportion of β does not decrease rapidly with the fall in working temperature in the $\alpha + \beta$ range. Figure 9 shows the β approach curves for the alloys Ti-15V, Ti-15V-3Al, Ti-4.5Fe, Ti-4.5Fe-4Al. Also included is that of the Ti-10-2-3 alloy which could be regarded as comparable to the Ti-15V-3Al with some of the V replaced by Fe. These curves were plotted using either binary/ternary phase diagram data[1] and Figs. 1 and 2 or metallographic measurements of phase fractions. They show that the steepness of the β-approach curve of the binary Ti-V and Ti-Fe alloys is beneficially reduced by the addition of Al. Further reduction in the slope of the β-approach curve is observed in Ti-10-2-3 alloy compared to the ternary alloys.

An important practical objective in relation to a metastable β alloy such as Ti-10-2-3 is to optimise the chemical composition towards achieving high strength, good ductility, fatigue strength and fracture toughness; to achieve this, it is essential that the amounts of the α stabiliser Al and β stabilisers, V and Fe, present should be such as to reduce the amount and stability of ω-phase while strengthening the α and β phases of the $\alpha + \beta$ microstructure. From the results obtained, in which about 3 and 4 wt% of Al are needed to suppress athermal ω in Ti-V-Al and Ti-Fe-Al based alloys respectively, it is questionable whether the 3 wt% of Al in the Ti-10-2-3 alloy is optimum for ω-phase suppression in an alloy with its V/Fe combination. The presence of athermal ω-phase in the alloy (Table 1) confirms this.

The data on solute partitioning between the α and β phases in the alloy indicates that there is an increasing V contribution to the strengthening of the β phase compared to that by Fe as the annealing temperature is lowered.

For example the V in the β of the alloy increases from 10 to 14 wt% between 800 and 700°C, while that of Fe in β increases from only 2 to 3 wt% in the same temperature range.

SUMMARY AND CONCLUSIONS

1. Experimental data for the Ti-rich portions of the Ti-V-Al and Ti-Fe-Al systems at 800 and 700°C compared to the solute partitioning in the Ti-10-2-3 alloy show that the partitioning ratio X_α/X_β of V or Fe between α and β is higher in the Ti-10-2-3 alloy than in the ternary alloys.

2. The greater β-stabilising effect of the Fe compared with V implies retention of more metastable, β phase in preference to α''-martensite when β phase is quenched; this is beneficial with respect to alloy strengthening on ageing.

3. The presence of Al in a β-stabilised alloy causes the proportion of β to α to decrease less rapidly with fall in annealing temperature in the α + β range.

4. For the development of quaternary alloys, the ternary constitutional and structural data are important in relation to optimising compositions and phase fractions of α and β phases while reducing ω formation and adequately strengthening the phases formed.

ACKNOWLEDGEMENTS

This work has been carried out with the support of Procurement Executive, Ministry of Defence.

REFERENCES

1. J L Murray, Phase Diagrams of Binary Titanium alloys, 1987, ASM, Metals Park.
2. J L Murray, Bull . of Alloy Phase Diagrams, 1981, 2(1), 48.
3. I I Kornilov and M A Volkova, Ti Alloys for New Techniques, Nauka Moscow, 1968, 78.
4. T Tsujimoto, J . Jap. Inst Met, 1968, 32, 970; Trans. Jap. Inst. Met., 1969, 10, 281.
5. K Hashimoto, H Doi and T Tsujimoto, J . Jap. Inst. Met, 1985, 49, 410.
6. A Seibold, Z. Metallk, 1981, 77 (1), 712.
7. C McCullough, J J Valencia, C G Levi and R Mehrabian, Acta. Met., 1989, 37, 5, 1321.
8. Wei Fuming and H M Flower, Mats Sci Tech, 1989, 5, 1172.
9. T Maeda, PhD thesis, University of London, 1988.
10. H M Flower, Mats Sci Tech, 1990, 6, 1182.
11. B Vandecastele, N Rizzi and J F Wadier, Proc 6th World Conf on Titanium, edited by P Lacombe, R Tricot and G Beranger, J De Physique, 1988, 1345.
12. U Zwicker, "Titan and Titanlegierungen", Springer-Verlag Publ, 1974.
13. F H Froes, J C Williams, C F Yolten, C H Hamilton and H E Rosenblum, Ti '80 Sci & Tech, edited by H Kimura and O Izumi, AIME Publ, 1980, 1025.
14. A Nwobu, H M Flower and D R F West, Mats Sci Tech, 1991, 7, 971.
15. E W Collings, "Alloying", ASM Publ 1988, 257.

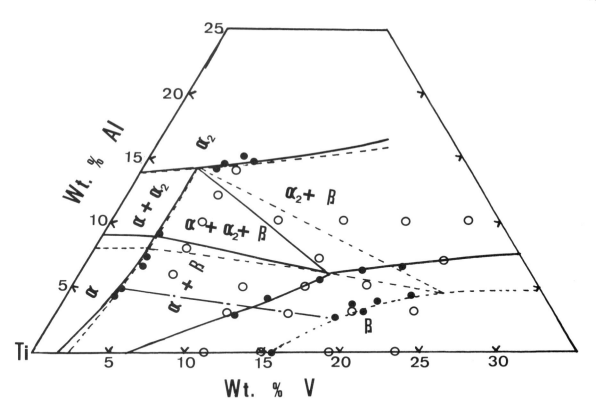

Fig. 1. Partial isothermal sections of Ti-V-Al system at 800 and 700°C.
o alloy compositions.　　　• terminal analysed compositions of phases.
————— 800°C　　　　　　　－－－－ 700°C.
—·—·— typical α + β tie-line.　　α$_2$ = Ti$_3$Al.

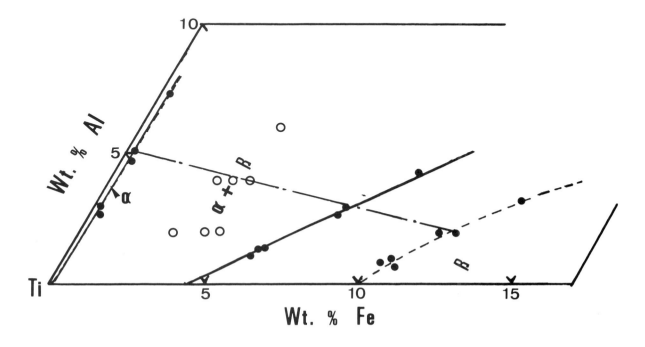

Fig. 2. Partial isothermal section of Ti-Fe-Al system at 800 and 700°C. Symbols etc as in Fig. 1.

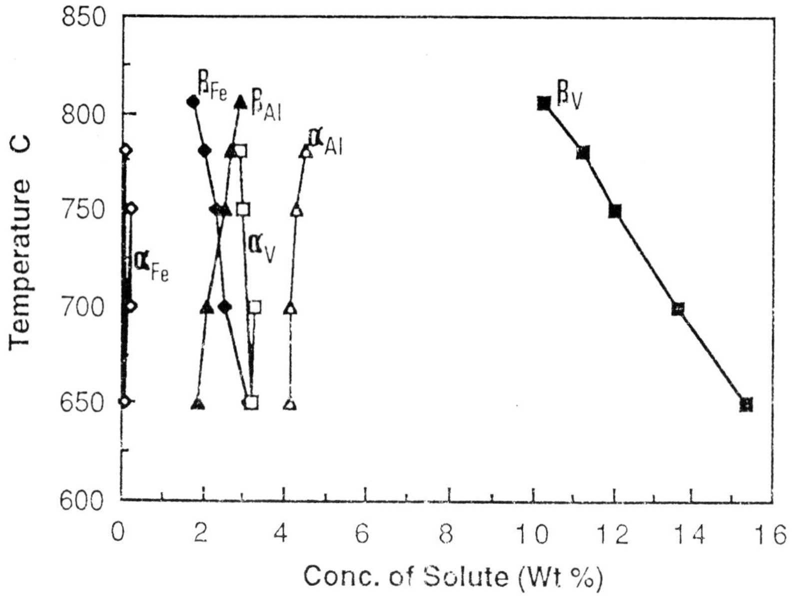

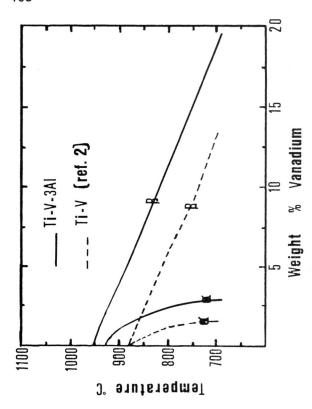

Fig. 3. Solute partitioning between α and β phases in Ti-10V-2Fe-3Al alloy.

Fig. 4. Isopleths showing effect of Al additions on the β transus in the Ti-V-Al system (based on Fig. 1 and Ref. 9).

Fig. 5. Isopleths showing effect of Al additions on the β transus in the Ti-Fe-Al system (based on Fig. 2 and other experimental data).

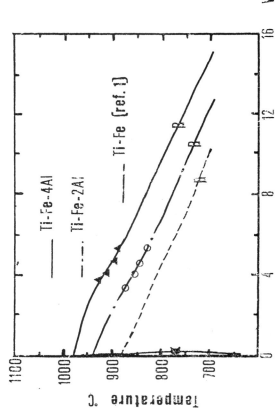

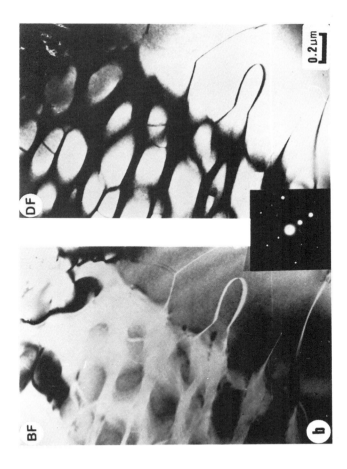

Fig. 6. Microstructures observed in Ti-6V-10Al alloy. (a) Furnace cooled from 1170°C to 800°C, annealed and water quenched (WQ); shows coarse α + β (transformed) structure - α dark, transformed β light (α₂ not seen by SEM), and (b) interface between α₂ and α region containing α₂ particles (by TEM, dark field g = (10$\bar{1}$0)α₂, zone normal near [1$\bar{2}$13] α). (c) 1170°C annealed, WQ and 800°C annealed, WQ; SEM shows α + β structure retaining the martensitic morphology produced by the initial quenching from 1170°C.

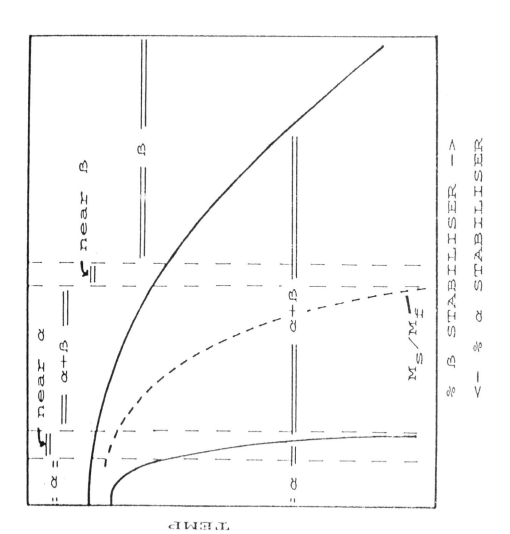

TEMP

% β STABILISER —>
<— % α STABILISER

Fig. 8. Semi-schematic isopleth showing general form of α/β relationships in Ti alloys containing α and β stabilisers. (Ref. 10).

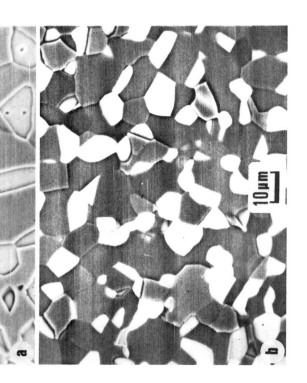

Fig. 7. Typical globular/equiaxed structures produced by hot working and annealing in the α + β range. (SEM, α light β dark/grey). (a) Ti-4.5Fe-2Al alloy annealed at 800°C (b) Ti-10V-2Fe-3Al alloy annealed at 700°C.

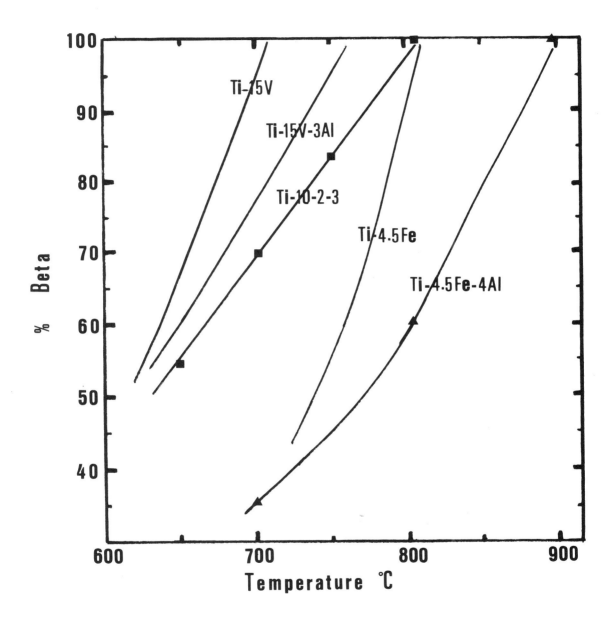

Fig. 9. β-approach curves for selected alloys from the Ti-V-Fe-Al system.

Table 1 Structure and hardness of β quenched alloys

ALLOY	STRUCTURE	HARDNESS HV
Ti-11V	$\alpha'' + \beta$	245
Ti-11V-3Al	α''	275
Ti-11V-7Al	α''	330
Ti-15V	$\beta + \omega$	328
Ti-15V-2Al	α''	235
Ti-15V-5Al	$\alpha'' + \beta$	270
Ti-15V-7Al	β	280
Ti-19V	$\beta + \omega$	290
Ti-19V-2Al	$\beta + \omega + \alpha''$	215
Ti-19V-3Al	β	245
Ti-19V-7Al	β	255
Ti-23V	β	225
Ti-23V-3Al	β	230
Ti-3.5Fe-2Al	$\alpha'' + \beta$	450
Ti-3.5Fe-4Al	$\alpha'' + \beta$	328
Ti-4Fe-2Al	$\alpha'' + \omega + \beta$	360
Ti-4Fe-3Al	$\alpha'' + \omega + \beta$	320
Ti-4Fe-4Al	$\alpha'' + \beta$	295
Ti-4.5Fe-2Al	$\alpha'' + \omega + \beta$	380
Ti-4.5Fe-4Al	$\alpha'' + \beta$	348
Ti-4.5Fe-6Al	β	355
Ti-5Fe-2Al	β	370
Ti-5Fe-4Al	β	355
Ti-10V-2Fe-3Al	$\alpha'' + \omega + \beta$	250

13
THERMOCHEMISTRY AND PHASE DIAGRAMS OF COMPOUND SEMICONDUCTORS

T. J. ANDERSON

Chemical Engineering Department, University of Florida, Gainesville, FL 32611,USA

ABSTRACT

Group III-V compounds and solid solutions are presently receiving intense investigation for electronic and optoelectronic device applications. Many of the processing steps in the fabrication of these devices involve the contact of a liquid or gas phase wlth the semiconductor at an elevated temperature. Such processes include bulk crystal growth, liquid phase epitaxy, organometallic chemical vapour deposition (OMCVD), molecular beam epitaxy, dopant diffusion and rapid thermal annealing. An equilibrium boundary condition is often specified in the analysis of these processes and therefore knowledge of the phase diagram is important.

Two types of compound semi conductor solid solutions are formed. One solid solution involves substitutional mixing of atoms from the same column of the periodic table on a sublattice. The second solid solution is produced by the point defect structure of the material and exhibits limited solution range. Experimental information about the behaviour of the first kind of solid solution is very limited. Rather, the solid solution behaviour is generally obtained by analysis of solid–liquid phase equilibria. A majority of binary phase diagrams and several tertiary systems have been determined and exerimental measurements of liquid phase component activities and enthalpies of mixing have been performed in many systems. A formalism for computing multicomponent compound semiconductor phase diagrams is presented, including a discussion of standard state selection. This formalism is illustrated for several group III-V Systems.

The point defect structure of compound semiconductor is extremely important as it controls the electrical behaviour of these materials. Indeed, our ability to control the point defect structure is the key link between processing and devices. Since processing often occurs at elevated temperature, a near equilibrium defect structure can be frozen in the material. Examples of the point defect structure of GaAs and AlSb are presented.

Equipped with reliable models for the behaviour of these two solid solutions, the influence of processing conditions on the properties of materials can be analyzed. One area where this is possible is in chemical vapour deposition of compound semiconductors. Examples of such applications are presented, including complex reaction equilibria of hydride VPE and Zn doping of InP during OMCVD.

14

INTERACTIONS OF METALS WITH CdTe

Q. Han, H. Cordes, J. Klingbeil and R. Schmid-Fetzer

Technical University Clausthal, Metallurgical Center, Robert-Koch-Str. 42,
D-3392 Clausthal-Zellerfeld, Germany

ABSTRACT

A number of metals are not stable as an electrical contact material on CdTe, leading to the problem of degradation or even device failure. Here the metal / CdTe interactions are viewed from the point of the ternary Cd-Te-metal phase diagram. For the examples Cd-Te-Ag and Cd-Te-W the phase diagrams are studied using a two step approach: (i) approximate thermodynamic calculation and (ii) experimental verification at 500°C. The rather stable CdTe phase dominates the equilibria which are complicated in the case of silver by extensive solid solubilities which is not the case for tungsten. The possible reactions in a solid state diffusion couple Ag/CdTe or W/CdTe are discussed on the basis of the phase diagrams. A Ag-14at.% Cd alloy is suggested to be much more suitable then pure Ag on CdTe. Tungsten was detected as a stable elemental contact material on CdTe.

1. INTRODUCTION

One of the major factors influencing the reliability of CdTe-based devices for many optoelectronic applications is the stability of the electrical contact to the semiconductor. During the formation of Schottky or ohmic contacts and subsequent processing steps and also during later exposure of the device to elevated temperatures a number of reactions may occur at the metal/CdTe interface. These reactions generally result in changes in the electrical and mechanical contact properties, leading to the problem of degradation or even failure.

From studies of thin metal films on CdTe one may conclude that CdTe is a rather reactive material to a number of metals. Non-abrupt interfaces and intermixing of Ag with CdTe has been observed by surface analytical methods [80Hum, 82Pat, 82Ebi] and there are also indications that thermal effects due to substrate heating during metal deposition increase this intermixing [80Hum]. The nature of this reaction has not been revealed. Oxide layers at the interface also play an important role in these interactions and for the resulting electrical properties as was found in these studies and [86Dha].

In this study these interactions shall be analyzed from the point of view of the ternary phase diagrams for the examples Cd-Te-Ag and Cd-Te-W. No data could be found for Cd-Te-W. One of the very few Cd-Te-metal systems where at least limited information is available is Cd-Te-Ag. The vertical section Ag-CdTe was studied essentially by X-ray diffraction and DTA [74Pan1]. It was found not to be pseudobinary with a three phase region CdTe+(Ag)+Ag2Te below 795°C. The solubility of silver in CdTe is low [74Pan1], with a maximum of $3.4 \cdot 10^{19}$ cm^{-3} at 793°C [74Pan2]. Silver diffused from the gas phase into CdTe at levels below 10^{17} cm^{-3} was studied by spectroscopic and electrical measurements [82Cha] and Ag was found on Cd sites acting as an acceptor. However, after storing at room temperature for about 100 days a silver complex forms which diminishes the doping properties of Ag and this complex vanishes reversibly after re-annealing. The Ag_2Te-CdTe section was found to be pseudobinary with a large solubility range of Ag_2Te [85Bon, 86Tri]. A critical review is given in [88Pet].

The purpose of this study is to construct isothermal sections at 500°C of the Cd-Te-Ag and Cd-Te-W systems and to discuss the materials compatibility accordingly.

2. METHODS

The following two step approach to the ternary system is taken, using theoretical and experimental methods :

(i) Simplified thermodynamic calculation at 25°C;

(ii) Experimental analysis at 500°C using bulk samples.

The approximate thermodynamic calculation is done as described in [88Sch] using an extended computer program [89Kli] and the following simplifications : (a) solid solubilities and ternary phases are disregarded, (b) the Gibbs energies of formation, ΔG, of the binary solid phases are estimated from their enthalpies of formation, ΔH, if no literature data are given. The estimation of ΔH is done using Miedema's model [89Mie], however, the parameters for the tellurides are not given in [89Mie] and are determined from an extrapolation that will be published elsewhere.

Using the results of this approximately calculated phase diagram, a number of alloy compositions are selected and prepared from the elements in evacuated silica capsules and annealed at 500°C for up to 27 days. Samples were analyzed by powder X-ray diffraction, metallography and electron microprobe analysis.

3. RESULTS FOR Cd-Te-Ag

3.1 Approximate Phase Diagram Calculation

The ΔG-data for the solid phases in the three binary edge systems of Cd-Te-Ag, taken as line compounds, are given in Fig. 1. The data for CdTe are from [74Mil]. The phase diagram calculated from these data is given in Fig. 2. Because of the dominating thermodynamic stability of CdTe the tie line CdTe-Ag governs the ternary equilibria and all the other tie lines originating from CdTe are merely a consequence of that. Even within a large error bar (± 10 kJ/g-atom) of the estimated ΔG-data no other tie line distribution is possible.

3.2 Experimental Determination

One of the approximations is, however, incorrect since the solubility of Cd in (Ag) is very large [86Mas], see Fig. 3. The other binary systems, Cd-Te [86Mas] and Ag-Te [78Mof, 86Mas], fulfil the assumption of negligible solid solubilities. The γ-phase of Ag-Te [78Mof] was not accepted by [86Mas]. The rules of heterogeneous equilibria require in this case that the single tie lines Ag-CdTe and Ag-Ag$_2$Te develop into two-phase regions (Ag)-CdTe and (Ag)-Ag$_2$Te. Consequently, the three phase field Ag+CdTe+Ag$_2$Te must shift from the Ag-vertex (Fig. 2) to some intermediate point of the (Ag) solution range (Fig.3) with a presumably low Te content. This expectation is verified by the experimental data given in Fig. 3 and also the quantitative location of the (Ag)-vertex at 14 ± 2 at.% Cd follows from these data. The data points in Fig. 3 denote the phases identified by arrows with the centre corresponding to the sample composition.

The existence of the three phase field (Ag) + CdTe + Ag$_2$Te on the Ag-CdTe section is also supported by data of [74Pan1] and the solubility of Cd in Ag$_2$Te is in accordance with [86Tri].

The other two phase fields and the liquid phase regions have been developed on the basis of the binary solubility data, taking into account the results from a polythermal space model of the liquidus surfaces that has been constructed using all the available experimental data. The isothermal sections at lower temperature develop from Fig. 3 by the decrease of the liquid phase fields, the occurrence of Ag$_5$Te$_3$ and the retrograde solubility of Ag$_2$Te [86Tri] and, to some extent, also of Cd in (Ag). The location of the (Ag)-vertex of the (Ag)+CdTe+Ag$_2$Te region might also vary but probably not too much. It should be pointed out that the principal feature of this phase diagram (centering of the tie lines at CdTe) essentially agrees with the approximate calculation of Fig. 2.

4. RESULTS FOR Cd-Te-W

The approximate calculation for Cd-Te-W also yields the CdTe-W tie line in favour of the hypothetical possibility Cd-WTe$_2$. The CdTe-WTe$_2$ tie line exists since there is no tie line to compete with. The calculated tie lines are plotted in Fig. 4. In the same figure the experimental data at 500°C are given which completely verify the calculations. The fragmentary W-Te binary phase diagram is from [86Mas, 78Mof]. No information could be found for W-Cd. Our own experimental data in this binary at 500°C and higher temperatures suggest a phase diagram as plotted in Fig. 3 with no compound formation and very limited mutual solubilities in the temperature range of interest. This information has, of course, been used also for the calculation of the ternary equilibria. In view of the small solubilities in liquid Cd and Te at 500°C, these degenerated regions are not distinguished from the Cd- and Te-corners in Fig. 4.

5. DISCUSSION

The interaction of Ag with CdTe in a solid state diffusion couple may be interpreted on the basis of Fig. 3 in two ways, leading to the same result as follows. In an Ag/CdTe diffusion couple the overall composition and the mean composition at the interface is located on the CdTe-Ag composition cut. This cut crosses the three phase field CdTe+Ag$_2$Te+(Ag) in Fig. 3 and, consequently, these three phases must occur in the reaction zone under the reasonable assumption of local equilibria.

From another point of view, at the fresh Ag/CdTe contact Cd from CdTe should dissolve in the pure Ag which acts as a sink for Cd because of its extensive solubility. The excess Te from the decomposed CdTe is now free to react with the available Ag to form Ag$_2$Te. Again, assuming local equilibria, the composition limits of (Ag), Ag$_2$Te and CdTe are given by the three phase field in Fig. 3. This figure is obviously also an important tool for the interpretation of the diffusion path of this interface reaction. A different explanation for the occurrence of Ag_2Te was given in [74Pan1]who suggested a driving force for the reaction $CdTe + 2Ag \rightarrow Ag_2Te + Cd$ above 1000K which fails to explain the interactions discussed in Fig. 3.

In addition, traces of Ag will diffuse into CdTe and form acceptor levels [82Cha]. From the above discussion it becomes clear, that pure silver is not a stable contact on CdTe. Especially the precipitation of an additional phase (Ag_2Te) and also the huge dissolution of Cd in (Ag) will change the state of matter, the microstructure and, accordingly, the properties at the interface. This may be a possible explanation for the degradation of Ag-contacts on CdTe.

The phase diagram in Fig. 3 is, however, not only an important tool for the understanding of the possible reactions but also offers a solution to this problem. If not pure Ag but the Ag-14at.%Cd phase is deposited onto CdTe the reactions and the precipitation of Ag_2Te will be avoided since this (Ag)-solution at the vertex of the three phase field is in equilibrium with CdTe; the small solubility of Te in (Ag) may be compensated similarly. The (Ag) solutions from 14 to 40 at.% Cd, which are also in two phase equilibrium with CdTe, may also be used as a non-reactive metallization, however, the higher Cd content might deteriorate other properties (electrical,...) of this contact. One should also note that the (Ag)-vertex shown in Fig. 3 for 500°C might be shifted at other temperatures.

Looking at Fig. 4 it is evident that tungsten is a very promising candidate for an inert metallization of CdTe since both phases are connected by a tie line. As in the case of silver the phase diagram addresses only one important part of the contact problem and additional questions (electrical properties, thermal expansion mismatch, adhesion etc.) must be studied separately. Or, as W. Hume-Rothery pointed out, the phase diagram is the beginning of wisdom and not the end of it.

REFERENCES

74Mil K.G. Mills, "Thermodynamic data for inorganic Sulphides, Selenides and Tellurides", London, Butterworths (1974).

74Pan1 O.É. Panchuk, I.E. Panchuk, D.P. Belotzkii and V.I. Grytsiv, V.I., "The cross section CdTe-Ag", Izvest. Akad. Nauk SSSR, Neorg. Mater. 10, 980-982 (1974).

74Pan2 O.É. Panchuk, V.I. Grytsiv and D.P. Belotzkii, "Solubility of Ag in n-CdTe", Izvest. Akad. Nauk SSSR, Neorg. Mater. 10, 581-584 (1974).

78Mof W.G. Moffat, "Handbook of binary phase diagrams", Genium Publishing Corporation, Schenectady, New York (1978).

80Hum T.P. Humphreys, M.H. Patterson and R.H. Williams, "Metal contacts to clean and oxidized cadmium telluride and indium phosphide surfaces", J. Vac. Sci. Technol. 17, 886-890 (1980).

82Cha J.P. Chamonal, E. Molva, J.L. Pautrat and L. Revoil, "Complex behaviour of Ag in CdTe", J. Cryst. Growth 59, 297-300 (1982).

82Ebi A. Ebina and T. Takahashi, "Studies of clean and adatom treated surfaces of II-VI compounds", J. Cryst. Growth 59, 51-64 (1982).

82Pat M.H. Patterson and R.H. Williams, "Metal-CdTe interfaces", J. Cryst. Growth 59, 281-288 (1982).

85Bon Z. Boncheva-Mladenova, V. Vassilev, T. Milenov and S. Aleksandrova, "Investigation on phase-diagram of the Ag_2Te-CdTe system", Thermochimica Acta 92, 591-594 (1985).

86Dha I.M. Dharmadasa, W.G. Herrenden-Harker and R.H. Williams, "Metals on cadmium telluride: Schottky barriers and interface reactions", Appl. Phys. Lett. 48, 1802-1804 (1986).

86Mas T.B. Massalski (ed.), "Binary alloy phase diagrams", Amercian Society for Metals, Metals Park, Ohio (1986).

86Tri L.I. Trishchuk, G.S. Oleinik and I.B. Mizetskaya, "Physicochemical study of interaction in the Ag_2Te-ZnTe and Ag_2Te-CdTe systems", Ukrain. Khim. Zhur. 52, 799-803 (1986).

88Pet G. Petzow and G. Effenberg (eds.), "Ternary Alloys", Vol. 1, VCH, Weinheim, 439-442 (1988).

88Sch R. Schmid-Fetzer, "Stability of Metal/GaAs-interfaces: a phase diagram survey", J. Electron. Mat. 17, 193-200 (1988).

89Kli J. Klingbeil and R. Schmid-Fetzer, "TerQuat", Unpublished Report, AG Elektronische Materialien, TU Clausthal (1989).

89Mie A.R. Miedema and A.K. Niessen, "Cohesion in metals", North-Holland Publishing Company, Amsterdam (1989).

117

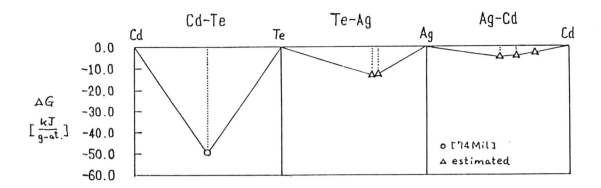

Fig. 1.The Gibbs energies of formation of the solid phases at room temperature in the three binary edge systems of CdTe–Ag. Composition scales are in at.%.

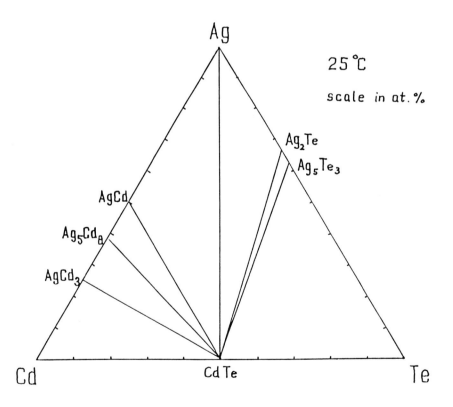

Fig. 2.Approximately calculated Cd–Te–Ag phase diagram at room temperature, disregarding solid solubilities.

118

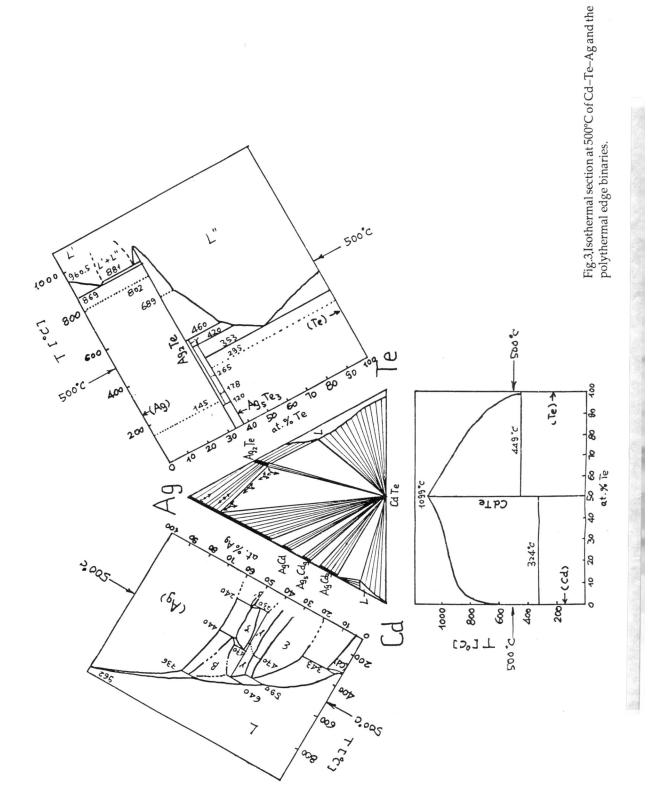

Fig.3.Isothermal section at 500°C of Cd–Te–Ag and the polythermal edge binaries.

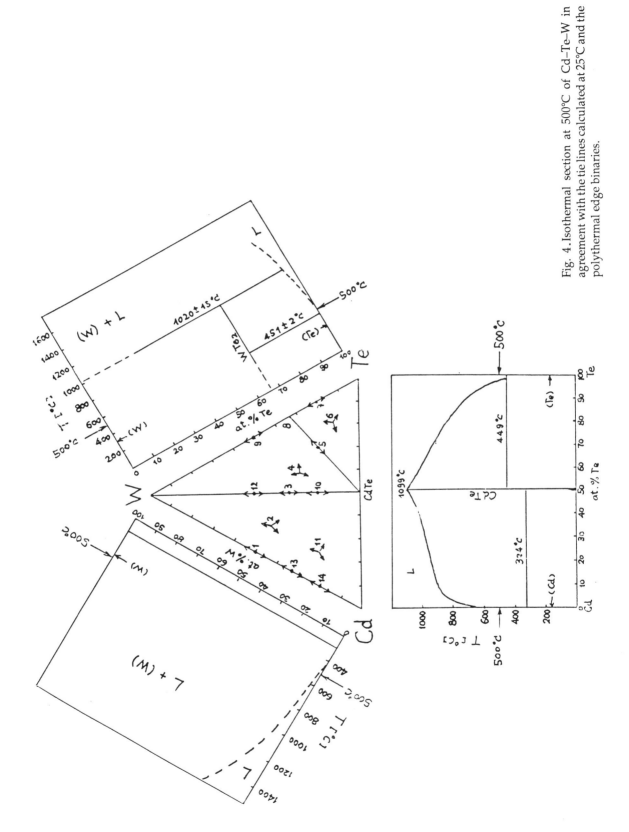

Fig. 4. Isothermal section at 500°C of Cd–Te–W in agreement with the tie lines calculated at 25°C and the polythermal edge binaries.

15

CAD COMPUTER DISPLAY AND APPLICATION OF (Cu-Sn)-X (X=Ag, As, Au, Pb, Zn) BINARY AND TERNARY PHASE DIAGRAMS TO FABRICATION AND PROCESSING PROBLEMS

M. R. NOTIS

Lehigh University, Bethlehem, PA 18015, USA

ABSTRACT

This paper demonstrates the use of CAD (computer aided design) systems to portray phase equilibria in the Cu-Sn binary and Cu-Sn-X (Ag, As, Au, Pb, Zn) ternary systems. the ternary diagrams produced are displayed in 3D using high resolution color graphics. They are continuously rotatable and can be displayed either as wire-frame or solid shaded models. Critical data can easily be extracted from the diagrams, smaller regions of the diagram can be magnified and displayed without loss of resolution, phase fractions can be determined from tie-line measurements, and constant phase fraction contour lines may be superimposed on the phase diagrams.

The Cu-Sn binary diagram includes recent additions proposed by V.C. Marcotte, and compares the revised diagram to the calculated diagram given by A.P. Miodownik. The Cu-Sn (and Cu-Zn) binary phase diagram is examined with respect to the tendency of different high temperature phases (β and δ) to undergo diffusive, massive, and martensitic phase transformations at lower temperatures.

The behavior of the (Cu-Sn)-Au system is of particular interest because of its relevance to the soldering of Au-Cu alloys with Pb-Sn solders. This paper examines the kinetics of reactions involving intermediate phase formation in the Cu-Sn-Au system in the solid-state. Diffusion couples made from Au-Cu alloys vs. pure Sn were annealed at 170°C for various times. Both binary and ternary intermediate phases were observed to form depending on the composition of the Au-Cu end member; two new ternary compounds were identified and an isothermal section of the phase diagram was generated. The spatial extent and number of intermediate phases formed in the reaction zone increased as the Au content of the Au-Cu end member was increased. The intermediate phases also displayed a shift from planar to non-planar morphologies as the Au content was increased. Growth constants for individual planar layer as well as the overall reaction zones were determined. The width of the layers showed a $t^{1/2}$ dependency indicating volume diffusion control of intermediate phase growth. The growth constants for the overall reaction zone in $AuCu_3$, $AuCu$, and Au_3Cu vs. pure Sn couples were 0.63, 2.3, and 3.4 μm/h$^{1/2}$ respectively. Electron probe microanalysis (EPMA) was used to map diffusion paths onto the 170°C isothermal section of the phase diagram. Using this path data and the diagram, an explanation is given to account for the shift from planar to nonplanar interface morphology in this system.

For the (Cu-Sn)-As ternary, the CAD system was used to display sixteen isothermal sections available from the literature, to contrast a quasibinary section existing between the compounds $CusAs_2$ and $SnAs$, to create a 3D liquidus surface, and to create phase regions of interest to the present study. CAD display was also employed to aid interpretation of EPMA composition analysis in the Cu-rich corner of the Cu-Sn-As system. Specifically, the CAD system was used to interpolate 3-phase equilibria data obtained experimentally and to verify consistency in the experimental results.

16
PURPLE PLAGUE AND THE Al-Au-Si SYSTEM

F.H.HAYES AND W.T.CHAO*

Materials Science Centre, University of Manchester and UMIST, Grosvenor Street, Manchester M1 7HS, UK
*present address: British Steel Technical, Swinden Laboratories, Rotherham, S60 3AR, UK

ABSTRACT

Experimental results are presented for the $Al-Al_2Au-Si$ portion of the Al-Au-Si ternary system which has been studied in relation to the formation of the purple plague phase, Al_2Au, in the presence of Si. Vertical sections for Al_2Au-Si, which is shown to be a pseudo-binary system, for $Al-Al_{0.5}Au_{0.25}Si_{0.25}$ and for $Al_{0.75}Si_{0.25}-Al_{0.75}Au_{0.25}$ are given plus a liquidus projection for the subternary system. Results are discussed in relation to the enhanced rate of Al_2Au formation in the presence of Si; a mechanism for this effect is proposed.

INTRODUCTION

The combined use of aluminium and gold in electronic devices plus exposure to temperatures of up to 300°C during processing can lead to the formation of intermetallic phases which promote mechanical failure through a mechanism involving Kirkendall voiding on the gold-rich side of the gold-aluminium interface[1,2]. The term 'purple plague' which is used to describe this effect came about as a result of the coloured appearence of the fracture surfaces due the presence of the purple phase Al_2Au. A great deal of research relating to this problem was carried out during the sixties and seventies[3] and many publications devoted to this subject appeared in the literature[4]. Attention has recently been focussed on the reappearence of purple plague as a source of failure in integrated circuit devices leading to renewed interest in this area[5].

A question that has remained unresolved for many years in connection with purple plague failures concerns the influence that the presence of silicon in the proximity of the gold-aluminium interface can have on the stablility and rate of formation of the damaging intermetallic phases. Selikson and Longo[6] noted significant differences both in colour and in the rates of formation of intermetallic phases resulting from reaction between the gold wire and the aluminium metallisation at the semiconductor chip compared with the normal purple phase which forms very slowly by reaction between the aluminium wire and the gold-plated post in integrated circuit devices. It was suggested that these differences arise from the presence of silicon in the former case. The view that silicon has an effect on the formation of Al-Au intermetallics was reinforced when the results of mechanical tests carried out on Al-wire bonded units heated for various times at 300°C were compared with the results of identical tests carried out on Au-wire bonded units. The gold wire units were found to have developed an extensive region of a black reaction product phase at and around the bonds on the semiconductor whereas the aluminium wire bonded units only developed the purple phase after hundreds to thousands of hours at the baking temperature. Furthermore, the percentage of bonds breaking at the gold wire-semiconductor region increased with time at temperature in contrast to aluminium wire with which the percentage of bonds failing remained constant. This led to the view that additions of silicon to aluminium-gold alloys might lower the invariant points leading to an enhanced reaction. Although no details were given concerning experimental techniques used in the investigation it is stated[6] that the Au-Si eutectic temperature was found to be lowered by 10 degrees by the addition of some 15 weight percent of aluminium. The Al-Si eutectic temperature was also found to be lowered by 10 degrees with the addition of 20 percent of gold. There is also the suggestion that the presence of silicon has a catalytic effect on the rate of formation of intermetallic phases between gold and aluminium. Commenting on these results, Prince[7] notes that it is difficult to reconcile the lowering of the Au-Si eutectic by 10 degrees with the likely equilibria in the $Al_2Au-AlAu-Si$ phase region of the Al-Au-Si ternary system and it may be anticipated that the reaction temperature will be far higher than the temperature of 353°C suggested by Selikson and Longo[6].

Subsequently, Philofsky[8,9] investigated the effect of silicon on intermetallic phase formation by means of electron probe microanalysis studies of butt-welded diffusion couples of Al-5wt%Si and pure Au wires annealed for various times at temperatures in the range 250-400°C. With the sample aged at 400°C for 100 min a layer 20μm thick was observed which was found to have the composition 23.1 at% Au, 62at% Al, 14.2 at% Si. It was on the basis of this observation that the existence of a stable ternary phase $AuSiAl_4$ was suggested which would lead in turn to enhanced intermetallic phase formation reaction between Au and Al in the vicinity of silicon.

Additional phase diagram information comes from the work of Lovosky and Politova[10] who determined the

1100°C isotherm by sampling the liquid phase for ten binary Al-Au alloys equilibrated with silicon and from Lovosky and Kolesnichenko[11] who determined the solubility of Al in Si for four different Al-Au ratios over the range 1000-1300°C.

In this paper we report some results of a study of the Al-Au-Si ternary system carried to establish the phase relationships and invariant reactions in the entire system and to investigate further reports of the existence of the ternary phase AuSiAl$_4$. Here we report our results for the Al-Al$_2$Au-Si portion of the system. Prior to any experimental work being carried out a series of computer calculations of the entire ternary phase diagram was performed using thermodynamic data for the edge binary systems taken from the literature.

INITIAL COMPUTER CALCULATIONS

Computer calculations using excess Gibbs energy coefficients given by Murray and McAlister for Al-Si[12], by Murray, Okamoto and Massalski for Al-Au[13], and by Ansara for Au-Si[14]. These calculations indicated the Al-Al$_2$Au-Si system to be a ternary eutectic system with the invariant reaction plane lying some 10 degrees below the Al-Si binary eutectic temperature. It was on the basis of these calculations that the vertical sections studied experimentally were selected.

EXPERIMENTAL

A total of thirty alloy compositions comprising the three vertical sections Al$_2$Au to Si, Al$_{0.5}$Au$_{0.25}$Si$_{0.25}$ to Al and Al$_{0.75}$Si$_{0.25}$ to Al$_{0.75}$Au$_{0.25}$ were studied using the Smith thermal analysis technique[15,16] employing a Kanthal wire wound furnace plus calibrated Pt/Pt-13%Rh thermocouples to locate phase boundary temperatures. Optical metallography, SEM-EDX, electron probe microanalysis and x-ray diffraction were used for phase identification.

Materials: Gold grain, minimum purity 99.99%, aluminium ingot 99.999% both supplied by Johnson-Matthey Ltd and silicon pieces, 99.9999% supplied by Aldrich Chemical Company Ltd were used to prepare alloy samples between 3 to 5 g in weight by completely melting weighed amounts of the pure components together in Al$_2$O$_3$ crucibles under a hydrogen atmosphere. The crucibles, 3-4 cm long, 12.0 +/- 0.06 mm o.d. and 1.23 +/- 0.20 mm wall thickness were made by slip-casting and firing Al$_2$O$_3$ powder type AMS-9 supplied by Mandoral Ltd. Immediately on melting, the liquid alloys were thoroughly stirred using the Al$_2$O$_3$ thermocouple sheath and then rapidly quenched into cold water to minimise segregation. The whole Al$_2$O$_3$ crucible assembly containing the quenched alloy and thermocouple assembly fitted into the rounded end of a supporting silica tube 16 mm o.d. and 1 mm wall thickness which in turn was placed inside the Smith thermal analysis furnace tube. Each solidified alloy sample was homogenised for several hours at a temperature some twenty degrees below the lowest reaction temperature for that composition immediately prior to to thermal analysis being carried out. Average heating rates of 1 degree per minute were used. As soon as a thermal event was detected indicating the presence of a phase boundary, heating was discontinued and, after an isothermal hold above the reaction temperature, the sample was cooled again under Smith control to confirm the existence of the boundary. A constant thermal response upon cycling around the temperature confirmed the presence of an equilibrium phase boundary; an isothermal reaction indicates the presence of an invariant reaction. Heating was then continued until the next thermal event was detected and the same procedure repeated until the sample was completely molten. Liquidus points were determined using cooling runs. The Pt/Pt-13% Rh thermocouples used were calibrated using the melting points of pure Ag, Cu, Zn, In, Al and the Ag-Cu eutectic point. The Smith thermal analysis equipment used in the present work operates satisfactorily up to temperatures in the region of 1200°C.

RESULTS

All of the thermal arrests observed in the three vertical sections of the portion of the Al-Au-Si system bounded by Al, Al$_2$Au and Si, plus phase field labels are shown in Figs. 1-3. In the Al$_2$Au-Si section ten alloy compositions were studied and the results clearly show this to be a pseudo-binary system with a single eutectic reaction at 994°C. The same reaction was found in a further experiment carried out with the 80% Si alloy on this section. The eutectic composition is placed at 16.5 at% Si. Optical metallographic examination of samples air-cooled from the liquid confirmed that at 15 at.% Si the primary phase is Al$_2$Au whereas at 25 at.% Si the primary phase was found to be pure silicon; in each case the matrix exhibited a eutectic microstructure. The melting point of the congruently melting Al$_2$Au phase was found to be 1061°C which is one degree higher than the assessed value[13].

The Al$_{0.5}$Au$_{0.25}$Si$_{0.25}$ to Al section contains three regions of primary crystallisation separated by monovariant points, two three-phase fields and an invariant reaction at 576°C. The two three phase fields are separated at 81 at.% Al on this section. The 576°C invariant reaction was observed at 96 at.% Au but not at 98 at.% Au. Thus the edge of this plane on this section is placed at 97 +/- 1 at.% Au.

The $Al_{0.75}Au_{0.25}$ to $Al_{0.75}Si_{0.25}$ section shows two regions of primary crystallisation separated at 21 at.% Si plus two three-phase fields. The 576°C invariant reaction was observed at all of the compositions studied in this section. The two three-phase fields intersect the reaction plane at 4.5 at.% Si.

DISCUSSION

The vertical sections presented above clearly show that within the Al-Au-Si system the Al_2Au-Si section is a pseudo-binary simple eutectic system and that the Al-Al_2Au-Si subternary system is a ternary eutectic system with a single invariant reaction involving formation of essentially the pure solid components from the liquid phase on cooling. In contrast to the views of Selikson and Longo[6] and of Philovsky[8,9] no evidence was found in this study for the existence of any stable solid phases other than the pure components Al, Al_2Au and Si. The present results also allow the construction of the polythermal liquidus projection shown in Fig. 4. The ternary eutectic point lies within one atomic percent and within one degree of the eutectic point in the Al-Si binary system. The 1100°C isothermal section shown in Fig. 4 is in good agreement with the data given by Lovosky et al.[10].

In the absence of any stable ternary phases such as $AuSiAl_4$ suggested by earlier workers, the explanation of the observation that the presence of silicon in the vicinity of Al-Au interconnections increases the rate of purple phase formation, must involve some other effect. One possibility is the effect that silicon might have on the stability of Al_2O_3 in the presence of metallic gold. The thin film of highly stable Al_2O_3 that is present on all aluminium surfaces with the possible exception of those prepared and maintained under UHV conditions, will exist on both the aluminium metallisation and on aluminium wires prior to the formation of any joints to gold. In the absence of any other elements this layer will be stable at the Al-Au interface after joining and will act as a barrier through which the Al and Au atoms will have to pass during subsequent inter-diffusion. Its presence therfore must influence to some extent the rate of formation of intermetallic phases at the interface and hence it must play a part in controlling the overall rate of the purple plague reaction. Thermodynamic calculations show that the introduction of a small amount of silicon into the system has the potential to modify this thin Al_2O_3 layer and hence indirectly to enhance the rate of purple plague formation. Under normal circumstances, due to the higher stability of Al_2O_3 compared to that of SiO_2, elemental silicon would not be able to reduce alumina. That is to say the chemical reaction represented by the following equation is not thermodynamically feasible under standard state conditions:

$$2/3.Al_2O_3(s) + Si(s) = 4/3.Al(s) + SiO_2(s)$$

The standard Gibbs energy change for this reaction is approximately +200kJ per mole of oxygen up to the melting point of aluminium. However, in the presence of a relatively large amount of gold and a small amount of silicon, reaction becomes thermodynamically feasible for two reasons. Firstly aluminium can dissolve to form a dilute solution in the gold, and secondly the SiO_2 formed can react further with Al_2O_3 to form a mixed oxide phase. Thus the gold acts as a sink for aluminium and silicon for oxygen. These two effects combined overcome the high stability of Al_2O_3 and modify the interfacial protective film. The process can be represented as follows:

$$(2/3+x).Al_2O_3(s) + Si(s) = 4/3.[Al]_{\{Au(s)\}} + SiO_2.xAl_2O_3(s)$$

Even a minute amount of reaction of this type is likely to be sufficient to cause a reduction in the barrier effect that pure alumina presents to the transfer of Al- and Au-atoms across the Al-Au interface and accelerate purple plague formation. Under normal bulk reaction conditions rates of reaction between such ceramic materials would approach zero at these temperatures. With interfacial reactions however where surface rather than bulk diffusion is involved reaction rates may not be negligibly small.

Thermodynamic calculations using the MTDATA system[17] with the SGTE unary, solution and substance databases together with interaction coefficients (taken from references(13) and (14)) demonstrate the feasibility of the above reaction within the temperature range of the experiments described in references (6),(8) and (9). Some additional support for this proposed mechanism comes from both the colour and composition of the reaction product observed by Philovsky[8],[9] which was said to be blackish and lies on the aluminium-rich side but close to the line joining Al_2Au and Si. This is exactly the sort of composition that would be expected to result from a reaction forming purple plague with silica containing a small amount of aluminium. Clearly further work is required to substantiate this proposed mechanism for the effect of silicon on enhanced purple plague formation. One thing is certain: no ternary $Au_xSi_yAl_z$ phases are involved in this reaction.

CONCLUSIONS

The reported enhancement of the purple plague phase, Al_2Au, by silicon is not as suggested in the literature the result of the existence of any stable ternary Al-Au-Si phases. The $Al-Al_2Au-Si$ subternary is a simple ternary eutectic system with a single invariant. The reported effect of silicon may arise from a reaction involving this element and alumina at the aluminium-gold interface which becomes thermodynamically feasible in the presence of gold.

REFERENCES

1. H.A.Lauffenberger and T.R.Myers, Holm Seminar on Electric Contact Phenomena, (1969), S61-68.

2. C.W.Horsting, IEEE Int. Reliability Physics Symposium, 155-158, (1972).

3. J.A.Cunningham, Solid State Electronics, (1965), 8, 735-745.

4. R.Schmidt, IRE Trans. Electron Devices, (1962), ED-9, 506.

5. P.K.Footner, B.P.Richards and R.B.Yates, GEC Journal of Research, (1986), 4, 174-180.

6. B.Selikson and T.A.Longo, Proc. IEEE (1964), 52, 1638-1641.

7. "Phase Diagrams of Ternary Gold Alloys", A.Prince, G.V.Raynor and D.S.Evans,pp116-118, Institute of Metals, 1990.

8. E.Philovsky, Proc.8th IEEE Int. Reliability Physics Symposium, Las Vegas, (1970), 177-185.

9. E. Philovsky, Solid State Electronics, (1970), 13, 1391-1399.

10. V.N.Lozovsky and N.F.Politva, Izvest. Akad. Nauk. SSSR, Metally, (1977), 3, 36-38.

11. V.N.Lozovsky and A.I.Kolesnichenko, Ivest. Akad. Nauk. SSSR, Neorg. Materialy, (1981), 17, 737-738.

12. J.L.Murray and A.J.McAlister, Bull. Alloy Phase Diagrams, (1984), 5, 74-84.

13. J.L.Murray, H.Okamoto and T.B.Massalski, Bull., Alloy Phase Diagrams, (1987), 8, 20-29.

14. I.Ansara, Private Communication.

15. C.S.Smith, Trans. AIME, (1940), 137, 236-245.

16. F.H.Hayes, Proc.Int.Conf. 'Calorimetry, Thermal Analysis and Thermodynamics', Basle, Switzerland, August 1989.

17. T.I.Barry, R.H.Davies, A.T.Dinsdale, J.A.Gisby, S.M.Hodson and N.J.Pugh, MTDATA Handbook: Documentation for the NPL Metallurgical and Thermochemical Databank, National Physical Laboratory, Teddington, Middlesex, UK (1989).

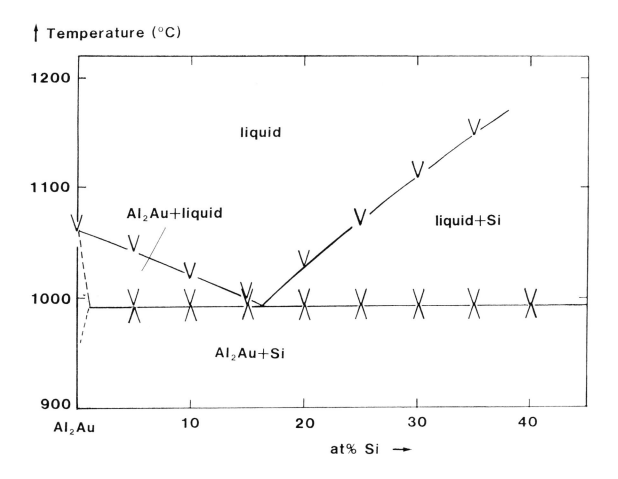

Fig. 1. Experimental results for the Al₂Au - Si section showing this to be a true pseudo-binary system.

126

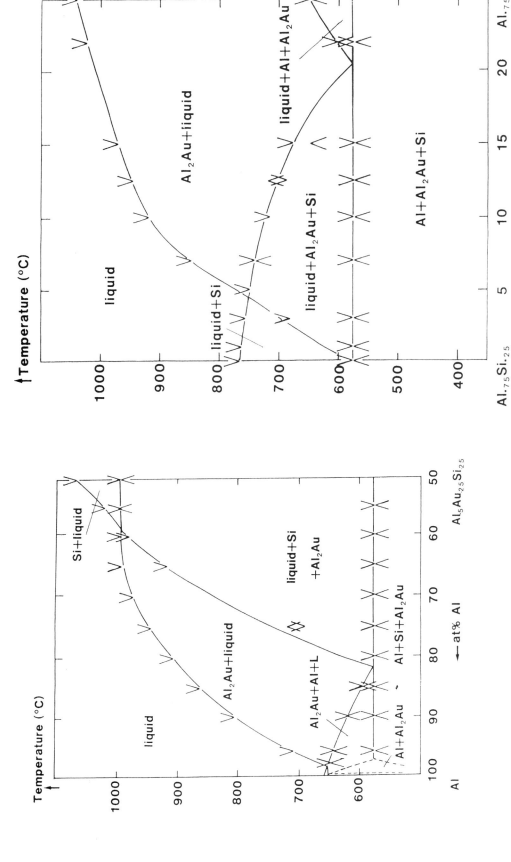

Fig. 3. Experimental results for the 75 at% Al section.

Fig. 2. Experimental results for the Al - Al$_{0.5}$Au$_{0.25}$Si$_{0.25}$ section.

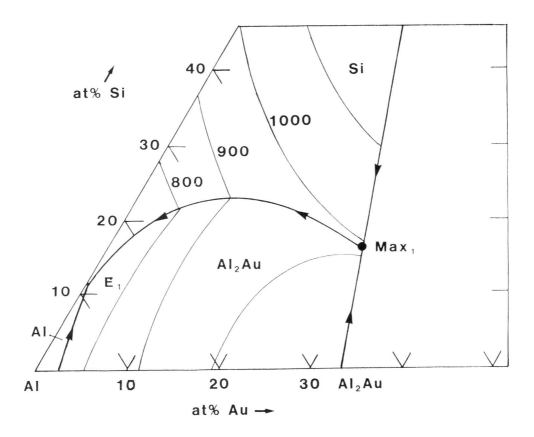

Fig. 4. Liquidus projection for the Al - Al₂Au - Si subternary system.

17

COMPUTER CALCULATED PHASE EQUILIBRIA: AN OVERVIEW OF INDUSTRIAL APPLICATIONS

T. G. CHART

Division of Materials Metrology, National Physical Laboratory, Teddington, Middlesex, UK

A knowledge of multicomponent phase equilibria is fundamental to materials science in terms of the extraction and refining of elements, their processing to marketable products and their end usage. These data are often unavailable, particularly for advanced and high-temperature materials, mainly due to the high costs involved and because of experimental difficulties. During the past decade the computation of phase equilibria from thermodynamic data has advanced significantly, in terms of both software and data, and it is now almost a routine matter to calculate equilibria for systems involving six or more components. Thus it is possible to calculate equilibria for real materials, and to provide meaningful predictions for a wide variety of alloys, molten salt systems, ceramics, slags and aqueous solutions. This technique is a rapid, reliable and cost-effective alternative to that of experimental determination, and is finding increasing industrial acceptance.[1]

This paper will illustrate briefly, using a series of examples, applications of calculated phase equilibria to processes of industrial relevance. Further information is given in references 2 and 3. The phase equilibria presented have been calculated using MTDATA. the NPL Databank for Metallurgical Thermochemistry, the applications of which have been described by Davies et al[4] and will be presented in a later paper as part of the present conference. Topics to be covered will include:

Phase equilibria for molten salt systems required to understand reactions involved in recovery and refinement of metals.

The solidification behaviour relevant to the modelling of segregation of impurities during continuous casting of steel.

Solid state reactions involving the embrittling sigma phase in multicomponent stainless steels.

The development of heat-resisting titanium-nitriding steels for nuclear reactor fuel claddings.

Production and control of superalloys with low inclusion levels.

The development of data for oxide systems of relevance to pyrometallurgy.

Phase equilibria for III-V electronic materials.

An example of the reliability with which multicomponent phase equilibria can be calculated from thermodynamic data for the binary sub-systems alone is shown in Fig. 1, a liquidus projection together with liquidus isotherms for selected temperatures for the $KCl-CaCl_2-ZnCl_2$ system. Data for this system are important to the recovery and separation of actinide metals. Superimposed on the diagram are experimental data made available subsequent to the calculation being performed. The level of agreement is remarkably good.

An example of calculated phase equilibria for a six-component system is shown in Fig. 2, and is relevant to titanium-stabilised stainless steels. This diagram quantifies in a convenient manner the proportions of the various phases present as a function of temperature. Due to the small amounts of carbides present a logarithmic ordinate has been used.

CONCLUSION

From the examples shown it will be evident that thermodynamic modelling, especially then coupled with available experimental data, allows a better understanding of many industrial processes and problems.

REFERENCES

1. L. Kaufman, "User Applications of Alloy Phase Diagrams", 1987, ASM International, Metals Park (Ohio).
2. "A Thermodynamic Approach to the Possibilities for Replacing Expensive Elements in High-Temperature Alloys", T. I. Barry, T. G. Chart, Commission of the European Communities Report, EUR 9564,1985.
3. "New Approach to Materials Design: Calculated Phase Equilibria for Composition and Structural Control", T. I. Barry, T. G. Chart, in "Research and Development of High Temperature Materials for Industry", ed E Bullock, 1989, Elsevier Applied Science, London and New York, p 565.
4. "Application of MTDATA to the Modelling of Multicomponent Equilibria", R. H. Davies, A. T. Dinsdale, T. G. Chart, T. I. Barry M. and H. Rand, in Proceedings of 6th International Conference on High Temperatures - Chemistry of Inorganic Materials, NIST, Gaithersburg, 1989.

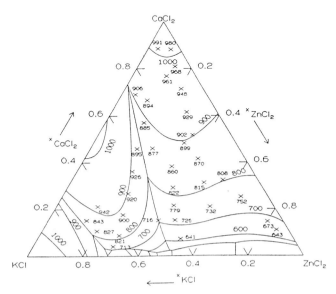

Fig. 1. Calculated liquidus projection and liquidus isotherms for the KCl-CaCl$_2$-ZnC12 system compared with experimental melting temperatures.

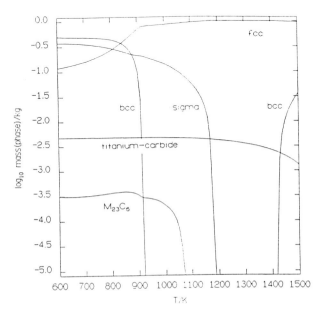

Fig. 2. Calculated equilibrium amounts of the phases present in a titanium stabilised stainless steel of composition in wt % 0.08 C, 18 Cr, 10 Ni, 1 Si and 0.4 Ti, balance Fe, as a function of temperature.

18

Thermo-Calc, A GENERAL TOOL FOR PHASE DIAGRAM CALCULATIONS AND MANIPULATIONS

B. SUNDMAN

Division of Physical Metallurgy, Royal Institute of Technology, Stockholm, Sweden

ABSTRACT

Various methods to define and calculate phase diagrams using the Thermo-Calc databank system are discussed and exemplified. The possibilities for manipulating the results of a calculation in the postprocessor are demonstrated. The use of the thermodynamic databank in order to generate alternative types of diagrams and to calculate single equilibria in order to complement the phase diagram information are shown.

INTRODUCTION

One of the most general computer programs for phase diagram calculations at present available is the Thermo-Calc system, developed at KTH in Sweden[1]. This program is a general thermodynamic workbench and in particular it allows the user to define any two-dimensional section through his system by combining various types of conditions. Examples of conditions are intensive state variables such as the temperature or the activity of a component or extensive variables like the volume or the fraction of a component. Many more general conditions can be used also, for example functions of state variables. This will be exemplified below.

USER INTERFACE

Thermo-Calc has a simple and easy-to-learn user interface and extensive on-line help. This makes it possible to learn the program by oneself after a short introduction. There is also a printed documentation set in three binders, one of which contains about 20 examples. In Fig. 1 the modular structure of Thermo-Calc is shown.

In Fig. 2 an extract from an on-line example is shown demonstrating some of the commands in the POLY-3 module. This example shows how one may obtain important information without calculating the whole phase diagram.

UNARY DIAGRAMS

Phase diagrams for unary systems i.e. systems with only one component, are not so common because they require that both the temperature and pressure properties are known. In Fig. 3 the unary diagram for pure iron is shown as a demonstration of the facilities of the post processor in POLY-3. All diagrams are plotted from one single calculation because in Thermo-Calc all information is saved during the calculation.

BINARY DIAGRAMS

This is the classical type of phase diagram with one composition axis and one temperature axis. There are several compilations of binary phase diagrams drawn by hand (or computer) but without use of thermodynamic calculations, the most recent published by NBS-ASM[2]. A user would normally not use a computerized data bank in order to find a binary phase diagram unless he also wants to obtain values for thermochemical data.

Binary phase diagrams must be assessed together with thermodynamic data in order that they can be used for the calculation of higher order systems. In this assessment one must select the models for the various phases and in Fig. 4 the scheme for assessment (of any system) is shown, taken from the Thesis by A. Fernandez Guillermet[3]. The choice of models for the phases is crucial for the extrapolation to higher order systems. As examples of recent binary assessments using sophisticated models the calculated binary Ti-C system from an assessment by H. Ohtani[4] is shown in Fig. 6 and the calculated Fe-O system from an assessment by Sundman[5] is shown in Fig. 6.

TERNARY DIAGRAMS

The problem of drawing phase diagrams instead of calculating them from thermodynamic data becomes immediately evident for ternary phase diagrams. The number of possible ways to present a ternary system is very large, isothermal sections, isopleths, isoactivity diagrams, liquidus surfaces etc. In very few cases one may find reliable experimental data for all possible sections. But with a computerized database one may from three binary systems make reliable extrapolations into the ternary system.

In Fig. 7 a liquidus surface of the system Fe-Cr-Ni is shown from the SGTE solution database[6]. Another type of ternary phase diagram is shown in Fig. 8 which is an isopleth calculation for the Al-Mg-Si system, also from the SGTE solution database. Finally Fig. 9 shows the calculated invariant equilibria in the system Al_2O_3-CaO-SiO_2 from an assessment by Hillert et al.[7].

MULTICOMPONENT PHASE DIAGRAMS

Binary and ternary phase diagrams are straightforward to calculate and understand but phase diagrams for systems with more components may be very difficult to interpret because they must be presented on two-dimensional paper and the number of lines will easily become very large. In Fig. 10 an isopleth calculation in a multicomonent alloy is shown. Such a diagram give a good overview of the system but is not very useful for alloy development because one does not know the amounts and compositions of the individual phases. Therefore one usually combines such diagrams with single equilibrium calculations as shown in Fig. 2 or with diagrams generated with just one axis variable as demonstrated in the next section.

PROPERTY DIAGRAMS

Often the user is satisfied with a calculation where only one variable at a time is varied for example the temperature or the amount of a component. One may then plot how some quantities such as the amounts of the phases or the activity of a component depend on the variable. Using the same steel as in Fig.10 one may calculate how this alloy varies at a fixed carbon content, here taken to be 0.8 %, as a function of temperature only. Various properties can then be plotted as shown if Figs. 11(a), (b) and (c).

With just one axis variable one may also use Thermo-Calc for various non-equilibrium calculations, the most important facility is the possibility to make solidification calculations assuming no back-diffusion in the solid, this is usually called a Scheil-Gulliver simulation. In Fig. 12 the fraction of solid phase versus temperature for solidification of a stainless steel is compared for an equilibrium solidification and a Scheil-Gulliver simulation. In Fig. 13 the change in Cr and Ni content in the liquid is shown for the same cases.

MODELS

Thermo-Calc can handle many models for describing non-ideality of phases. Examples are: the compound energy model[8], the two-sublattice ionic liquid model[9], cell model for slags[10], the associated model[11], and the Cluster Variation Method[12]. The magnetic transition is also modelled according to Inden[13] and aqueous solutions can be treated with the Pitzer model[14].

DATABASES

In order to obtain realistic results Thermo-Calc must have access to a database of assessed thermodynamic data. Such assessment work is important in order to cover many new materials where the database today is very limited. In Europe the Scientific Group Thermodata Europe (SGTE) is carrying out an ambitious effort to obtain a consistent and general thermodynamic database[6]. In Thermo-Calc one may also use a number of specialized databases like the Fe-base database for steels, the IRSID slag database and the Fe-base database for steels.

CONCLUSIONS

With a general tool like Thermo-Calc one may easily obtain any type of diagram. The possibility of performing many different types of calculation within the framework of a single easy-to-learn program makes it very valuable in order to understand what may happen if for example the composition or the heat treatment temperature of a material is changed.

Thermo-Calc has been used by steel industries in their research and development since 1985 and the response from the non-expert user community has been very important to improve the user-friendliness of the system. It is available on many different operating systems like VMS, UNIX, MS-DOS, OS/2 and NSO/VE.

REFERENCES

1. J-O. Andersson, B. Jansson and B. Sundman, Calphad Vol 9, (1985), p 153.
2. T. B. Massalski, J. L. Murray, L. H. Bennet, H. Baker, Phase Diagrams for Binary Alloys, ASM, Metals Park (1986).
3. A. Fernandez Guillermet, Thesis, Royal Institute of Technology (1988).
4. H. Othani, unpublished research.
5. B. Sundman, TRITA MAC 0344, Royal Institute of Technology (1990).
6. I. Ansara and B. Sundman, The Scientific Group Thermodata Europe, Proc. Conf. CODATA, Ottawa, (1986).
7. M. Hillert, X. Wang and B. Sundman, TRITA MAC 407 (1989).
8. M. Hillert, B. Jansson and B. Sundman, Z. Metallkde, (1988) 79, p 81.
9. M. Hillert, B. Jansson, B. Sundman and J. Ågren, Met. Trans A (1985)16A, p 261.
10. M. L. Kapoor and M. G. Frohberg, Arch. Eisenhuttenw (1974) 45, p 663.
11. F. Sommer, Z. Metallkde (1982) 73, p72.
12. B. Sundman and T. Mohri, Z. Metallkde (1990), 81, p 251.
13. G. Inden, Z. Metallkde (1975) 66, p 725.
14. K. S. Pitzer, J. Phys. Chem. (1973) 77, p 268.

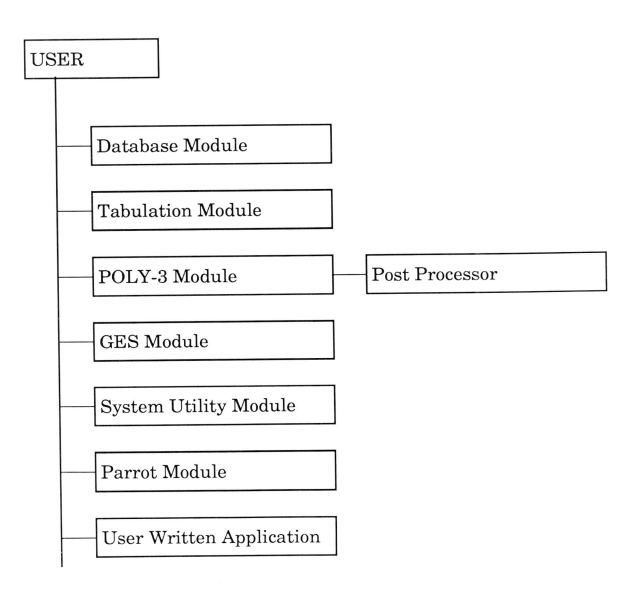

Fig. 1. Modular structure.

```
POLY_3:l-c
   ... the command in full is LIST_CONDITIONS
P=100000, N=1, W(CR)=1.5E-2, W(MN)=5E-3, W(SI)=3E-3, W(C)=3E-3,
W(NB)=1E-3
FIX PHASES FCC_A1
POLY_3:sh t
   ... the command in full is SHOW_VALUE
T=1008.4836
```

POLY_3:@@ *This is the maximum temperature for soft annealing in this steel.*
POLY_3:@@ *Now check which elements can change this temperature most by a*
POLY_3:@@ *small change. The change of the calculated temperature for a small*
POLY_3:@@ *change of the amount of a component acn be calculated as a derivative*
POLY_3:@@ *using the dot "." between the calculated variable and the condition*
POLY_3:**sh t.w(mn) t.w(cr) t.w(nb) t.w(c) t.w(si)**

```
   ... the command in full is SHOW_VALUE
T.W(MN)=-3246.8566
T.W(CR)=53.047067
T.W(NB)=-157.25418
T.W(C)=1167.58
T.W(SI)=1891.1811
```

POLY_3:@@ *A negative value means the temperature will decrease if the*
POLY_3:@@ *amount is increased. Assume we want to increase the Mn content*
POLY_3:**s-c w(mn)**

```
   ... the command in full is SET_CONDITION
Value /.005/: .006
POLY_3:c-e
   ... the command in full is COMPUTE_EQUILIBRIUM
 8 ITS,  CPU TIME USED   7 SECONDS
POLY_3:sh t
   ... the command in full is SHOW_VALUE
T=1005.2206
```

POLY_3:@@ *The temperature decreased about 3 degrees. According to the derivatives*
POLY_3:@@ *calculated above, one could increase the temperature with the same*
POLY_3:@@ *amount by increasing the amount of Si 1.5 times of the change in Mn*
POLY_3:**s-c w(si)**

```
   ... the command in full is SET_CONDITION
Value /.003/: .0045
POLY_3:c-e
   ... the command in full is COMPUTE_EQUILIBRIUM
 4 ITS,  CPU TIME USED   4 SECONDS
POLY_3:sh t
   ... the command in full is SHOW_VALUE
T=1008.0588
```

Fig. 2. Example.

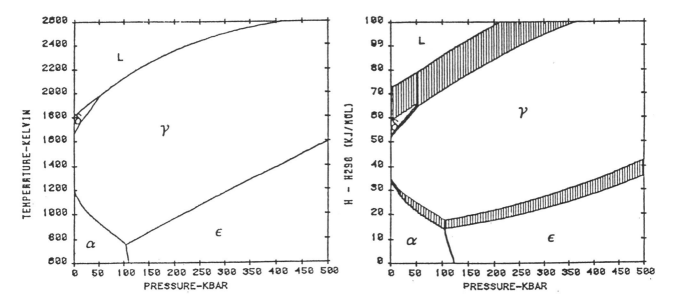

Fig. 3a Phase diagram for Fe,
temperature vs. pressure

Fig. 3b Phase diagram for Fe,
enthalpy vs. pressure

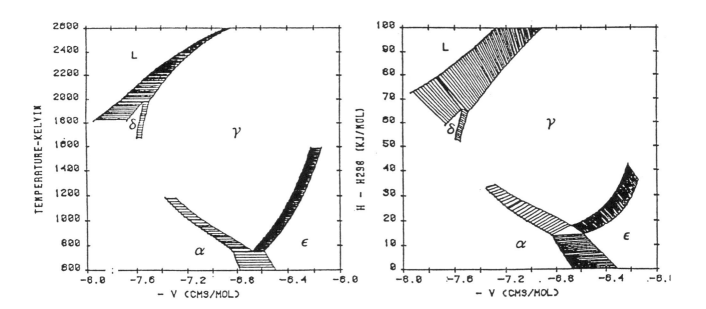

Fig. 3c Phase diagram for Fe,
temperature vs. volume

Fig. 3d Phase diagram for Fe,
enthalpy vs. pressure

Fig. 3. Unary Fe.

136

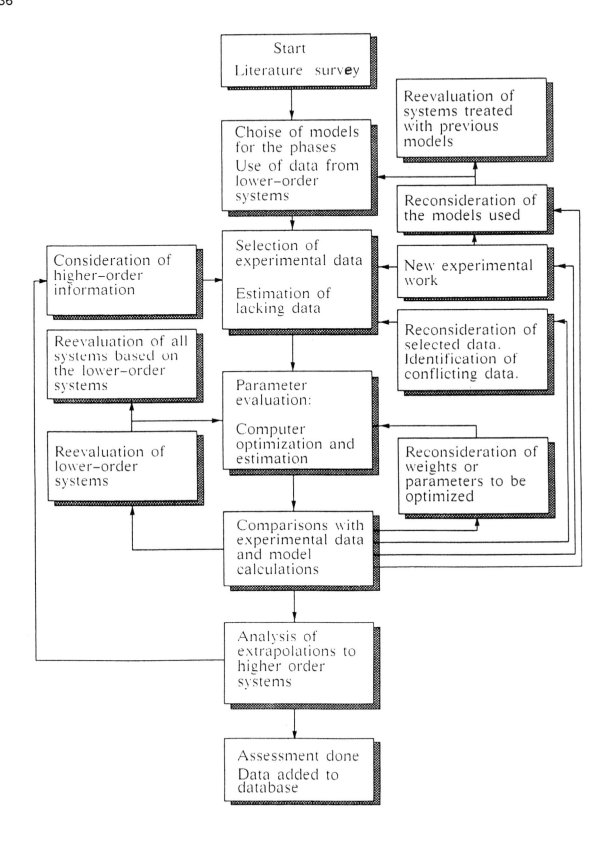

Fig. 4. Assessment scheme.

137

THERMO-CALC (90. 6.18: 4.54) :

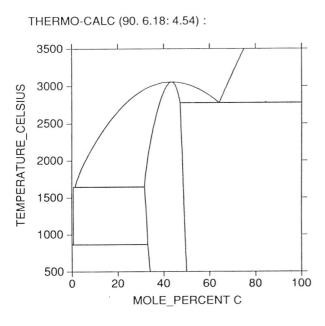

Fig. 5. Ti-C system.

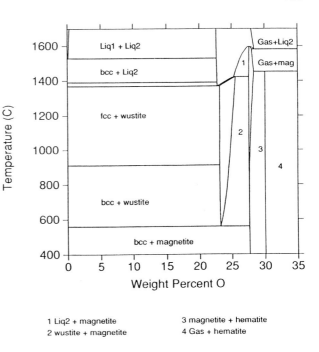

1 Liq2 + magnetite   3 magnetite + hematite
2 wustite + magnetite   4 Gas + hematite

Fig. 6. Fe-O system.

THERMO-CALC (89.10.15:15.11) :

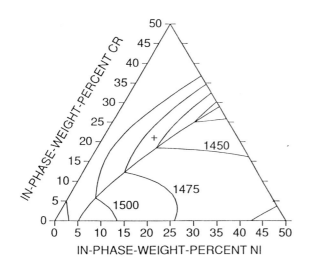

Fig. 7. Fe-Cr-Ni isotherm.

THERMO-CALC (90. 6.18: 4.37) :

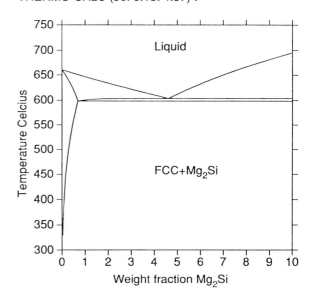

Fig. 8. Al-Mg-Si isopleth.

138

THERMO-CALC (90. 1.22:12.17) :Tick marks are degree C
Z-AXIS = 1000. + 100.0 * Z

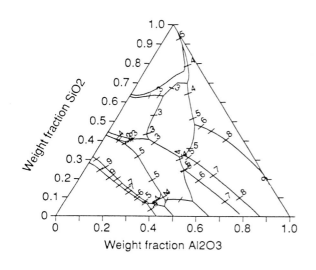

Fig. 9. Invariant equilibria in Al_2O_3-CaO-SiO_2.

Fig. 10. Isopleth.

THERMO-CALC (90. 6.18: 4.25) :

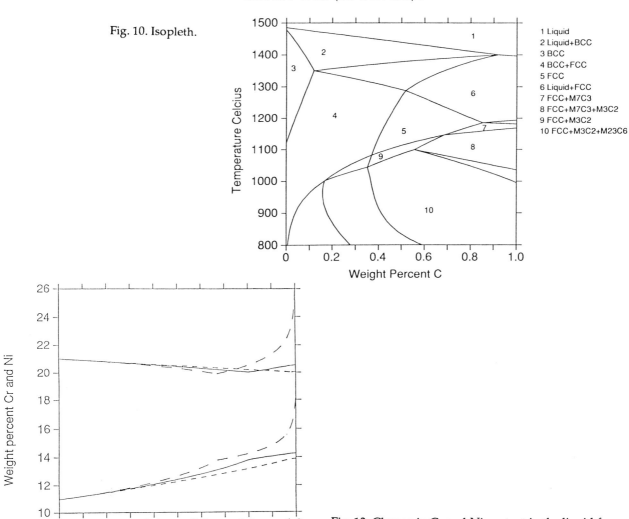

1 Liquid
2 Liquid+BCC
3 BCC
4 BCC+FCC
5 FCC
6 Liquid+FCC
7 FCC+M7C3
8 FCC+M7C3+M3C2
9 FCC+M3C2
10 FCC+M3C2+M23C6

Fig. 13. Change in Cr and Ni content in the liquid for same cases as Fig. 12.

THERMO-CALC (90. 6.18: 4.43) :

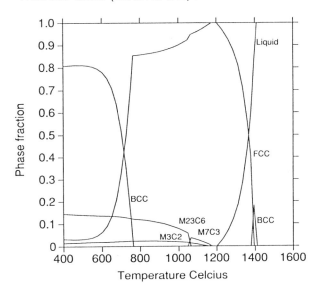

Fig. 11(a) phase fractions vs T.

THERMO-CALC (90. 6.18: 4.58) :

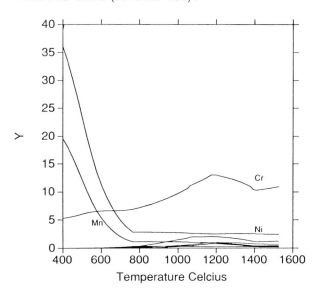

Fig. 11(b) composition in fcc vs T.

THERMO-CALC (90. 6.18: 5. 1) :

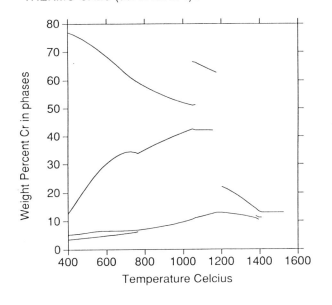

Fig. 11(c) fraction of Cr vs T.

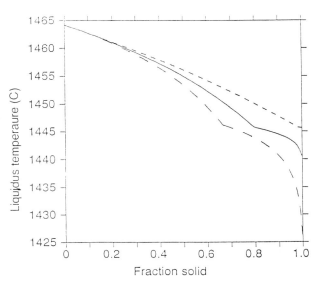

Fig. 12 Liquidus T vs solid fraction.

19

MTDATA — THE NPL DATABANK FOR METALLURGICAL THERMOCHEMISTRY

R. H. Davies, A. T. Dinsdale, S. M. Hodson, J. A. Gisby, N. J. Pugh,
T. I. Barry and T. G. Chart

Division of Materials Metrology, National Physical Laboratory, Teddington, Middlesex TW11 0LW, UK

ABSTRACT

The MTDATA databank is used as a tool in the analysis of diverse problems in chemistry, metallurgy and material science. Examples of areas of application include: pyro- and hydro-metallurgy, alloy development, process metallurgy, molten salt chemistry, hot and aqueous corrosion, pollution control, chemical vapour deposition, coating, etching and crystal growth.

Practical problems in these areas generally involve many components, deliberate or accidental, as well as the product phases: gaseous, liquid and crystalline. The principle behind the operation of MTDATA is that mathematical models incorporated in the software enable equilibria in multicomponent systems to be calculated on the basis of critically assessed data for the thermodynamic properties of simpler systems. Thus, using the facilities within MTDATA, it is possible to explore compositions and conditions of temperature and pressure for which no direct experimental data exist. Because MTDATA allows the user to monitor the results of equilibrium calculations in great detail and the effect of the thermodynamic data on these results, it is also a very useful educational tool and is already being used as such in a number of universities.

INTRODUCTION

The principles of calculation of phase equilibria, using as a basis critically assessed thermodynamic data for the components of the system and their binary and ternary combinations are well established. The review by Bale and Eriksson (90BAL/ERI) contains many examples. The purpose of this paper is to illustrate the use of MTDATA, the NPL Databank for Metallurgical Thermochemistry.

MTDATA is a single computer program comprising over 150000 lines of Fortran 77 source code. About 1000 lines of code are specific to the computer/operating system hosting the program. Normally the databank is operated interactively via a command language which allows the use of prompting, menus and access to the on-line help system at all times. Keywords are used for entering options in full or as unambiguous abbreviations. All user input is buffered so that more than one command sequence may be entered on a line of input. Those less familiar with MTDATA can enter the same set of commands on separate lines with prompting, menus and help being provided for each item as required. Groups of commands may be combined and stored in a file for subsequent execution as an MTDATA macro. Macros allow considerable customisation of the user interface with the options of inserting user input break points, extra prompting and explanatory text. The following example of user input is all that is required to compute and plot a ternary diagram for the Fe-Cr-Mo system at 1200K.

define system "Fe,Cr,Mo" source sgte_solution ! set temperature 1200 ! compute !

In the sequence above, there are three commands, define, set and compute, each of which is terminated by an exclamation mark. If the default database is the SGTE solution database (87ANS/SUN), the use of unambiguous abbreviations reduces the above to:

d sy "Fe,Cr,Mo" ! se t 1200 ! co !

The MTDATA databank comprises a number of modules for data retrieval and calculation, the TERNARY module being used in the above example. Each module will now be described and examples of use given. These examples have been selected to illustrate the way in which MTDATA works and are not necessarily taken from any specific practical application.

UTILITY

Data, in the form of assessed parameters for various models that describe the thermodynamic properties of phases, are stored in databases created and maintained by means of MTDATA's integral data management system. This allows large collections of data to be made readily available to the calculation modules of the databank. The UTILITY module is used for creating, loading and listing databases, which may include a user's own database or major data compilations such as the SGTE databases. The listing of the contents of a user's own database is achieved as follows.

list database my_data format name !

Only the names of the datasets would be listed such as Fe<LIQUID>, C,Fe<LIQUID> and Fe:C<CEMENTITE:3:1>. Complete data from non-proprietory databases could be listed simply by using the keyword data in the place of name. Any data listed by the UTILITY or THERMOTAB modules can also be loaded back into a database. These facilities in conjunction with macros allow a data manipulation environment to be built suitable for amending a single number in a dataset or managing large solution databases.

UTILITY also contains a facility for plotting graphics files that have been created by other modules and customised by the user to meet any requirement for publication, simplification or the highlighting of particular features.

ACCESS

The ACCESS module is used to retrieve thermodynamic data for systems of components. Some modules have direct database interfaces and can recover data directly; however, it is often more appropriate to use the ACCESS module, especially when a system is being searched for across more than one database. Databases available with MTDATA include the Scientific Group Thermodata Europe (SGTE) databases for pure substances, solutions and recommended unary values (89DIN, 90DIN) for the elements and key compounds.

Specifically the ACCESS module allows the user to define a system and a list of databases from which to retrieve data. For example:

define system "Ca,K,Zn,Cl" source my_data sgte_sub !

Here a search for data would be carried out in the private database MY_DATA and also in the SGTE Substance Database. As the system definition is in terms of elements as components all species containing these elements would be recovered. Systems can also be defined in terms of non-elemental components such as "KCl,CaCl2,ZnCl2". Calculations for this system would be confined to the plane determined by the three components, whereas the definition in terms of elements would allow the proportion of chlorine to vary independently of the other components. Similar considerations apply when the model for a phase includes charged species. A facility is included that allows the user to determine whether charge is formally a component or present in species but constrained to zero in phases.

The ACCESS module allows items of data (for phases or their unary constituents) to be rejected outright or substituted, under user control, by data from a database lower on the search list. Data may then be saved into a text file for use by other modules.

THERMOTAB

THERMOTAB is used for tabulating and plotting thermodynamic functions for individual substances or a chemical equation. The user enters chemical equations such as:

define equation "CO2<g> = CO<g> + 1/2 O2<g>" !

Equations are automatically balanced, if required, so the above equation could have been entered as "CO2<g> + CO<g> + O2<g> =" with THERMOTAB being left to insert the correct species multipliers. Data are automatically recovered from the chosen database and may be displayed in terms of the actual data stored in the database or tabulated and plotted to show the variation of the thermodynamic properties with temperature for chemical equations or individual substances. Users are at liberty to select from one of a set of standard table formats or, as illustrated in Table 1, to choose their own format. This feature enables easy comparison with established compilations which use a variety of tabulated thermodynamic functions. Plotting is equally flexible.

It is possible to select data from more than one database for different species in a THERMOTAB equation. This facility makes THERMOTAB very useful for the comparison of data from different sources as indicated by the following example, in which the data for substances 1 and 2 in the equation are derived from adata and bdata respectively.

define equation "CO<g> = CO<g> " ! use adata 1 bdata 2 !

Although THERMOTAB was originally designed to use data for pure condensed substances, ideal gaseous or aqueous species, the module's direct database interface allows all database data to be displayed. For example:

define equation "Fe,Ni:C<CEMENTITE:3:1>" ! list data 1 !

The equation keyword is used here to identify an interaction parameter for cementite, which is modelled in terms of two sublattices with site ratios of 3:1.

UNARY

The development of a thermodynamic database that can be used in the calculation of equilibria in multicomponent systems requires that all the data are consistent. This is achieved by building the data from a foundation, preferably internationally agreed, for pure components and their binary combinations. The components include the elements and important compounds such as salts and oxides.

Figure 1 shows the Gibbs energy calculated by UNARY as a function of temperature for three stable phases of titanium, liquid, hcp and bcc, as well as one unstable phase, fcc. The Gibbs energy curves are plotted relative to the bcc phase and cross at temperatures corresponding to the transition temperatures. The Gibbs energy of the fcc phase is important because it makes a major contribution to the solubility that will be predicted for titanium in that phase, as illustrated by Fig. 2.

BINARY

The fcc or γ-phase of pure iron is stable over the temperature range 1184.8K to 1667.5K. Titanium dissolves in the fcc phase but only to a limited extent. Figure 2 shows the use of BINARY to calculate the shape of the γ-loop and to compare the result with experimental data that have been assembled in a datafile. Once the user has retrieved the data, calculations are initiated by moving the cursor to a suspected two-phase field and pressing a defined key. Thereafter the calculations are completely automatic, except where phase fields are disconnected. The data used in the calculation of Fig. 2 have been used in the prediction of phase equilibria in multi-component steels (89BAR/CHA, 89HOW).

GPLOT

In the Fe-Cr system the σ-phase is stable over a limited range of temperature and composition. Figure 3 shows that, in the middle of the composition range, the Gibbs energy of the σ-phase is lowest, whereas, on either side, a two-phase combination of the σ and bcc phases is stable. GPLOT is used together with UNARY, BINARY and MULTIPHASE in the computer aided assessment of data. By arrangement with Dr H-L Lukas, the LUKAS optimiser (77LUK/HEN, 82LUK/WEI) is also available from within MTDATA and can exchange data with BINARY and GPLOT.

MULTIPHASE

The principal calculation module of MTDATA is MULTIPHASE. As the name suggests, this can solve equilibrium problems involving many phases and components. Given the composition, temperature and pressure (or volume) of the system, MULTIPHASE will calculate the equilibrium state in terms of the distribution of components and species between the stable phases. The volume or pressure will also be determined. Output may be tabular or graphical, the latter allowing a large number of plots to be drawn for the same set of equilibrium calculations. For the purposes of preparing plots users may either step one of the independent variables such as temperature or composition or specify arbitrary start and final compositions and the number of steps between them.

MULTIPHASE calculates equilibria using an extremely robust Gibbs energy minimisation procedure developed by Hodson which requires no initial guessing (88DIN/HOD). In the example illustrated in Fig. 4a-d equilibria have been calculated across the ternary system $CaO\text{-}FeO\text{-}Al_2O_3\text{-}SiO_2$ between a start and final composition. In order to display the often very extensive results of such calculations MULTIPHASE offers many plotting options. The example illustrates just four of these.

Figure 4a shows the masses of the phases present and Fig. 4b the distribution of SiO_2 between the phases. The composition of the liquid phase in terms of the components (the user has elected to confine the plot to the range over which the liquid is stable) is plotted in Fig. 4c. The final diagram of the set shows the species present in the liquid, some of which, for example Ca_2SiO_4, are associates. This type of diagram can be used in data assessment since it enables the user to monitor which constituents of the solution and therefore which interactions are making a significant contribution to the Gibbs energy. For purposes of data assessment and exploration of metastable phenomena it is possible to suppress individual substances or phases from the calculation. Hayes et al (90HAY/HET) have used this facility to model metastable phenomena in steels. The use of MTDATA in modelling the austenite - ferrite transformation and comparison with more approximate methods is described by Reed and Bhadeshia (91REE/BHA). Applications to the modelling of processing of ceramics have been described by Rand and Argent (89RAN/ARG) and to pyrometallurgical extraction from sulphide ores by Taylor and Dinsdale (90TAY/DIN).

Other variables that can be plotted include the activity of components and stoichiometric substances, the Gibbs energy relative to definable reference states, partial pressures, overall pressure and volume. The scale can be logarithmic or linear and its range can be chosen automatically or manually. Amounts can be expressed in terms of moles or mass. To simplify very crowded plots the user can focus on the compounds of a chosen component. Normally the abscissa is the same as the independent variable but a dependent variable can be selected.

In addition to the default Stage-1 algorithm described above, the Stage-2 process, similar to other established chemical equilibrium programs (90BAL/ERI), may be invoked to give extra accuracy in small species amounts when each thermodynamic component has a well established and independent activity in the system. Instead of the results being available to a constant tolerance in species amount, Stage-2 determines the speciation to a tolerance on a logarithmic scale.

If some information about the equilibrium state is available it may be useful to evaluate equilibria applying this information as a constraint. For example the oxygen partial pressure, the pH or the mole fraction of iron in the liquid phase may have been measured. It is almost always best to express the problems posed to MULTIPHASE in terms that match as closely as possible the experimental or actual conditions. This ensures that the problem has a solution and that the results are pertinent to the real situation.

TERNARY

Based on the MULTIPHASE Stage-1 Gibbs energy minimisation algorithm, the TERNARY module calculates and plots isothermal ternary phase diagrams via a simple graphical interface. The diagram is plotted automatically as equilibria are calculated. Facilities are provided, by the graphical interface, for labelling phase fields semi-automatically. Diagrams may be replotted with magnification, component rotation and experimental data overlaid.

Figure 5 shows a diagram calculated by TERNARY for the $FeO\text{-}Fe_2O_3\text{-}SiO_2$ system. In this system iron is present in two valence states and the equilibria are not confined within the system $FeO\text{-}Fe_2O_3\text{-}SiO_2$ because iron and oxygen are implied by making the amounts respectively of Fe_2O_3 or FeO negative. The diagram shows that at compositions along the boundaries of the region corresponding to the liquid phase, the slag is in equilibrium with variously bcc iron, cristobalite, oxygen gas (0.1 Pa) and wüstite and spinel (magnetite) solid solutions. The compositions of the coexisting phases correspond with the positions of the ends of the tie-lines.

TERNARY also allows diagrams for reciprocal systems to be calculated and plotted. Figure 6 illustrates this for the NaCl-KCl-NaI-KI system. In systems of this type there are only three independent components but symmetry demands that the diagram should be plotted on a square rather than a triangle. The model for the solid solutions allows independent mixing on two sublattices. In this particular case, because all the ions are univalent, the same simple model can also be used for the liquid phase. The parameters were assessed or validated from published information. The diagram shows the existence of a miscibility gap emmanating from the NaCl-NaI and KCl-KI systems. The predictions for the reciprocal system bear all the features indicated by experimental studies (35RAD). If altervalent ions were involved the variable site two-sublattice and other models could be used, thereby making it easy to compare the ability of the various models to represent the overall phase equilibria.

APPLICATION

The APPLICATION module provides an interface that allows user supplied Fortran 77 code to be linked with MTDATA. A simple example which has been put to use is the calculation of a composition surface for a range of slags for which the proportion of crystallisation by mass is set to a particular mass fraction. Also included as an application module is a facility for calculating the phases present as a function of temperature across a multicomponent system. Figure 7 shows an example for the join between two compositions in the CaO-FeO-Al$_2$O$_3$-SiO$_2$ system. The presence of ferric iron has been ignored. The diagram gives a general impression of the phase equilibria but not of the proportions or compositions of the coexisting phases. MULTIPHASE should be used for detailed calculations.

A set of "tools" is provided to facilitate the writing of applications. The applications may involve extensive computation, effectively using MTDATA as a set of subroutines to solve equilibrium or non-equilibrium problems.

COPLOT

COPLOT calculates and plots predominance area or Pourbaix diagrams of the type much used by those concerned with hot and aqueous corrosion (87BAR/DIN) and with hydrometallurgy. The application of MTDATA to the redox chemistry of H$_2$S in contact with water is described by Kelsall and Thompson (91KEL/THO). An example of a Pourbaix diagram is given in Fig. 8. Two components, iron and sulphur, are present in an aqueous environment. The diagram shows the dominant phases, containing respectively iron and sulphur, found within the range of applied pH and E$_h$. COPLOT has been written so as to allow the components whose behaviour is being investigated to be set by amount relative to 1kg of water and a gas phase of variable volume. This enables the module to be used to explore very rapidly a wide range of conditions, for example where sulphur and/or iron is present in solution at very low concentrations, as is important for corrosion avoidance. COPLOT displays only the solid phases or "predominant" aqueous or gaseous species, thus, once the most important variables have been established, MULTIPHASE should normally be used to obtain more detailed information.

FITANDPLOT

FITANDPLOT is a generalised multirange fitting module which integrates with other MTDATA modules. The user determines the function and manipulates the closeness of the fit either by setting a tolerance or fixing the splitting points between ranges. Errors in the input data can be exposed by means of a difference method and, if desired, semi-automatically corrected as indicated in Fig. 9.

User Aspects and Availability

Most of the modules of MTDATA are fully documented and equipped with online help. The user guide incorporates information on the data structures, the thermodynamic models and the principles of operation of the modules, particularly MULTIPHASE. The cooperation of users in the improvement and development of MTDATA is encouraged through the thriving MTDATA Users Group.

MTDATA is available for use on personal computers based on Intel 80386/486 processors with the MS-DOS operating system and DEC VAX computers under VMS. A UNIX version is planned.

ACKNOWLEDGEMENTS

The authors gratefully acknowledge the contributions to MTDATA made by NPL staff and others over the 20 years since the first primitive version of what is now THERMOTAB went on line. Sponsors of Mineral Industry Research Organisation Project RC54 are thanked for support of work on oxide systems.

REFERENCES

35RAD V. P. Radishchev, Zh. Oshch. Khim., 1935, 5, 465.

77LUK/HEN H.-L. Lukas, E. Th. Henig and B. Zimmermann, CALPHAD, 1977, 1, 225.

82LUK/WEI H.-L. Lukas, J. Weiss and E. Th. Henig, CALPHAD, 1982, 6, 229.

87ANS/SUN I. Ansara and B. Sundman, CODATA Report "Computer Handling and Dissemination of Data, 1987, 154.

87BAR/DIN T. I. Barry and A. T. Dinsdale, Mater. Sci. Technol., 1987, 3, 501-511.

88DIN/HOD A. T. Dinsdale, S. M. Hodson, T. I. Barry and J. R. Taylor, Proc. Conf. 27th Annual Conference of Metallurgists, CIM, Montreal, 1988, 11, 59

89BAR/CHA T. I. Barry and T, G, Chart, New approach to materials design: calculated phase equilibria for composition and structure control. "R & D of high temperature materials for industry", editor E. Bullock (1989) Elsevier, Barking, 565.

89DIN A. T. Dinsdale, NPL Report DMA(A)195, September 1989.

89HOW A. A. Howe, "The Segregation and Phase Distribution during Solidification of Carbon, Alloy and Stainless Steels", 1989, ECSC Final Report 7210.CF/801.

89RAN/ARG B. Rand, and B. B. Argent, Seramik Sains Malaysia, 1989.

90BAL/ERI C. W. Bale and G. Eriksson, Canad. Metal. Quart., 1990, 29, 105-132.

90DAV/DIN R. H. Davies, A. T. Dinsdale, T. G. Chart, T. I. Barry and M. H. Rand, High Temp. Science, 1990, 26, 251.

90DIN A. T. Dinsdale, "Binary Alloy Phase Diagrams, 2nd edition", 1990, ASM International, editor T. H. Massalski, Vol 1, T13.

90HAY/HET F. H. Hayes, M. G. Hetherington and R. D. Longbottom, Materials Sci. Technol. 1990, 6, 263.

90TAY/DIN J. R. Taylor and A. T. Dinsdale, Proc. Conf. "User Aspects of Phase Diagrams", 25-27 June 1990, Petten, Institute of Metals, "Application of the calculation of phase equilibria to the pyrometallurgical extraction from sulphide ores."

91KEL/THO G. H. Kelsall and I. Thompson, J Appl. Electrochem. in press.

91REE/BHA R. C. Reed and H. K. D. H. Bhadeshia, submitted to Materials Sci. Technol.

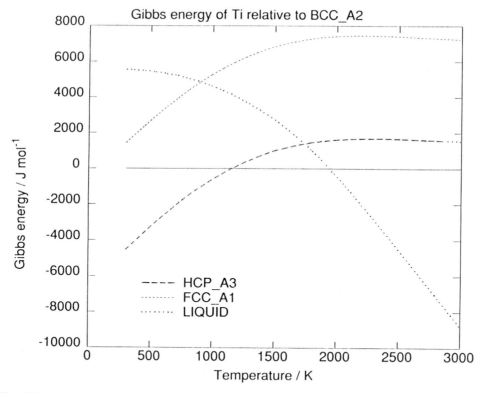

Fig. 1. The Gibbs energy of the liquid, hcp and fcc phases of titanium relative to the bcc phase as a function of temperature.

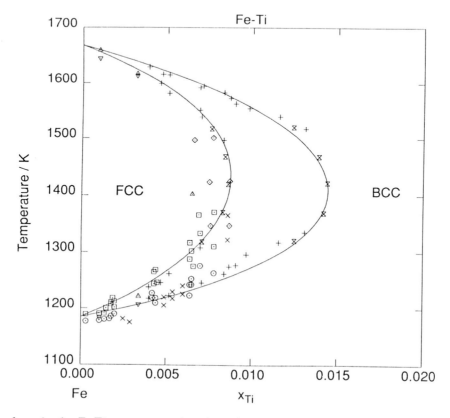

Fig. 2. The γ-loop in the Fe-Ti system as a function of temperature. A comparison of the calculated phase boundaries with the experimental observations on which they were based.

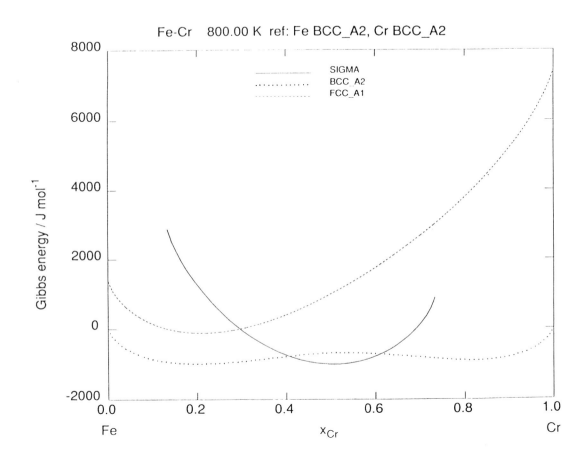

Fig. 3. The Gibbs energies of the phases in the Cr-Fe system as a function of composition. The stable phase combination is that of lowest Gibbs energy.

148

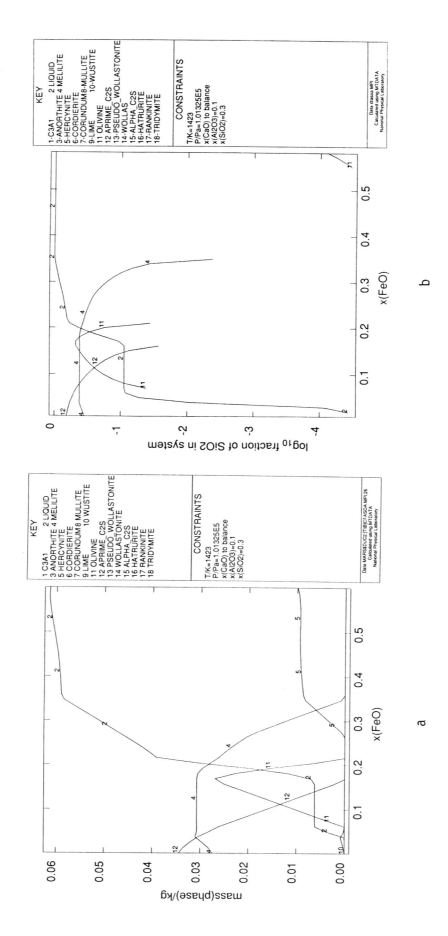

Fig. 4. Equilibria in the CaO-FeO-Al₂O₃-SiO₂ system as a function of composition by weight over the range 0.4 CaO, 0.1 FeO, 0.1 Al₂O₃, 0.4 SiO₂ to 0.1 CaO, 0.7 FeO, 0.1 Al₂O₃, 0-1 SiO₂ (a) the mass of phases, (b) the distribution of SiO₂ between the phases, (c) the composition of the liquid phase, (d) the speciation of the liquid.

149

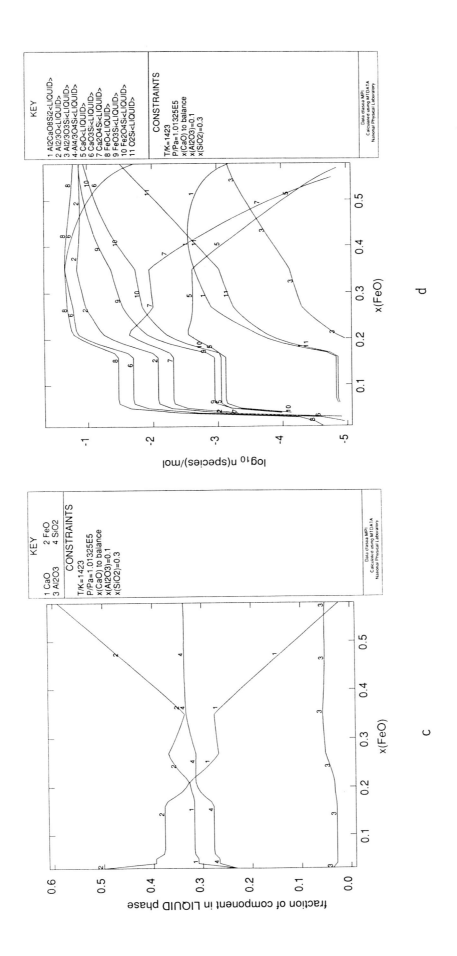

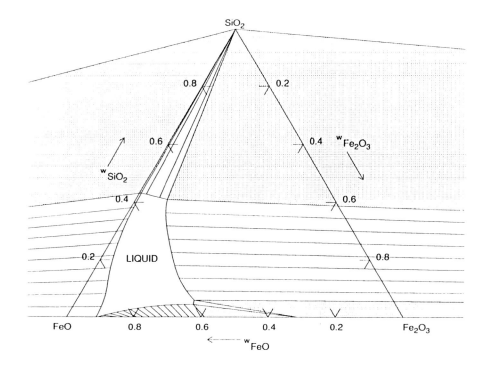

Fig. 5. Phase diagram for the FeO-Fe_2O_3-SiO_2 system at 1673 K, with an imposed pressure of 0.1 Pa. The ends of tie-lines and the corners of tie-triangles correspond to the compositions of coexisting phases.

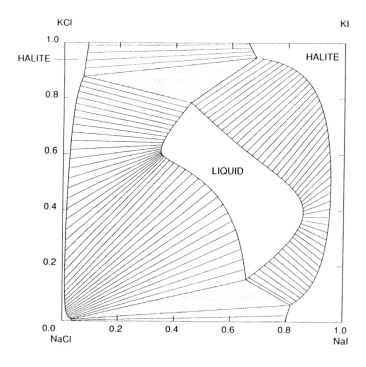

Fig. 6. Phase equilibria in the reciprocal system NaCl-KCl-NaI-KI at 825 K as calculated using the *reciprocal* option of the TERNARY module.

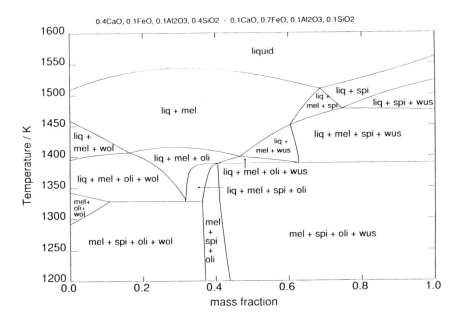

Fig. 7. Phase boundaries for the system CaO-FeO-Al₂O₃-SiO₂ as a function of temperature and a range of composition from 0.4 CaO, 0.1 FeO, 0.1 Al₂O₃, 0.4 SiO₂ to 0.1 CaO, 0.7 FeO, 0.1 Al₂O₃, 0.1 SiO₂. This diagram reveals the coexisting phases but not their compositions, which in general do not lie in the plane of the diagram. Compare with Fig. 5 in which this information is available as a function of a single variable. Note that as in Fig. 5 ferric ironwould also be present in the liquid and some crystalline phases.

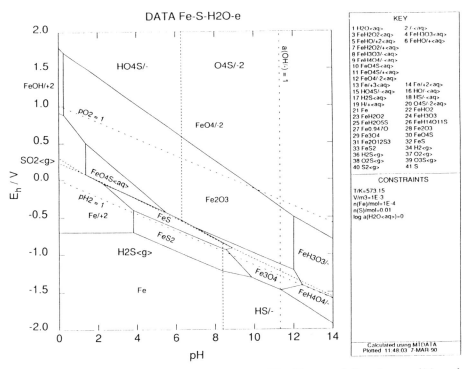

Fig. 8. Predominance area diagram for the Fe-S-H₂O system. The diagram delineates conditions for coexistence of aqueous species with each other and with condensed phases. The amounts of iron and sulphur in the coexisting substances have been set by the user to be 0.0001 and 0.01 moles per kg of water respectively. The behaviour of the sulphur (closely dashed lines, large labels) is largely independent of the iron, whereas that of the iron (solid lines smaller labels) is dependent on the chemistry of the sulphur. Note that SO₂ and H₂S have areas of predominance in the gas phase for the gas volume chosen of 0.001m³. The diagram also shows the 1 atm lines for H₂ and O₂ and the condition where the activity of the hydroxyl ion exceeds unity. Normally areas are labelled by numbers; the diagram has been edited and replotted by the UTILITY module.

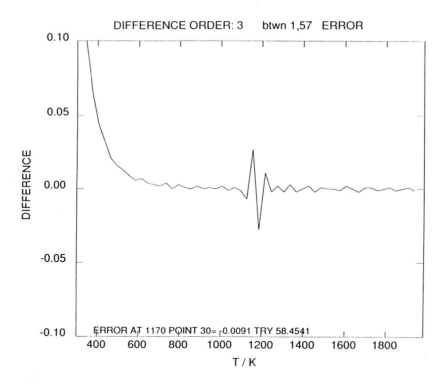

Fig. 9. A check of a set of data in which the user has entered 58.445 instead of 58.454 for the point at 1170 K. FITANDPLOT highlights the error and suggests an alternative value.

Table 1 Illustration of a user controlled format generated by THERMOTAB showing various thermodynamic properties as a function of temperature

CO<g>

T	Cp	S	H-H0	G-H0	-(G-H0)/T	-(G-H298)/T	G-Hser	Beta
K	J/K mol	J/K mol	J/mol	J/mol	J/K mol	J/K mol	J/mol	-G/RTln10
298.15	29.141	197.56	8673.4	-50228.	168.46	197.56	-1.69430E+05	29.683
500.00	29.818	212.73	14602.	-91761.	183.52	200.87	-2.10963E+05	22.039
1000.0	33.211	234.44	30365.	-2.04072E+05	204.07	212.75	-3.23274E+05	16.886
1500.0	35.201	248.32	47520.	-3.24962E+05	216.64	222.42	-4.44165E+05	15.467
2000.0	36.274	258.61	65418.	-4.51806E+05	225.90	230.24	-5.71008E+05	14.913

21

USER ASPECTS OF PHASE EQUILIBRIA IN HIGH T_C CERAMICS SUPERCONDUCTORS

R. S. ROTH

Ceramics Division, National Institute of Standards and Technology, Gaithersburg, MD 20899, USA

ABSTRACT

The systems of most user interest in the area of high T_c ceramic superconductors include the three component alkaline earth rare earth cuprates, e.g. $BaO-Y_2O_3-CuO$ and the four component alkaline earth bismuth/thallium cuprates, e.g. $SrO-CaO-Bi_2O_3-CuO$. In order to understand the equilibria and metastable formation of the various phases it is necessary to understand the reactions of all of the boundary binary (and ternary) systems. Most of our current knowledge of phase equilibria in these systems is confined to reactions in air, there is little data for any but the superconducting phases themselves in variable P_{O_2} conditions and even less data for CO_2 free atmospheric conditions (CO_2 - 5 ppm). The user oriented needs for these diagrams are generally two-fold: to grow large single crystals and to help in processing of high J_c (current carrying) bulk ceramics. A review of the literature together with new experimental data is used to construct possible phase diagrams for the various systems. Binary systems of interest include $BaO-Y_2O_3$, $Y_2O_3-CuO_x$, $BaO-CuO_x$, Bi_2O_3-CuO, $SrO-CuO$, $CaO-CuO$, $SrO-Bi_2O_3$, $CaO-Bi_2O_3$ and $SrO-CaO$. The ternary systems reviewed are $BaO-Y_2O_3-CuO$, $SrO-CaO-CuO$, $SrO-Bi_2O_3-CuO$, $CaO-Bi_2O_3-CuO$ and $SrO-CaO-Bi_2O_3$. Emphasis is placed on the role of P_{O_2} and P_{CO_2} on the equilibrium conditions of the various phases.

INTRODUCTION

Since the first discovery of high T_c superconductivity in oxides by Bednorz and Müller [1] and the consequent findings of still higher T_c by Wu et al [2] Cava et al [3] and Maeda et al [4], many attempts have been made to study the phase equilibria in various oxide systems in both cuprates and bismuthates. The reasons for the studies varied widely with the investigators' interests and expertise as did the methods used for the investigation and the interpretation of the results obtained. The obvious first user interest in phase equilibria in such systems is to determine the phase or phases responsible for the interesting properties, in this case high T_c. The second and most intense effort comes during the attempt to determine how to optimize the properties by chemical substitution. In most cases the original such studies were "cookbook", "table-top", type experiments which can only loosely be classified as phase equilibria studies. However, the intense world-wide interest in these phases quickly resulted in many more sophisticated experimental and theoretical studies. As the amount of information quickly grew beyond the ability of any individual to absorb all the published data, the types of information available and the user aspects of the studies became more specialized. Although, most engineering applications will require individualized special studies, the needs for accurate phase equilibria data are mostly in the fields of synthesis, first of all in single phase ceramics and secondly in crystal growth. However, special studies are badly needed for the optimization of high J_c, which is related to the ability of the specimen to carry a high current and still exhibit useful superconducting properties. This is apparently related to the formation, during synthesis of "pinning-sites" which may (or may not) be due to the emplacement of impurity atoms or micro-domains of secondary phases.

The oxide systems of interest in this field of high T_c superconductors are relatively limited, and presently confined mostly to cuprates, but also include a few bismuthates. This discussion will be limited to portions of the following systems:

1. $(Ba, Sr, Ca)O: (RE,Y)_2O_3: CuO_x$

2. $(Ba, Sr, Ca)O: (Bi, Tl, Pb)O_x:CuO_x$

Often still other oxides are added for property evaluation. These two complex systems include the superconducting phases with ideal formulas:

a. $La_{2-x}(Ba,Sr)_xCuO_4$
b. $(Ba,Sr)_{2-x}(RE,Y)_{1+x}Cu_3O_{6+y}$
c. $(Ba,K)(Bi,Pb)O_{3\pm x}$
d. $A_2Ca_{n-1}B_2Cu_nO_{2n+4}$ (A=Sr,Ba; B-Bi,Tl,Pb)
e. $Ba_2Ca_{n-1}TlCu_nO_{2n+3}$
f. $Pb_2Sr_2(Y,Ca)Cu_3O_{8+\delta}$

Few, if any, details of the phase equilibria are yet known for either the Tl or Pb containing systems. Experimental difficulties in these systems are exacerbated by low temperature volatility and problems of oxidation/reduction.

2. BINARY OXIDE SYSTEMS

The high T_c oxide phases are all very complex, containing at least three cations. Nevertheless, binary phase equilibria is the first step needed to understand the equilibria in the more complex 3, 4, and 5 component systems. Little is known of the relevant thermodynamic data needed to calculate the possible equilibria in these systems. Most of the available phase diagrams represent the results of interpretation of various experimental data derived from DTA, TGA, SEM, room temperature and H.T. x-ray diffraction and quench data, or often even less sophisticated techniques. The phase diagrams are mostly temperature/composition diagrams at constant P_{O_2}, usually air. There are a very few diagrams also available of variable P_{O_2}, generally plotted against temperature for constant cation ratios. There are essentially no complete phase diagrams constructed for all the relevant details of temperature - composition - P_{O_2} space. Even the more sophisticated phase equilibria papers seldom even mention the important aspects of specimen purity and possible contamination by container materials and atmospheric gases, to say nothing of the possible precision and accuracy of temperature measurements. For the purpose of discussion in a limited time and space this paper will be limited to the systems involved in the two superconductors, "$Ba_2YCu_3O_{6+x}$" and "$Sr_2CaBi_2Cu_2O_{8+\delta}$". Therefore, the "binary" systems involved in the production of these two phases include:

1. $BaO-Y_2O_3$
2. $Y_2O_3-CuO_x$
3. $BaO-CuO_x$
4. Bi_2O_3-CuO
5. $SrO-CuO$
6. $CaO-CuO$
7. $SrO-Bi_2O_3$
8. $CaO-Bi_2O_3$
9. $SrO-CaO$

None of the systems involving BaO or CuO are strictly binary in air as CuO is reduced to Cu_2O at $\sim 1026°C$ and some of the compounds can be oxidized, containing formally Cu^{+3} ions or peroxide molecules. In addition, most of these phases are extremely sensitive to the partial pressure of CO_2 in the environment and several of the observed phases are apparently structurally oxycarbonates.

2.1 $BaO-Y_2O_3$
In spite of the published phase diagram [5] the binary system $BaO-Y_2O_3$ probably contains only two compounds, BaY_2O_4 and $Ba_3Y_4O_9$, Fig. 1. All the other phases probably cannot be synthesized in atmospheres containing less than ~ 5 ppm CO_2. It cannot be emphasized too much that an understanding of the phase relations in the ternary system $BaO-Y_2O_3-CuO$ (and its binary boundary systems) must include a study in carefully controlled mixed gases with known P_{O_2} and P_{CO_2}. All gases, especially the oxygen, must be purified by passing the gas through a CO_2 getter like "Askerite".

2.2 $Y_2O_3-CuO_x$
A phase diagram for the system $Y_2O_3-CuO_x$ (Fig. 2) was also published in [5] for equilibrium in air. This is the only portion of the system that does not seem to be affected by CO_2, although, it is quite sensitive to P_{O_2}. The compound $Y_2Cu_2O_5$ can be reduced to $YCuO_2$ by heating in Ar or N_2 at $\sim 1000°C$. The latter phase has the dellafossite structure and shows poly-typic stacking faults which are probably quite sensitive to impurities in the gas and the actual P_{O_2} of the atmosphere. In addition, $YCuO_2$ can be oxidized by heating in O_2 at $\sim 500°C$, to at least $YCuO_{2.5}$, still with the dellafossite structure, with a large decrease in the c-axis and increase in the a-axis. No phase diagrams have been published for any but the air "isobar" for this binary system.

2.3 BaO-CuO$_x$

The system BaO-CuO$_x$, in air, was also reported [5] but no details were previously published of data for other isobars. The compound BaCu$_2$O$_2$ is known to occur at low P$_{O_2}$ and is an important impurity phase found in processing of the Ba$_2$YCu$_3$O$_{6+x}$ superconductor, especially at low P$_{O_2}$. However, this phase is also known to form, even in air, from specimens which have partially melted, either stably or metastably.

Apparently, the ternary eutectic liquid which forms in air, containing ~ 2-3% $^1/_2$Y$_2$O$_3$, has lost enough oxygen so that a BaCu$_2$O$_{2+x}$ phase crystallizes metastably from the melt. This phase is much less susceptible to reaction with atmospheric moisture than is pure BaCu$_2$O$_2$ and the d-spacings (amount of oxidation) are dependent on the temperature from which the melt is quenched.

BaCuO$_2$ is the only phase known to exist in air in this binary system. However, it has a wide P$_{O_2}$ stability range and is stable in both pure oxygen and "tank" Ar or N$_2$. It seems likely that the actual chemical formula is BaCuO$_{2\pm x}$ and even the exact cation/cation ratio is in doubt as neutron diffraction data does not confirm the reported X-ray diffraction structure. The compound Ba$_2$CuO$_3$ is known to form in one atmosphere O$_2$ but its exact P$_{O_2}$ stability limits have not been investigated. This phase also reacts quickly with atmospheric moisture and must be handled under "dry" conditions [6]. The compound Ba$_2$Cu$_3$O$_{5.5+x}$ has been found to form readily in one atmosphere of oxygen and although it is apparently quite stable even at 80:20 N$_2$/O$_2$, it will not form in the presence of CO$_2$ and is quickly decomposed with the formation of BaCO$_3$ when heated in air. The phase equilibria diagram of the system BaO$_2$-CuO$_x$ is shown in Fig. 3 for P$_{O_2}$ ~ 1.00. The Ba$_2$Cu$_3$O$_{5.5+x}$ phase begins to decompose above 800°C for all pressures between 1 and ~ 5 atmospheres of O$_2$. The 1:1 phase BaCuO$_{2\pm x}$ does not form at 700°C but is stable at 800°C under the same conditions.

Other phases higher in Ba content than 2:1 have been proposed for this system [7] but not confirmed. However, x-ray diffraction of specimens prepared in O$_2$ between BaO$_2$ and Ba$_2$CuO$_3$ show patterns not characteristic of pure BaO$_2$. It has been found that pure BaO$_2$ begins to lose oxygen at ~ 600°C and at 800°C in pure O$_2$ reaches a formula near BaO$_{1.75}$. A specimen of BaO$_2$ quenched from 800°C in a sealed Pt tube shows that the tetragonal BaO$_2$ lattice (CaC$_2$ structure type) has contracted so that the c-axis has changed from 6.851Å to 6.666Å with very little change in the a direction.

2.4 Bi$_2$O$_3$-CuO

The phase equilibrium diagram for the system Bi$_2$O$_3$-CuO was published [8] and redrawn in "Phase Diagrams for Ceramists" (PDFC) as Fig. 6392 [9]. It apparently contains only one compound Bi$_2$CuO$_4$ and only solidus data are given in the original paper. No further attempts to define equilibria in this system have been made as the melting relations in this binary have little effect on the phase relations of the quaternary system of interest.

2.5 SrO-CuO

The phase equilibria have been investigated in this binary system and details previously reported [10,11] are shown in Fig. 4. The compound Sr$_2$CuO$_3$ melts incongruently at ~ 1085°C. The third compound identified as Sr$_{14}$Cu$_{24}$O$_{41}$ [12] also melts incongruently at ~ 955°C. However, the difference in this incongruent melting temperature and that of the binary eutectic has not been found experimentally. From a practical point of view the actual melting temperatures are grossly affected by reaction with container materials, and Pt, Au, Ag/Pd alloys, Al$_2$O$_3$, etc., all give different experimental results.

2.6 CaO-CuO

Phase relations in the system CaO-CuO were previously reported [13] and a recent reinvestigation [14] Figure 5 shows the compound Ca$_2$CuO$_3$ melting incongruently at ~ 1030°C and the 1:2 phase CaCu$_2$O$_3$ existing only from ~ 985°C to 1018°C. The phase reported as Ca$_{1-x}$CuO$_2$ with the NaCuO$_2$ structure [15] was found to be single phase at a CaO:CuO ratio of 45.3:54.7 in one atmosphere of oxygen at 700°C. The exact composition of this phase should vary with change in the synthesis conditions of P$_{O_2}$ and temperature.

2.7 SrO-Bi$_2$O$_3$

The system SrO-Bi$_2$O$_3$ was previously reported [16] and redrawn as Fig. 6428 in PDFC [9]. The system has been completely reinvestigated and reported in [11] and [17] Fig. 6. The results of each investigation are fairly similar. The rhombohedral solid solution of SrO in Bi$_2$O$_3$ (the Sillen-phase) was found to have a maximum in the melting curve at about Sr$_{0.5}$Bi$_{1.5}$O$_{2.75}$ and ~ 960°C. "SrBi$_2$O$_4$" was found to have a non-quenchable phase transition at ~ 825°C, melting incongruently at ~ 940°C. The body centered tetragonal phase with a minimum at 765°C and a composition Sr$_{0.9}$Bi$_{1.1}$O$_{2.55}$ was found to extend to be essentially the high temperature polymorph of Sr$_2$Bi$_2$O$_5$ with a phase transition at about 925°C, melting incongruently at 985°C. The rhombohedral compound Sr$_3$Bi$_2$O$_6$ is the most stable phase in the binary system, melting incongruently to SrO plus liquid at ~ 1210°C. A new phase was also found at Sr$_6$Bi$_2$O$_9$ which decomposes in the solid state at ~ 965°C to Sr$_3$Bi$_2$O$_6$ plus SrO.

2.8 CaO-Bi₂O₃

The system $CaO-Bi_2O_3$ was previously reported [18] and redrawn as Fig. 6380 in PDFC [9]. This system has been completely reinvestigated and reported in [14] Fig. 7. The diagram shows considerable discrepancies with the original. The two phases previously reported as $Ca_7Bi_{10}O_{22}$ and $Ca_6Bi_5O_{13.5}$ are shown to actually be $Ca_4Bi_6O_{13}$ and $Ca_2Bi_2O_5$. These compositions have been confirmed by powder x-ray diffraction and single crystal structure determination [19]. Single crystals were grown mainly from salt solutions using Na_5K_5Cl as a flux. The rhombohedral Sillen-phase in this system has a maximum solid solution at about 840°C and 22% CaO, transforming to a face-centered cubic (fcc) solid solution which exhibits some superstructure (due to oxygen/vacancy ordering) near the minimum eutectoid and maximum CaO content. Two phases occur at low temperature probably due to Ca/Bi ordering. They have been tentatively assigned the compositions 5:14 and 1:2 (as was done in [18]) although the patterns are too complex to identify the exact composition. Both are triclinic when annealed at ~650°C, although the 1:2 phase apparently shows a somewhat simpler x-ray diffraction pattern at 700°C than at 650°C, indicating a possible phase transformation. $CaBi_2O_4$ transforms at about 725°C to the rhombohedral solid solution plus $Ca_4Bi_6O_{13}$, which again transforms to the "fcc" phase plus $Ca_4Bi_6O_{13}$ at about 775°C and again to the "body centered cubic" (bcc) phase at about 825°C. This "bcc" phase also shows superstructure (presumably due to oxygen/vacancy ordering) at the lower temperatures. $Ca_2Bi_2O_5$ is triclinic but transforms to an unknown symmetry polymorph. This transition was found to occur at ~885°C on the low CaO side of the compound and at ~925°C on the high CaO side. $Ca_2Bi_2O_5$ decomposes to the "bcc" solid solution plus CaO at ~955°C. The maximum temperature "bcc" composition at ~45% CaO melts incongruently at about 965°C.

2.9 SrO-CaO

An attempt has been made to determine the phase equilibria diagram for the system SrO-CaO but much of the detail still remains to be determined. $SrCO_3$ and $CaCO_3$ when heated together in air to about 1400°C form a single phase solid solution. However, when reheated at lower temperatures this solid solution is decomposed (unmixed) to two phases as shown in the previously unpublished Fig. 8. The appearance of this exsolution curve is unusual and may perhaps be influenced by the partial pressure of CO_2 in the atmosphere or a deviation of the unit cell dimensions from Vegard's law. Attempts to measure solidus temperatures of the solid solutions have not been successful. Experiments were performed in Mo or W containers with reducing atmospheres and melting always occurred at temperatures near 1600°C due to formation of alkaline earth molybdates and tungstates and reduction of the alkaline earth oxides. It is obvious that a containerless melt in oxidizing atmosphere (skull-melting) is the method needed to determine these melting relations. Unfortunately this technique has not been available during the course of this study.

3. TERNARY OXIDE SYSTEMS

The ternary systems of interest here include the systems $BaO-Y_2O_3-CuO$, $SrO-CaO-CuO$, $SrO-Bi_2O_3-CuO$, $CaO-Bi_2O_3-CuO$, and $SrO-CaO-Bi_2O_3$. Most of these systems are not really true ternary systems. At least copper oxide reduces to form Cu_2O at higher temperature and many of the phases of interest oxidize to include more oxygen than indicated by the components listed. In addition the alkaline earth oxides and phases formed with these oxides are very sensitive to the CO_2 content of both the starting materials and the surrounding atmosphere. A further experimental complication is caused by the great reactivity of many of the components with their container materials. The ternary liquids are especially corrosive to almost all noble metals, Au, Pt, Ag and Pd and incorporate most container oxides into the crystalline phases during solidification. During interpretation of experimental data special care must be taken to investigate and account for all of these complications. Unfortunately this is seldom done and even more seldom mentioned in the literature.

3.1 BaO-Y₂O₃-CuOₓ

The ternary system $BaO-Y_2O_3-CuO_x$ has probably been the most heavily studied oxide system in the shortest period of time, during the recorded history of phase equilibria studies in ceramic systems. This system has been investigated and reported by so many workers throughout the world that it is almost impossible to review all of the data and "sort-out-the-wheat-from-the-chaff". Unfortunately much of the published information has been accumulated by scientists with little or no experience or expertise in experimental investigation of ceramic oxide systems. The rush for publication in this field has resulted in the reporting of conclusions without the detailed description of the experimental conditions which are needed by a reviewer to gauge the reliability of the results. Most of the work has been done in air and very little has been reported on the equilibrium conditions at various partial pressures of oxygen with carefully controlled low CO_2 contents. This data is badly needed and only recently has any such information been forthcoming for any of the phases other than the superconductor $Ba_2YCu_3O_{6+x}$.

The first phase equilibria diagrams reported for this systems [5, 7, 20] were all done in air. These, and many subsequent published diagrams for instance [21] all agree on several major points but disagree strongly on others. The only areas of agreement are with respect to the existence of the two ternary phases BaY_2CuO_5 and $Ba_2YCu_3O_{6+x}$ and the pseudo-binary joins shown as solid lines in Fig. 9. There has been no attempt made to confirm the phase relations in the higher BaO portion of this system at a $P_{O_2}=0.2(O_2/N_2=20/80)$ with controlled $CO_2 \leq 5$ppm.

The major culprit responsible for the conflicting data in the high BaO portion of the system is a phase generally referred to as "the other perovskite" with a composition variously labelled 3:1:2, 4:1:2, 4:1:3, 8:1:4, etc. There is more or less general agreement that this phase may have a variable cation composition and that it certainly can be oxidized and reduced by heat treatment. However, there is no agreement on the conclusion of whether or not it must contain CO_2 in order to be a stable phase.

A recent paper [22] has given data on the appearance of the phase equilibria diagram at various P_{O_2} especially at the low end of the scale. However, these authors believe the "3:1:2" to be a stable phase in the system and show it occurring below $\sim 10^{-4}$ P_{O_2} where the 2:1:3 superconductor phase is no longer stable. Although this paper disagrees with the conclusion that "3:1:2" is only stable in the presence of CO_2 no experimental evidence is given for this conclusion. No diagrams have previously been published for the equilibrium conditions occurring at various temperatures for one atmosphere of O_2. However, the occurrence of both Ba_2CuO_3 and $Ba_2Cu_3O_{5+x}$ which do not form in air must cause considerable changes in the diagram. In addition two more ternary phases, previously reported only for very high O_2 pressure [23, 24] are now known to actually be stable at one atmosphere of O_2 [25]. These are $Ba_2YCu_4O_8$ and $Ba_4Y_2Cu_7O_{14+x}$. The 2:1:4 phase is known to be stable at 800°C but decompose at ≤ 850°C. The true stability limits of the 4:2:7 phase have not been delineated, Fig. 10.

Neither of these phases, nor the 2:1:3 phase can be formed at 700°C in O_2. Indeed, an equilibrium pseudo-binary join apparently exists between Y_2O_3 and $Ba_2Cu_3O_{5+x}$ at 700°C and one atmosphere O_2, Fig. 11. The possible nonbinary join between $Ba_2YCu_3O_{6+x}$ and CuO is shown in Fig. 12.

Although this system has been very extensively studied it still needs much more work under extremely carefully controlled atmospheric conditions before the necessary information can be obtained. Various methods have been hypothesized for obtaining higher J_c bulk ceramics in this system, including using the decomposition of 2:1:4 to 2:1:3 plus CuO to fix pinning sites. However, the details of the phase equilibria for this reaction have not yet been fixed.

3.2 SrO-CaO-CuO$_x$

Phase relations in the system SrO-CaO-CuO$_x$ were the earliest to be studied as a boundary system in the four component Bi_2O_3 superconductors. A diagram illustrating equilibria in air at ~ 950°C was previously published [13], Fig. 13. Extensive solid solution is shown for the three SrO-CuO$_x$ phases. A complete solid solution exists between Sr_2CuO_3 and Ca_2CuO_3. Ca^{+2} substitutes for Sr^{+2} in $SrCuO_2$ up to ~ 75% and in "$Sr_{14}Cu_{24}O_{41}$" up to ~ 50%. A new ternary phase was found with a very small homogeneity range $(Sr_xCa_{1-x})CuO_2$ x $\sim 0.15 \pm 0.02$. This phase has a small tetragonal unit cell and is essentially the end member of the homologous series $(Tl, Bi)_2(Ba, Sr)_2(Ca, Sr)_{n-1}Cu_nO_{4+2n}$ with all the (Tl, Bi) layers removed. It should be noticed that compositions in two phase solid solution regions do not contain equivalent alkaline earth ratios in each phase. The phase with the higher alkaline earth content will always contain more CaO than the phase in equilibrium with it containing more CuO. Later experimental work (unpublished) has indicated that the "$(Sr, Ca)_{14}Cu_{24}O_{41}$" solid solution probably deviates from the 14:24 ratio as the CaO content increases. At a given P_{O_2} (air ~ 0.2), more alkaline earth ions are needed as the Ca^{+2} content increases in order to allow full oxidation to 41 oxygen ions.

It is important to realize that in order to reach equilibrium in these CuO containing systems temperatures near 950°C must be used for solid state reactions. The four component superconductors melt below 900°C. Therefore, most solid state reactions used to synthesize the Bi^{+2} and Cu^{+2} containing superconductors never reach equilibria and contain excess unreacted CuO until after some melting occurs.

3.3. SrO-Bi$_2$O$_3$-CuO$_x$

The SrO-Bi$_2$O$_3$-CuO$_x$ system is probably the most complex of the four boundary ternary systems of the Bi_2O_3 superconductors, as well as the most difficult to reach equilibria. As melting occurs near 900°C in many of the compositions, the solid state equilibria must be approached at ~ 875°C. This is very near the decomposition temperature of $SrCO_3$ in air so that problems involved with the incorporation of CO_2 are also present in this system, although these problems are probably not as severe as for those systems involving BaO. The phase diagram for this system [17, 26], Fig. 14, shows four ternary phases $Sr_2Bi_2CuO_6$, $Sr_8Bi_4Cu_5O_{19+x}$, $Sr_3Bi_2Cu_2O_8$ and a solid solution region labelled the Raveauphase which corresponds to the general formula $Sr_{1.8-x}Bi_{2.2+x}Cu_{1\pm x/2}O_2(0.0\leq x \leq 0.15)$ (see enlargement of diagram, Fig. 15). Superconductivity was originally reported by Raveau [27] in this phase but the compositions of the phase are very variable. Superconductivity apparently only occurs at high SrO contents and possibly only from metastable compositions (negative values of x). It should be noted that the single layer Bi^{+3} superconductor which should have the ideal composition $Sr_2Bi_2CuO_6$ actually has more Bi^{+3} and less Sr^{+2} starting at $Sr_{1.8}Bi_{2.2}CaO_{6+x}$ and the actual $Sr_2Bi_2CuO_6$ phase has a different structure [26]. The compound $Sr_8Bi_4Cu_5O_{19+x}$ can only be made single phase in O_2 (or with Al^3 impurities) [17] and superconductivity in the Raveau phase is influenced by treatment in O_2 [28].

Note that CuO is in equilibrium with the maximum amount of SrO in the rhombohdral Sillen phase. Therefore the low temperature melting relations of the Bi_2O_3-CuO binary system do not affect the melting relations within the ternary system above this join.

3.4 CaO-Bi₂O₃-CuO

The CaO-Bi_2O_3-CuO system [29], Fig. 16, is the simplest of the ternary boundary systems of the Bi_2O_3-super-conductors. At ~ 800°C to the solidus temperatures no ternary phases are formed and $Ca_2Bi_2O_5$ exists in equilibrium with CuO and Ca_2CuO_3. At lower temperatures in O_2 where the phase $Ca_{1-x}CuO_2$ is stable (~ 700°C) there is some possibility that the pseudo-binary joins shift to equilibria with the 2:3 phases, but this has not been definitely proven. Note that again CuO is in equilibrium with the maximum amount of CaO in the Sillen rhombohedral solid solution so that the low temperature melting of the Bi_2O_3-CuO binary system does not affect the melting relations of this ternary.

3.5 SrO-CaO-Bi₂O₃

The system SrO-CaO-Bi_2O_3 [29], Fig. 17, shows considerable solid solution and two new monoclinic phases. Note that although CaO can substitute for SrO in several of the phases there is no sign of substitution of SrO in the CaO-Bi_2O_3 binary phases. It is apparently easy to substitute a small ion, Ca^{+2}, for a larger ion, Sr^{+2}, but the larger ion cannot fit into the structure of the phase containing the smaller ion. The crystal structure of these binary and ternary alkaline earth bismuth phases are currently under investigation to help to elucidate this phenomena and throw some light on the possible structural and compositional variations which might be predicted for the superconducting four component phase(s).

In order to determine the melting conditions in this ternary system it is necessary to establish non-binary cuts through the ternary. One such diagram, previously unpublished, is shown in Fig. 18 for the join Bi_2O_3-$(Sr_{0.5}Ca_{0.5})O$. Note that the rhombohedral Sillen phase exists all the way to the liquidus, with a maximum near 25% alkaline earth, as was found for the SrO-Bi_2O_3 binary. The C-centered monoclinic phase near 47% alkaline earth melts incongruently, while the primitive monoclinic phase near 67% alkaline earth decomposes in the solid state. The $Sr_3Bi_2O_6$ rhombohedral phase accepts 50% CaO substituted for the SrO and is still the most stable compound in the system.

4. THE QUATERNARY SYSTEM SrO-CaO-Bi₂O₃-CuO

Many papers have been written discussing the synthesis of the quaternary phases with the ideal formulas of "$Sr_2CaBi_2Cu_2O_8$" and "$Sr_2Ca_2Bi_2Cu_3O_{10}$", with two and three "$(Ca,Sr)CuO_2$" layers respectively. However, no detailed studies have been reported showing the phase relations of this four-component system. General agreement seems to exist that the two-layer phase can be synthesized essentially single phase at the 3:3:3:4 composition but not at the 2:1:2:2 composition. All workers agree that the three-layer phase is more stable with additional PbO but there is no general agreement as to whether this phase can be formed single phase at some composition in the quaternary system. Several workers [30, 31] have listed the phases which were found in equilibria with the two-layer (~80K) superconductor although not enough details have been published yet for general agreement. The first of these [30] indicates that the "2:1:2:2" phase forms pseudo-binary joins at 850°C with CaO, $(Sr,Ca)_2CuO_3$, $(Sr,Ca)_{14}Cu_{24}O_{41}$, CuO, Liquid, the "Raveau-Solid Solution", $Sr_6Bi_2O_9$ and both ternary SrO-CaO-Bi_2O_3 monoclinic solid solutions. In the second study [31] the joins found in equilibrium with "2:1:2:2" at 800°C (with and without Li_2-CO_3 added as a "mineralizer") were CuO, Ca_2CuO_3, the "Raveau Solid Solution" and both of the SrO-CaO-Bi_2O_3 monoclinic phases. There are many other binary and ternary phases in the quaternary system and the complete equilibria tetrahedra for this system have yet to be determined.

Three psuedo-ternary planes of $(Sr,Ca)O$-Bi_2O_3CuO are currently being examined in this laboratory, at the Sr/Ca ratios of 2:1, 1:1 and 1:2. The equilibria are being studied by first forming single phase alkaline earth bismuth oxide and alkaline earth copper oxide end members and then mixing two or more phases to determine the quaternary compatibility tetrahedra. This is necessary in order to avoid metastable melting of the Bi_2O_3 (m.p ~ 825°C) and ensure that no metastable CuO exists in the specimens. It is hoped that the single phase region of the two-layer phase can thus be mapped and a determination made as to the stability of the three layer phase.

ACKNOWLEDGEMENTS

This paper summarizes that status of our knowledge of phase equilibria diagrams pertaining to high T_c

superconductor systems drawn mostly from studies made at the National Institute of Standards and Technology (formerly The National Bureau of Standards). Persons who have contributed substantially to these studies include, C. J. Rawn, B. P. Burton, F. Beech, J. D. Whitler, J. J. Ritter, N. M. Hwang, and S. P. Bayrakci. We wish to than L. A. Bendersky for electron diffraction data, J. P. Cline for high temperature x-ray diffraction patterns, and M. D. Hill for DTGA, as well as S. K. Peart and T. R. Green for the computer-graphic representations of the phase equilibria diagrams.

REFERENCES

1. J. G. Bednorz and K. A. Müller, Z. Phys. B-Condensed Matter 64, 189 (1986).
2. M. K. Wu, J. R. Asburn, C. J. Torng, P. H. Hor, R. L. Meng, L. Gao, Z. J. Huang, Y. Q. Wang and C. W. Chu, Phys. Rev. Lett. 58, 908 (1987).
3. R. J. Cava, B. Batlog, R.B. VanDover, D. W. Murphy, S. A. Sunshine, T. Siegrist, J. R. Remeika, E. A. Rietman, S. Zahurak and G. P. Espinosa, Phys. Rev. Lett. 58 1676 (1987).
4. H. Maeda, Y. Tanaka, M. Fukutomi and T. Asana, Jap. J. Appl. Phys. 27 L209 (1988).
5. R. S. Roth, K. L. Davis and J. R. Dennis, Ad. Ceram. Mat. 2 [3B], 303 (1987).
6. W. K. Ng-Wong, K. L. Davis and R. S. Roth, J. Am. Ceram. Soc. 71 [2], C64 (1988).
7. K. G. Frase and D. R. Clarke, Ad. Ceram. Mat. 2 [3B], 295 (1987).
8. B. G. Kakhan, V. B. Lazarev and I. S. Shaplygin, Zh. Neorg. Kim 24 1663 (1979). Russ J. Inorg. Chem. (Engl. Transl.) 24 922 (1979).
9. Phase Diagrams for Ceramists, Vol. 6, R. S. Roth, J. R. Dennis and H. F. McMurdie, Eds., Amer. Ceram. Soc. Westerville, OH (1987).
10. R. S. Roth, C. J. Rawn, J. D. Whitler, C. K. Chiang and W. K. Wong-Ng, J. Amer. Ceram. Soc. 72 395 (1989).
11. N. M. Hwang, R. S. Roth and C. J. Rawn, J. Amer. Ceram. Soc. (accepted for publication) (1990).
12. T. Siegrist, L. F. Schneemeyer, S. A. Sunshine, J. V. Waszczak and R. S. Roth, Mat. Res. Bull. 23, 1429 (1988).
13. R. S. Roth, C. J. Rawn, J. J. Ritter and B. P. Burton, J. Amer. Ceram. Soc. 72 [8] 1545 (1989).
14. N. M. Hwang, R. S. Roth, C. J. Rawn and J. J. Ritter, J. Amer. Ceram. Soc., (to be submitted) (1990).
15. T. Siegrist, R. S. Roth, C. J. Rawn J. J. and Ritter, Chem. of Mater. (accepted for publication) (1990).
16. R. Guillermo, P. Conflant, J. C. Bovin and D. Thomas, Rev. Chim. Min. 15 153 (1978).
17. R. S. Roth, C. J. Rawn, B. P. Burton and F. Beech, J. Res. Natl. Inst. Stand. Technol. 95, [3] 1 (1990).
18 P. Conflant, J. C. Bovin and D. Thomas, J. Solid State Chem. 18 133 (1976).
19. J. B. Parise, C. C. Torardi, M. H. Whangbo, C. J. Rawn, R.S. Roth and B. P. Burton, Chem. of Mater. (Submitted for publication) (1990).
20. G. Wang, S. J. Hwu, S. N. Song, J. B. Ketterson, L. D. Marks, K. R. Poeppelmeier and T. O. Mason, Ad. Ceram. Mat. 2 [3B] 313 (1987).
21. D. M. De Leeuw, C. A. H. A. Mutsaers, C. Langereis, H. C. A. Smoorenburg and P. J. Rommers, Physica C. 152 [1] 39 (1988).
22. B. T. Ahn, V. Y. Lee, R. Beyers, T. M. Gür and R. A. Huggins, Physica C 167, 529 (1990).
23. J. Karpinski, E. Kaldis, E. Jilek and S. Rusiecki, Nature (London) 336, 660 (1988).
24. D. E. Morris, J. H. Nickel, J. Y. T. Wei, N. G. Asmar, J. S. Scott, U. M. Scheven, C. T. Hultgren and A. G. Markelz, Phys. Rev. B 39 7347 (1989).
25. D. M. Pooke, R. G. Buckley, M. R. Presland and J. L. Tallon, Phys. Rev. B. 41 6616 (1990).
26. R. S. Roth, C. J. Rawn and L. A. Bendersky, J. Mater. Res., 5 46 (1990).
27. C. Michel, M. Hervieu, M. M. Borel, A. Grandin, F. Deslandes, T. Provost and B. Raveau, Zeit. Phys. B. 68 421 (1987).
28. B. C. Chakoumakos, J. D. Budai, B. C. Sales, and E. Sonder, High Temp. Superconductors, ed. J. B. Torrance,

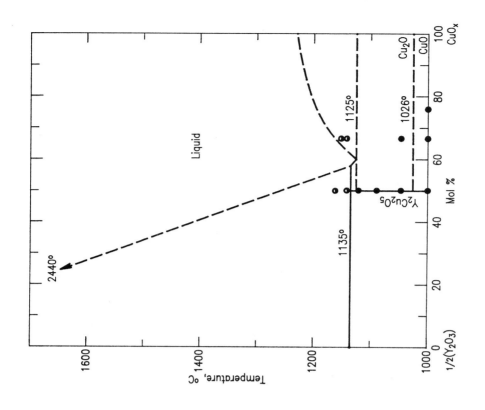

Fig. 2. Phase equilibria in the system Y_2O_3-CuO_x (after ref. 5).

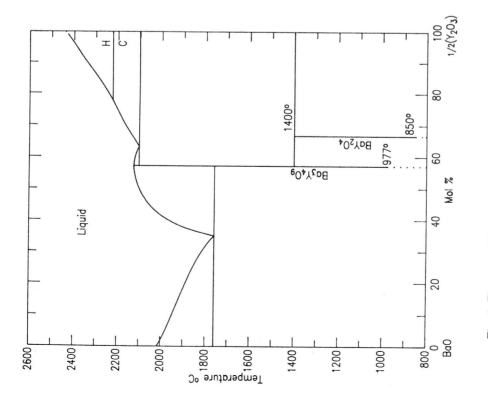

Fig. 1. Phase equilibria in the system BaO-Y_2O_3, as calculated by J.D. Anderson (ref. 5); modified by assuming "$Ba_2Y_2O_5$" and "$Ba_4Y_2O_7$" are really oxycarbonates and BaY_2O_4 and $Ba_3Y_4O_9$ are probably stable at temperatures well below the originally reported values.

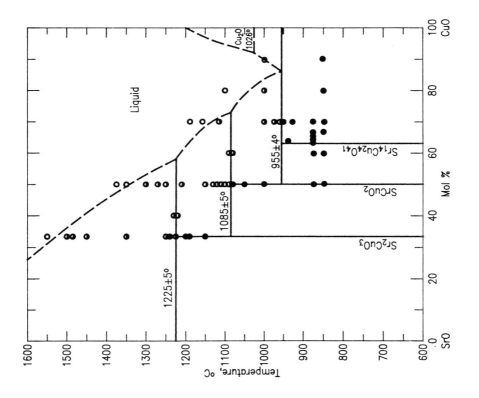

Fig. 4. Phase equilibria in the system SrO-CuO$_x$ (after refs. 10, 11, 17).

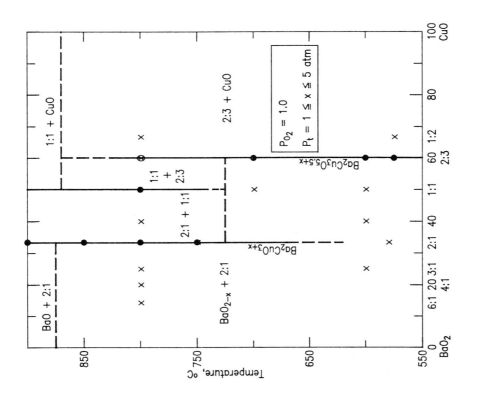

Fig. 3. Phase equilibria in the low temperature portion of the system BaO$_2$-CuO(P$_{O2}$= 1.0, P$_{total}$ = 1≤ x ≤ 5 atm, P$_{O2}$ ≤ 5ppm).

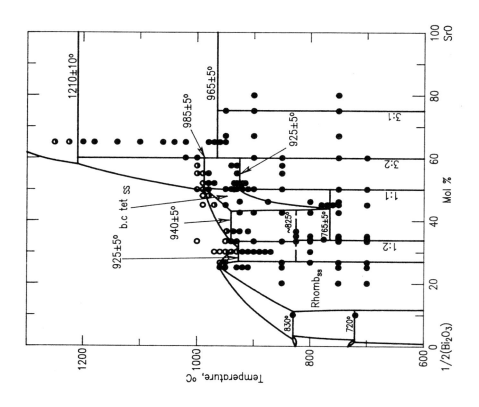

Fig. 6. Phase equilibria in the system SrO-Bi₂O₃ (after refs. 11, 17).

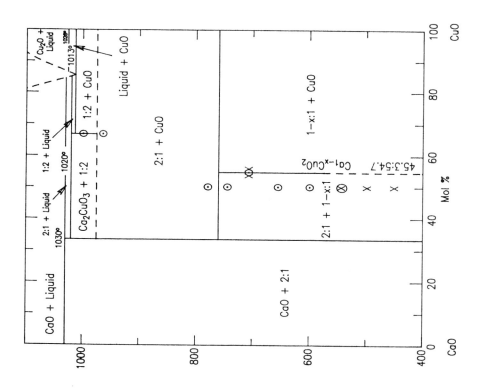

Fig. 5. Phase equilibria in the system CaO-CuO (after refs. 13, 14).

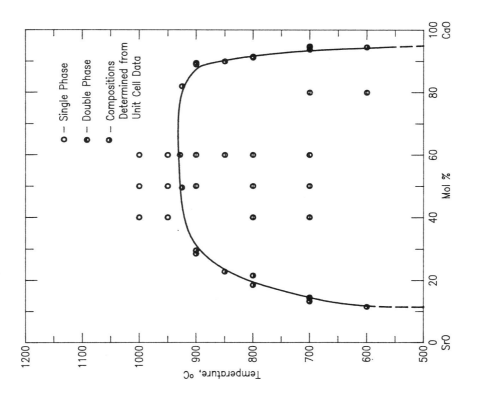

Fig. 8. Phase equilibria in the low temperature portion of the system SrO-CaO.

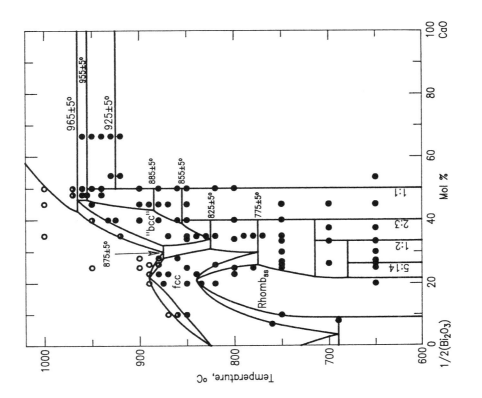

Fig. 7. Phase equilibria in the system CaO-Bi₂O₃ (after ref. 14).

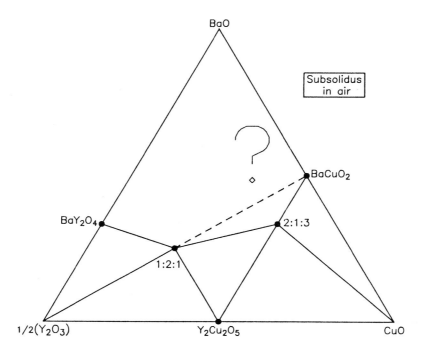

Fig. 9. Generally agreed phase relations in air for subsolidus conditions in the system BaO-Y$_2$O$_3$- CuO.

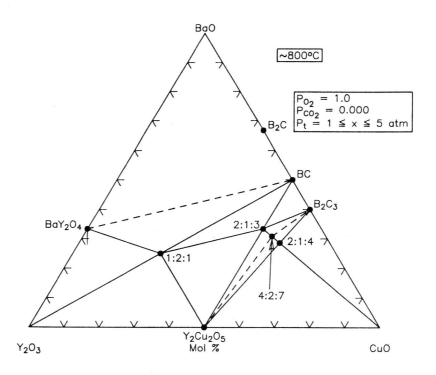

Fig. 10. Probable phase relations in oxygen at 800°C for the system BaO-Y$_2$O$_3$-CuO(P_{O2} =1.000, P$_{total}$ = 1 ≤ x ≤ 5 atm, P$_{CO2}$ ≤0.000).

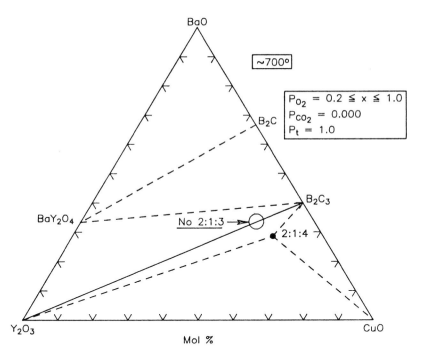

Fig. 11. Probable phase relations at ~ 700°C in the system BaO-Y$_2$O$_3$-CuO(P$_{O2}$ = 0.2 ⩽ x ⩽,P$_{total}$ = 1.0 atm, P$_{CO2}$ =0.000).

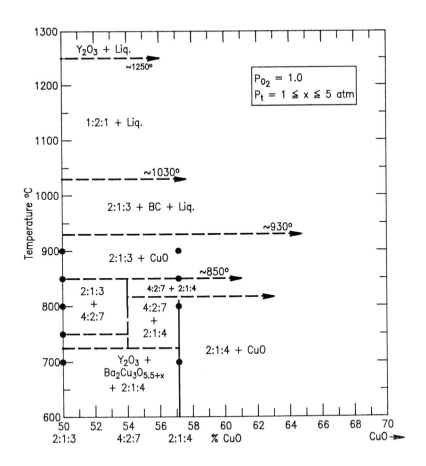

Fig. 12. Probable phase relations in the non-binary system Ba$_2$YCu$_3$O$_{6+x}$ -CuO (P$_{O2}$ = 1.0, P$_{total}$ = 1 ⩽ x ⩽ 5 atm).

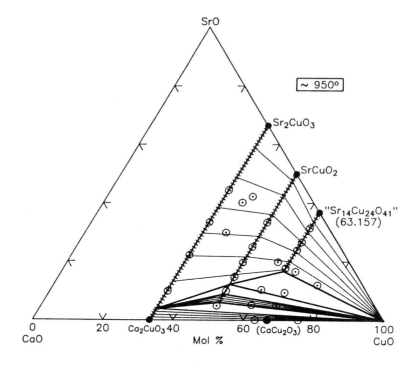

Fig. 13. Phase relations in the system SrO–CaO–CuO (after ref. 13).

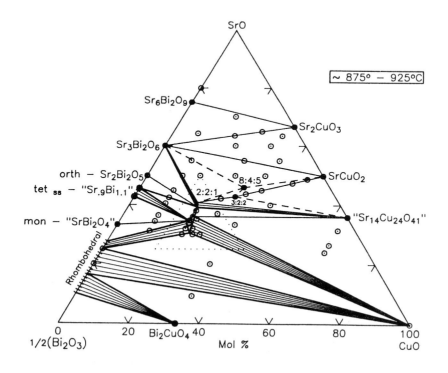

Fig. 14. Subsolidus phase relations in air in the system SrO–Bi$_2$O$_3$–CuO near 875°–925°C (after ref. 17, 26).

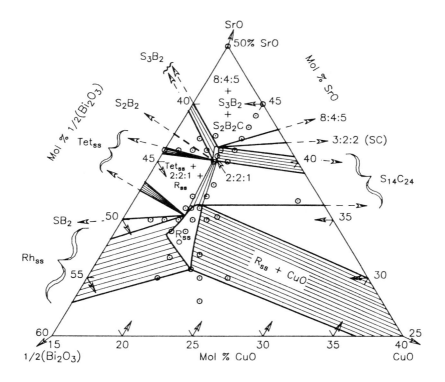

Fig. 15. Enlargement of a portion of Fig. 14 showing phase relations near $Sr_2Bi_2CuO_6$ and the "Raveau-type" solid solution.

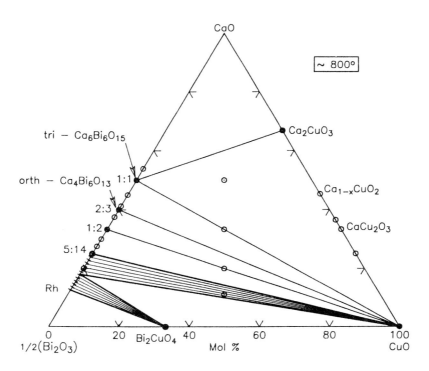

Fig. 16. Phase relations in air in the system $CaO-Bi_2O_3-CuO$ near 800°C and below (after ref. 29).

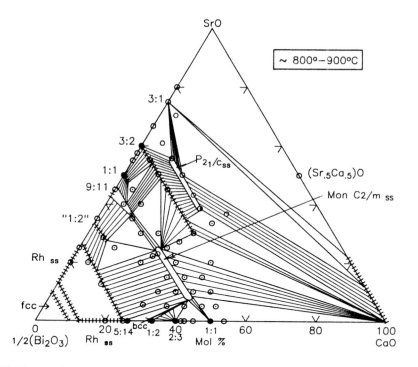

Fig. 17. Phase relations in air in the system SrO-CaO-Bi₂O₃ near 800°-900°C (after ref. 29).

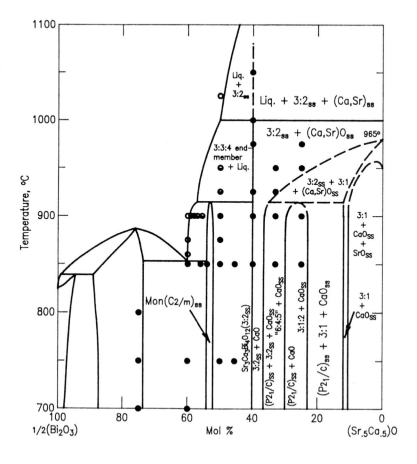

Fig. 18. phase relations in air in the system (Sr₀.₅ Ca₀.₅) O - Bi₂ O₃.

22

PHASE DIAGRAMS OF Fe-Nd-B AND RELATED SYSTEMS FOR THE OPTIMIZATION OF HARDMAGNETIC PROPERTIES

B. GRIEB, E.-TH. HENIG, G. SCHNEIDER*, G. KNOCH, G. PETZOW AND D. DE MOOIJ[†]

Max-Planck-Institut für Metallforschung, Institut für Werkstoffwissenschaft, PML, Heisenbergstrabe 5, 7000 Stuttgart 80, Germany

* Robert Bosch GmbH, Postfach 50, 7000 Stuttgart, Germany

[†] Philips Research Laboratories, P. 0. Box 80000, 5600 JA Eindhoven, The Netherlands

ABSTRACT

Permanent magnets based on Fe-Nd-B show high magnetization and high coercivity at room temperature but unfavourable properties at higher temperatures. This behaviour has to be improved. The hardmagnetic properties are limited by nucleation of severed domains, therefore the surface and interfaces of the grains constituting the material are the controlling factor. For any optimization a detailed knowledge of the phase relations and the microstructure development is needed. The parameters thus varied are gross composition plus additives, the liquid-phase sintering process parameters and the subsequent heat treatment.

Phase relations and listed parameters in the systems Fe-Nd-B, Fe-Dy(Tb)-B, Fe-Nd-Al, Fe-Nd-C and Fe-Nd-Dy(Tb)-B were investigated.

INTRODUCTION

The magnetic properties of Fe-Nd-B based magnets are determined by both the intrinsic physical properties of the occurring phases (anisotropy field (H_A), Magnetization (M_S), Curie-Temperature (T_C) and the extrinsic properties of their microstructures. Consequently both of these properties must be improved for an enhancement of the material. Therefore, a knowledge of the phase relations and the development of the microstructure due to certain sintering parameters must be available for the manufacturers. In addition, the influence of additives and contaminations in the grain boundary areas is known to have a major importance.

One of the possibilities to improve the intrinsic magnetic properties of the $Fe_{14}Nd_2B$ (ϕ) phase is to replace the Nd partially by other rare-earth elements (RE). The ϕ phase formed by Tb or Dy, Fe and B have more than three times the coercivities of ϕ with Nd, Fe and B [86Pin]. On the other hand magnetizations are much lowered by the heavy rare-earth element [85Liv]. Therefore improved magnetic properties should be attained with a mixture of Nd and Tb or Dy. The ternary systems are the base required for the investigation of the quaternary systems.

The coercivity in Fe-Nd-B magnets can be increased by the addition of Al_2O_3 or Al [88Gri]. This is due to increased wettability [89Kno] and the formation of new phases which are stabilized by Al.

The $Fe_{14}Nd_2C$ phase shows intrinsic magnetic properties which are comparable to those of $Fe_{14}Nd_2B$ [88Bus]; therefore the prospect to produce magnets out of this material is of interest.

It is impossible so far to produce hard magnets from Fe-Nd-C by a sintering process due to the phase relations and an inhibited formation of $Fe_{14}Nd_2C$.

EXPERIMENTAL

Samples were prepared from the elements with purities > 99.9 %. The elements or prealloys were melted in an arc furnace under an Ar atmosphere. Thermal investigations (differential thermal analysis (DTA) and annealings) were made in alumina or silica crucibles. Compositions of phases were determined by electron probe microanalysis (EPMA, EDX).

As-cast and DTA specimens were examined by standard metallography (different domain size revealed by the Kerr effect in polarized light distinguishes the ferromagnetic phases) and X-ray diffractometry using monochromated Cu-Kα radiation.

Magnetic properties were measured with a vibrating sample magnetometer. The magnets were prepared by a standard sintering process (1050°C / 1 h) followed by a heat treatment at about 600°C for 1 h.

Fe-Nd-B

Phase Relations

The sequence of phase fields in the Fe-rich corner of the Fe-Nd-B system is shown as the liquidus projection (Fig. 1) and the reaction scheme (Fig. 2). The critical tie line $L \rightleftharpoons \phi + Fe_2B$ and a ternary eutectic $L \rightleftharpoons Fe + \phi + Fe_2B$ represent stable reactions though they contradict earlier reports in which Fe coexists with $Fe_4Nd_{1.1}B_4$ (η) [84Sta, 85Mat]. The isothermal section at 1000°C is shown in Fig. 3. A two-phase field (ϕ +Nd-rich liquid) exists between the maximum temperature of ϕ formation (1180°C) and the temperature of the ternary eutectic at which the residual liquid solidifies (655°C). Four vertical sections have been constructed from DTA data [86Sch]. The section Nd: B=2: 1 is shown in Fig. 4a. In specimens with a composition near ϕ metastable solidification was observed if they have been superheated. The short range order of ϕ clusters in the liquid is destroyed at a certain temperature above the decomposition. From this temperature the crystallization of ϕ is suppressed for kinetic reasons. Then the primary crystallization field of Fe is expanded to lower Fe compositions and to lower temperatures. A secondary metastable crystallization of a high temperature phase (χ) was observed. This phase decomposes at 1105°C into Fe and ϕ. In the related system Fe-Dy(Tb)-B quenching of this χ phase from high temperatures succeeded [87Gri]. It was found to be of the $Fe_{17}RE_2$ (2/17) type containing some B. The further solidification is analogous to the stable system. The vertical section at a ratio Nd:B=2:1 of the metastable system is shown in Fig. 4b.

Microstructure

Microstructure refinement of the sintered body is the basis for an optimization of sintered magnets. The important characteristic features are on a micrometer scale: phase distribution, grain-size distribution, contiguity, porosity, and the degree of alignment of the grains; on a nanometer scale: the microstructure of the Nd-rich grain boundary regions and the structure of the ϕ grain surfaces. Large potential lies in the microstructural improvement of the ternary Fe-Nd-B magnet, suggested by the comparison of the calculated nucleation field of ϕ with the measured coercive fields of real Fe-Nd-B magnets [87Durl]. In a commercial sintered magnet with the composition $Fe_{77}Nd_{15}B_8$ [84Sag] in principle three phases occur. These are:

- ϕ, responsible for the good magnetic properties,

- a Nd-rich phase, leading to liquid above 655°C and good densification by liquid phase sintering as well as a magnetic decoupling between neighboring ϕ grains,

- η of the composition given as $Fe_4Nd_{1-x}B_4$ with x = 0.1 [85Bez]. The Curie temperature is 13 K. This phase is therefore paramagnetic in the temperature range of interest for magnet applications.

Additional phases visible in the microstructure are Nd_2O_3, oxygen stabilized Fe-Nd phases and pores. A sketch of such a real microstructure is shown in Fig. 5a. Quantitative analysis of the phases yields volume percentages of 82 to 85% for ϕ, 10% for (Nd, pores, Nd_2O_3) and 5 to 8% for η. The ideal microstructure is two-phase with the rotund grain being perfectly isolated by the non-ferromagnetic Nd-rich phase. Other non-ferromagnetic phases such as pores, oxides and the η phase should be avoided (Fig. 5b).

Two-Phase Magnet - Avoidance of the Non-Ferromagnetic $Fe_4Nd_{1.1}B_4$ (η) Phase

Non-ferromagnetic inclusions reduce the remanence as well as the coercive field in a permanent magnet. They cause large stray fields [87Dur2] inducing nucleation of reverse domains in the neighborhood of these nonmagnetic inclusions. This demagnetization effect is more severe at higher temperatures as shown in reference [87Dur2]. At sintering temperatures of 1000 to 1080°C the composition $Fe_{77}Nd_{15}B_8$ lies inside the three phase field $L + \phi + \eta$ (Fig. 3). This causes the occurrence of a certain amount of large η grains in the magnet. As can be seen from the isothermal section, Fig. 3, there exists a two-phase field $L + \phi$ at the sintering temperature of about 1050°C. If the magnet composition is located in this two-phase field, only ϕ grains surrounded by the Nd-rich liquid phase at the sintering temperature exists. Large η grains do not occur. Upon cooling some particles crystallize out of the liquid and finally η is a part of the eutectic. This amounts to less than 0.5 vol. % of very small η particles. Two effects are attained with this optimized magnet:

1. The magnetic properties of the sintered magnet have been improved. The improved temperature dependence compared with a "three-phase" magnet can be seen in Fig. 6.

2. The corrosion resistance of the "two-phase" magnet is superior to that of the "three-phase" magnet [87Men].

Fe-Nd-(DY,Tb)-B

Phase Relations
The stable formation mechanism of the ϕ phase with either Tb or Dy is different from the one with Nd. With Tb or Dy ϕ forms peritectically via the reaction $L + Fe_{17}(Tb$ or $Dy)_2 \rightleftharpoons \phi$ whereas with Nd the reaction is $Fe + L \rightleftharpoons \phi$. In the $Fe_{17}(Tb$ or $Dy)_2$ phase (2/17) a certain amount of boron is dissolved [89Gril, 89Gri2]. The formation temperature of $Fe_{14}Tb_2B$ is 1218°C and that of $Fe_{14}Dy_2B$ is 1214°C [89Gri2].

Constitution along $Fe_{14}(Nd_{1-x}(Tb$ or $Dy)_x)_2B$
The results of DTA and metallographic analyses along the temperature concentration cut in the quaternary system have to be divided into a stable and a metastable system [89Gri3]. The different results are due to different cooling behaviours with the heating temperature of the sample attained. If the samples were heated only 20 to 30 K above the decomposition temperature of the ϕ phase a sort of short range order in the liquid still remains. Cooling from these temperatures lead to the stable system where the ϕ phase is formed at higher temperatures. Superheating of more than $\sim$ 30 K leads to the metastable behaviour. The formation of ϕ is suppressed. The same effect was observed in the ternary system [86Sch, 89Gril, 86Gri].

$Fe_{14}(Nd_{1-x}(Tb$ or $Dy_x)_2B$ Stable System - Heating just above the ϕ Formation Temperature
In Figs. 7a and b the formation temperatures of ϕ along $Fe_{14}(Nd_{1-x}(Tb$ or $Dy_x)_2B$ are shown. The results in the two systems are quite similar. Effects and phase fields above the formation temperature are not measured because cooling starts just above ϕ. Two ranges of different slope can be distinguished: (i) one range from $Fe_{14}Nd_2B$ to 20% substitution of Nd with about 1 K/at.% and (ii) one range from 20% Nd to complete substitution of Nd by Tb with 0. 31 K/at. % and by Dy with 0. 25 K/at. %, respectively.

$Fe_{14}(Nd_{1-x}(Tb$ or $Dy_x)_2B$ Metastable Systems - Superheating above the Liquidus
This system is the relevant one for the alloying process by melting together the elements. Figures 8a and b show the results of the DTA measurement of the superheated samples. No qualitative differences in the two quaternary systems are observed. The phases cristallizing primarily from the liquid are iron on the Nd-rich side and the 2/17 phases on the Tb- or Dy-rich sides. The temperatures of primary cristallization of iron have insignificantly different courses. Starting at $Fe_{14}Nd_2B$ in the Fe-Nd-Dy-B system this temperature decreases faster and the solidus of iron runs into the solidus of the $Fe_{17}Dy_2$ phase at an amount of 60 to 70 % Dy of rare-earth element, which then becomes the primary phase. In the Fe-Nd-Tb-B system the courses meet at about 90 % Tb. Strong undercooling effects are observable for the 2/17 phase formation in the Fe-Nd-Dy-B system especially at a higher Dy amount of rare-earth metal. On the Nd-rich side of the systems the secondary phase is the 2/17 phase. The primary crystallites (iron) are surrounded by a seam of Fe and ϕ which is the decomposition product of $Fe_{17}Nd_2B$ with some boron dissolved [86Gri2]. The two-phase regions of L and 2/17 exist over the whole range of substitution. Therefore the formation mechanism of ϕ from L and the 2/17 phase is the formation mechanism in the metastable Fe-Nd-B system, too. The preparation of the ternary $Fe_{17}Nd_2B$ phase via quenching failed, because the temperature range in which this phase exists is very small. Subsequent to the formation of 2/17, ϕ crystallizes from L + 2/17. Undercooling mechanism prevent an exact and reproducible determination of the metastable formation temperatures, scattering up to 25 K was observed. In both quaternary systems the three-phase field between the two two-phase fields L + Fe and L + 2/17 is not resolved, only the areas with different primary crystallization (Fe or 2/17) could be demarcated. The formation of the ϕ phase in the stable Fe-Tb-B and Fe-Dy-B systems is contrary to the Fe-Nd-B stable system but the same in all the metastable systems.

Superposition of the stable and the metastable systems:
In the stable systems the formation mechanism of apparently has to change from $L + Fe \rightleftharpoons \phi$ on the $Fe_{14}Nd_2B$ side to $L + 2/17 \rightleftharpoons \phi$ on the $Fe_{14}(Tb$ or $Dy)_2B$ sides. Up to a certain degree of substitution of Nd by Tb or Dy the first reaction is the stable one, beyond the second one. This transition can be settled at about a 20% portion of Tb or Dy, Fig. 9.

Magnetic Properties
The addition of Dy to magnet compositions, which are different from the stoichiometric composition of ϕ, leads, in addition to the change of the formation mechanism, to other phases which are stabilized by Dy. The change in the formation mechanism leads to different primary phases which is relevant for the alloying process. These phases are either Fe or $Fe_{17}RE_2$, which are ferromagnetic and therefore of unfavourable influence on magnetic properties. They are present in the as-cast alloy. Additional phases, such as $FeRE_2$, $FeRE_3$, stabilized due to the Dy substitution, are ferromagnetic in character. They can influence the magnetic properties of the magnet after sintering because they are equilibrium phases at this temperature. The beneficial influence of Dy on the intrinsic properties are overlaid. Detailed description of these relationships are given in the Poster and the pertinent publication titled

172

"Influence of Dy on Constitution and Magnetic Properties of Fe-Nd-B-Based Magnet Compositions" by K. Fitz, B. Grieb, E.-Th. Henig, and G. Petzow.

Fe-Nd-Al

Phase Relations

In this ternary system at lower Al contents (< 30 at.%) several binary and ternary phases are in equilibrium (Fig. 10). Iron can be replaced by Al in the Fe-rich phases $(Fe,Al)_{17}Nd_2$, δ and μ. The two new phases, called δ and μ, cover the following homogeneity ranges below 600°C (Fig. 11):

(δ) $Fe_{67.5-x}Al_xNd_{32.5}$ $(7 < x < 25)$ and (μ) $Fe_{63.5-x}Al_xNd_{36.5}$ $(2.5 < x < 5)$.

Both, δ and μ are observed in Fe-Nd-B-Al magnets. The phases are formed by peritectic reactions: δ at $\approx$900°C and μ at $\sim$ 750°C. The equilibria at temperatures above 700°C and at higher Nd contents (> 70 at.%) are more complicated due to the influence of binary Nd - Al compounds; they have not been studied in detail. However, this does not influence the relationships of the ternary phases δ and μ.

The δ phase is tetragonal with antiferromagnetic character and shows a high susceptibility of 0.05. This structure type was observed before in La-Ga-Co [85Sic]. The magnetic influence of δ on magnets is unclear.

The structure of μ is just under investigation: it is not cubic and reveals a very complex X-ray diffraction pattern. Its magnetic behaviour is ferromagnetic with a large anisotropy field $\mu_0H_A > 8$ T combined with a magnetization of about 0.85 T. Compared to $Fe_{14}Nd_2B$, the coercivity of μ is larger but the magnetization is only half because of the high Nd amount.

The μ phase is observed at the surface of the $Fe_{14}Nd_2B$ grains. It forms by peritectic reaction and nucleates at ϕ. This growth mechanism smoothes the surface of ϕ. The occurrence of μ in the magnetic material and its large anisotropy field is a precondition for the explanation of the large increase in coercivity by the Al addition.

MAGNETIC PROPERTIES

The influence on magnetic properties (Fig. 11) seems to be due to microstructural modifications. The occurrence of new phases and improved wettability (Fig. 12) are given [90Kno]. Similar relationships are observed by the addition of Ga instead of Al.

Fe-Nd-C

Phase Relations

The compound $Fe_{14}Nd_2C$ is normally not found in as-cast or DTA (dT/dt = 5 K/min, cooled from liquid) alloys [87Sta]. The phases observed in a DTA sample with a gross composition corresponding to ϕ are primary Fe, secondary $Fe_{17}Nd_2C_x$, and phases in the grain boundary region. No $Fe_{14}Nd_2C$ phase is observed. This is due to the slow reaction kinetics and nucleation difficulties of that phase; therefore, long homogenization times are necessary ($\approx$30 d) [88Moo].

In stable equilibrium the $Fe_{14}Nd_2C$ phase should be observed at higher (70 to 80 at.%) Nd amounts as a primary phase (Fig. 13).

The formation of the $Fe_{14}Nd_2C$ phase occurs by peritectic reaction as in all Fe-RE-B systems, but the phase relations in the Fe-Nd-C system are not the same as observed in Fe-Nd-B; new phases occur.

Figure 14 shows one isothermal section of the system in the temperature range between 600 and 700 C. In this range the λ phase, which is ferromagnetic, is stable.

This phase is ferromagnetic with uniaxial anisotropy (domain structure). The Fe:Nd ratio is $\approx$3:2; the C content less than 10 at.%. The phase is observed in DTA samples at compositions between $Fe_{14}Nd_2C$ and Nd with small C amounts.

The influence of this phase with eutectic morphology and ferromagnetic properties on nucleation hardened magnetic material is unfavourable and must be avoided to guarantee the isolation of ϕ grains.

The binary and ternary carbides of Nd are very corrosive. Alloys are destroyed by the occurrence of Nd_2C_3, NdC_2 and $FeNd_4C_6$.

To produce hard magnetic material based on Fe, Nd and C by liquid sintering, several circumstances have to be considered:

(1) At temperatures below 900°C a liquid phase occurs only at low C contents occurs.

(2) The λ and the $Fe_{17}Nd_2C_x$ phases are ferromagnetic and influence the properties of the magnet material. $Fe_{17}Nd_2C_x$ is "easy plane" material; the properties of λ are unknown.

(3) Binary and ternary carbides of Nd are very corrosive.

(4) Nucleation centres of $Fe_{14}Nd_2C$ have to be formed and the diffusion rate must be increased. The long time required for the formation of the hard magnetic phase has to be shorten.

In summary, the phase relations and the nucleation difficulties are very unfavourable for production of a magnet material by a sintering process, but it may be possible by other processing routes to produce hard magnets of Fe-Nd-C (melt spun process or bonded magnets).

REFERENCES

[83Sag] M. Sagawa, S. Fujimara, N. Togawa, H. Yamamoto, and Y. Matsuura, J. Appl. Phys. 55 (1984) 2083.

[84Sta] H. H. Stadelmaier, N. A. El-Masry, N. C. Liu, and S. F. Cheng, Mat. Lett. 2 (1984) 411.

[85Bez] A. Bezinge, H. F. Braun, J. Muller, and K. Yvon, Solid State Com. 55 (1985) 131.

[85Liv] J. D. Livingston, "Iron-Rare Earth Permanent Magnets", 8th. Int. Workshop on Rare Earth Magnets and Their Applications, Dayton, Ohio (1985) Paper No. VI-1.

[85Mat] Y. Matsuura, S. Hirosawa, H. Yamamoto, S. Fujimura, M. Sagawa, and K. Osamura, Japan. J. Appl. Phys. 24 (1985) L635.

[85Sic] O. M. Sichevich, R. V. Lapunova, A. N. Sobolev, Yu. N. Grin', and Ya. P. Yarmolyuk, Kristallographia 30 (1985) 1077.

[86Gri] B. Grieb, Konstitutionelle Untersuchungen zu den Homogenitätsbereichen von $(Nd_xSE_{1-x})_2Fe_{14}B$'', Diploma Thesis, Univ. Stuttgart, 1986.

[86Pin] P. Pinkerton, J. Magn. Mat., 54-57 (1986) 579.

[86Sch] G. Schneider, E.-Th. Henig, G. Petzow, and H. H. Stadelmaier, Z. Metallkde. 77 (1986) 755.

[87Dur1] K.-D. Durst and H. Kronmüller, J. Magn. Magn. Mat. 68 (1987) 63.

[87Dur2] K.-D. Durst, H. Kronmüller, and G. Schneider, Proceedings of the 5th Int. Symp. on Magn. Anisotropy and Coercivity in Rare-Earth Transition Metal Alloys, Bad Soden, FRG (1987) 209.

[87Gri] B. Grieb, E.-Th. Henig, G. Schneider, and G. Petzow, Proceedings of the 5th Int. Symp. on Magn. Anisotropy and Coercivity in Rare-Earth Transition Metal Alloys, Bad Soden, FRG (1987) 395.

[87Men] R. van Mens, G. W. Turk, and M. Brouha, Proceedings of the 9th Int. Workshop on Rare-Earth Magnets and Their Applications, Bad Soden, FRG (1987) 395.

[87Sta] H. H. Stadelmaier N. A. El-Masry and N. C. Liu, Proc. 5th Int. Symp. on Magnetic Anisotropy and Coercivity in RE-TM Alloys, Part II, Bad Soden, FRG, Sept. 3 (1987) 389.

[88Bus] K. H. J. Buschow, Ferromagnetic Materials, Vol. 4, Ed. by E. P. Wohlfarth and K.H.J. Buschow, Elsevier Publishers B.V., Amsterdam (1988).

[88Gri] B. Grieb, K. G. Knoch, E.-Th. Henig and G. Petzow, J. Magn. Mag. Mat. 80 (1988) 75.

[88Moo] D.B. de Mooij and K.H.J. Buschow, J. Less-Common Metals 142 (1988) 349.

[89Gri1] B. Grieb, E.-Th. Henig, G. Schneider, and G. Petzow, Z. Metallkde. 80 (1989) 95.

[89Gri2] B. Grieb, G. Müller, E.-Th. Henig, G. Petzow, and H. H. Stadelmaier, Z. Metallkde., 80 (1989) 807.

[89Gri3] B. Grieb, G. Schneider, E.-Th. Henig, and G. Petzow, Z. Metallkde. 80 (1989) 515.

[89Kno] K. G. Knoch, G. Schneider, J. Fidler, E.-Th. Henig, and H. Kronmüeller, IEEE Trans. Mag. 25 (1989) 3426.

[90Kno] K. G. Knoch, B. Grieb, E.-Th. Henig, H. Kronmüller, and G. Petzow, to be published in IEEE 1990.

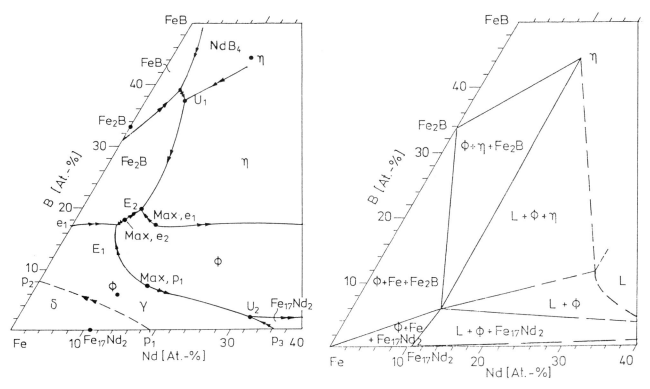

Fig. 1. Fe-Nd-B, liquidus projection [86Sch].

Fig. 3. Fe-Nd-B, 1000°C isothermal section [86Sch].

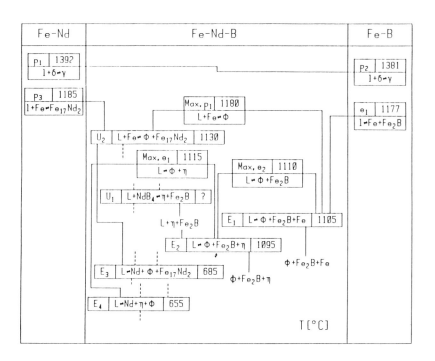

Fig. 2. Fe-Nd-B, reaction scheme [86Sch].

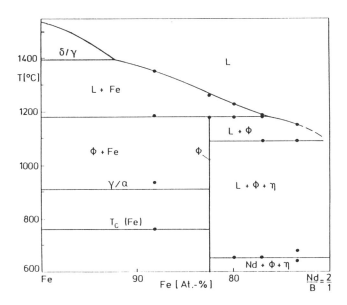

Fig. 4a. Fe-Nd-B, vertical section, Nd: B = 2:1, - stable [86Sch] .

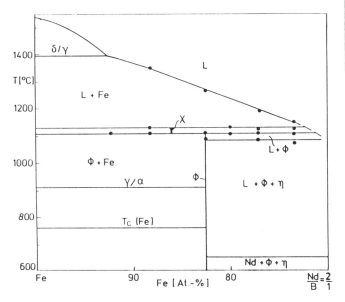

Fig. 4b. Fe-Nd-B, vertical section, Nd: B = 2:1, - metastable [86Sch].

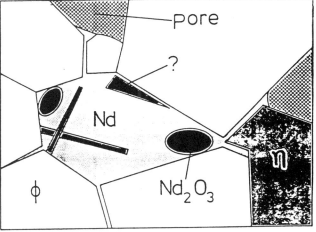

Fig. 5(a) Sketch of a s intered Fe-Nd-B magnet microstructure, real [86Sch].

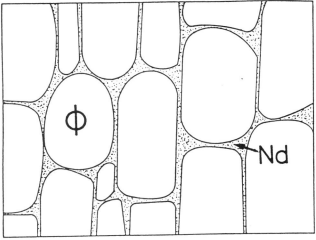

Fig. 5(b) Sketch of a sintered Fe-Nd-B magnet microstructure, ideal [86Sch].

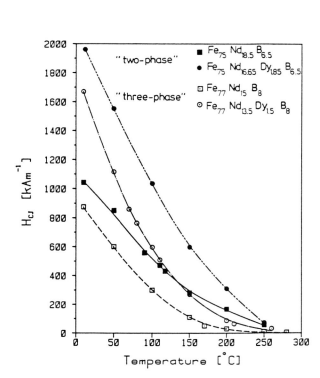

Fig. 6. Temperature dependence of the coercive field of different magnet compositions.

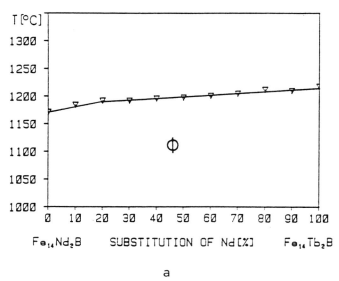

a

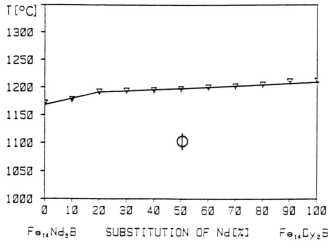

b

Fig. 7. Formation temperatures of the stable Φ phases along (a) $Fe_{14}(Nd_{1-x}Tb_x)_2B$ and (b) $Fe_{14}(Nd_{1-x}Dy_x)_2B$ stable systems [89Gri3]. Effects above Φ are not detected because cooling starts just above the Φ formation.

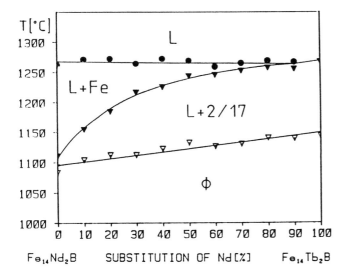

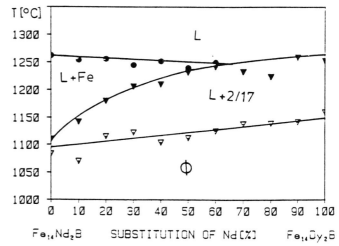

Fig. 8. Temperature-composition sections along the different Φ phases (a) $Fe_{14}(Nd_{1-x}Tb_x)_2B$ and (b) $Fe_{14}(Nd_{1-x}Dy_x)_2B$ ($0 \leqslant x \leqslant 1$) - superheated samples.

b

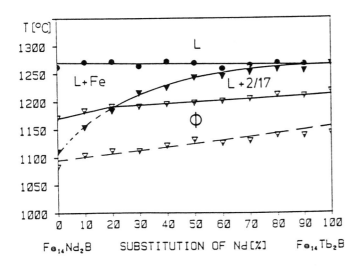

Fig. 9. Superposition of the stable (Fig. 7a) and the metastable (Fig. 8a) system $Fe14(Nd_{1-x}Tb_x)_2B$ ($0 \leqslant x \leqslant 1$)

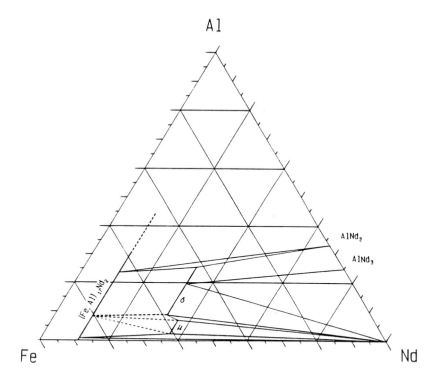

Fig. 10. Fe-Nd-Al, 580°C isothermal section (≡ RT).

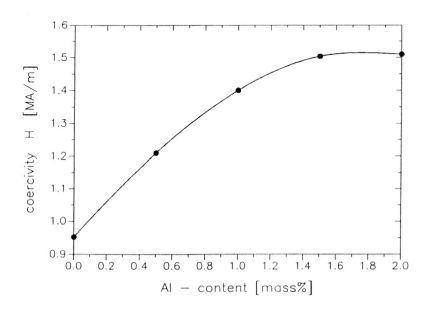

Fig. 11. Coercivity HcJ vs. Al amount (added before melting).

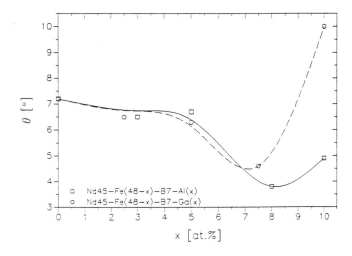

Fig. 12. Concentration dependence of the equilibrium wetting angle Θ (x) for the Al and Ga addition.

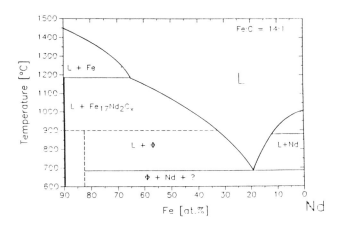

Fig. 13. Fe-Nd-C, vertical section along a constant Fe: C ratio of 14: 1.

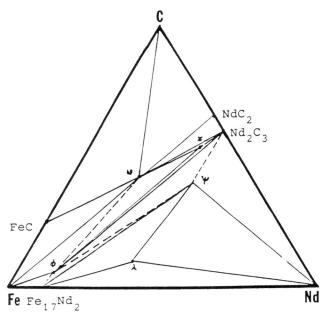

Fig. 14. Fe-Nd-C, isothermal section in the temperature range between 600 and 700°C.

23

CORRELATION BETWEEN PHASE DIAGRAMS AND ELECTRICAL PROPERTIES OF Cu-ALLOYS AND FILMS DEPOSITED VIA EVAPORATION IN VACUUM

N. R. BOCHVAR AND E. V. LYSOVA

Baikov Institute for Metallurgy, Soviet Academy of Sciences, Moscow

The aim of this research is to compare the electrical properties of some Cu-based alloys containing additions of Mn, Sn and Ge with the properties of vacuum evaporated films on the basis of a phase and "composition - property" diagrams investigation. The influence of the concentration of the alloying elements on the electrical resistivity of the films was also investigated.

DTA, microscopic and X-ray analyses, Auger spectroscopy and electrical resistivity measurements were used. The purity of the materials used for the production of the alloys and the films was better than 99.96%. The alloys were arc-melted in a vacuum furnace under a He-atmosphere. Ingots were rolled to a deformation about 80%, then annealed for 100h at a temperature 50°C below the solidus temperature in the quartz vacuum ampules and then air-cooled.

Electrical resistivity ($\wp$) is a microstructure dependent property and due to the dependence on the phase diagram type (or metallochemical interaction of components) it is possible to obtain a considerable difference in both $\wp$ and the electrical resistivity temperature coefficient α. The main influence on the electrical properties is exerted by elements which form broad ranges of solid solutions. Such behaviour results from solvent lattice distortion, and from solvent plus alloying element chemical interaction. In some Cu-based systems, where the alloying elements exhibit small ranges of solid solubility in the Cu-based solid solution (solubility ~ 0.1 - 0.01%), the $\wp$ and α values show negligible changes in their dependence on alloying (20-40% compared with Cu). In the case of broader ranges of solid solubility the $\wp$ and α values change by more than a factor of ten. In Fig. 1 $\wp$ and α vs composition curves for some Cu-based binary systems are presented [1]. For displaying systems with different alloying element solubility in Cu, parts of the Cu-Zr and of the Cu-Ge phase diagrams are also presented.

The investigation of alloys with elements that are able to form wide ranges of solid solution with Cu (Mn, Sn, Ge, Ga, S) is of particular interest in this research. In this case it is possible to vary the concentration of the alloying elements and thus to obtain different combinations of electrical properties.

For more complicated systems such as ternary systems the observed dependence remains. In Fig. 2 some vertical sections for the Cu-Mn-Ge, Cu-Mn-Sn, Cu-Ge-Sn ternary systems and corresponding $\wp$ and α vs composition curves are presented. In these systems broad areas of Cu-based solid solutions occur. Two ternary compounds were found in the Cu-Mn-Ge system. In the Cu-corner of the Cu-Mn-Sn and Cu-Ge-Sn phase diagrams only phases corresponding to the binary systems are found. It should be noted that in the Cu-Mn-Ge and Cu-Mn-Sn systems α possesses both positive and negative values. In Fig. 3 $\wp$ and α dependencies for further sections in these systems are displayed [2-4].

The significant difference between the electrical properties of the alloys and the evaporated films depends both on the size factor and the phase composition as well as the structure state difference between the alloys and the films. The size factor for the films of any composition reveals itself in the thickness dependence of $\wp$ for thicknesses less than 0.3 mkm. If the film thickness is more then 0.3 mkm $\wp$ is practically independent of the thickness. The phase composition and the structural state of the films are determined by the preparation process and by the heat treatment condition.

In this investigation we used the evaporation method for forming films with concentration gradients of the component varying with thickness. To achieve this the different abilities for evaporation of the alloying elements and the Cu-based alloys that depends on the saturated vapour pressure were used. Thus the high vapour pressure component forms a layer on the support, and the low vapour pressure component forms a surface layer. Therefore the intermediate layer is impoverished in the alloying elements. The film structure mentioned above consisting of three layers was obtained in vacuum $1.33 \cdot 10^{-4}$ Pa by total evaporation of a certain amount of alloy from the tungsten electrical resistance heating vapourizer. The mean condensation rate was 0.005 mkm/s. The film thickness was 0.5-0.6 mkm.

It is important to note that there is a significant difference between the $\wp$-values and those for the alloys films: $\wp$ values for the alloys are greater than $\wp$-values for the films. In Table 1 $\wp$ and α values for several different alloys and vacuum evaporated films are displayed.

It should be noted that $\wp$ for the alloys changes considerably more than $\wp$ for the films and $\wp$ values for the films are rather similar to each other and independent of the amounts of the alloying elements present. α demonstrates the same dependence: α for the film change less than α-values for the alloys and possess positive values.

These differences in properties testify to the presence of a film structure with a certain concentration gradient. The higher the difference between the Cu and the alloying element saturated vapour pressure the more pronounced the concentration gradient is and hence there occurs a greater difference between the electrical properties of the films and those of the alloys.

Using the Auger spectroscopy method it is possible to confirm this suggestion which is to the presence of several layers in a film. The film with the thickness of 0.1 mkm was evaporated on to a Ta foil and was etched by Ar ions (residual gases pressure 5.10^{-1} Pa). In Fig.4 concentration profiles for the films produced by vacuum evaporation of the Cu-5%Sn-2%Mn and Cu-5%Sn-3%Ge alloys are presented. In the first film the surface is enriched by Sn, in the second by both Sn and Ge. Note that Ge concentrates only in a thin surface layer (~ 0.01 mkm). The Sn concentration has its maximum value at the surface and decreases in the direction of the support. Mn condenses near the support, the layer being about 0.004 mkm. The distribution of the elements in the films confirms the theoretical suggestion that is connected with the fact that the element which possesses the higher saturated vapour pressure (here Mn) in comparison with the base of the alloy condenses at the beginning of evaporation process and enriches the support zone. Ge possesses a lower saturated vapour pressure than Cu, condenses at the end of the evaporation process and distributes within the surface zone of the film [5]. Sn possesses a saturated vapour pressure similar to that of Cu (slightly lower). Hence it distributes throughout the film.

To determine the phase composition of each fraction the alloy was evaporated in three stages, using NaCl crystals as the support. The film thickness at each stage was about 0.05 mkm. Electron diffraction structure analysis shows that during the initial evaporation stage the Cu-5%Sn-2%Mn film consists of Cu solid solution and Mn oxide (Lattice parameters a=0.366 nm and a=0.443 nm correspondingly). The intermediate layer consists almost entirely of pure Cu (a=0.362 nm). The surface layer consists of a mixture of the $Cu_{31}Sn_8$ and Cu_3Sn phases. Annealing at temperatures up to 200°C has no effect on the distribution of elements within the film. It confirms that the structures obtained are rather stable. The layer/fraction thicknesses near the support and near the surface are very small and rather similar to each other in the difference in the amounts of the alloying elements. Therefore, the influence on the electrical properties has an intermediate fraction impoverished by the alloying elements. At the same time those fractions, enriched by the alloying elements (near the support and near the surface) determine the technological properties of the films, for example, its adhesion to the support, its resistance to corrosion and others.

Therefore, in determining the required properties of the films it is possible to select the type and the amount of the alloying with a high accuracy.

It should be noted, that using of other types of film structure forming processes (for example, magnetron sputtering) will lead to different relationships between the composition and the electrical properties, but this is not the theme of the discussion here.

REFERENCES

1. M. E. Dritz, E. B. Lysova, N. R.Bochvar, A. M. Sasov and B. I. Popov, "The principles of Cu-alloys alloying for film material creation with special properties", Metall. i Metallov. Tsvet. Splav.:Sbornic Statei,Nauka, 1982, p.172-177.

2. M. E. Dritz, N. P. Leonova, N. P. Bochvar, V. V. Baranchikov and E. V. Kulbachevskay, "Phase equilibria in Cu-rich alloys of the Cu-Mn-Ge system", Izv. Akad. Nauk SSSR Metally, 1985, N 4, p.205-209

3. N. P. Leonova, N. R. Bochvar, E. V. Lysova, "Phase equilibria in Cu-rich alloys of the Cu-Mn-Sn system", Izv. Akad. Nauk SSSR Metally, 1987, N 4, p.203-205.

4. N. R. Bochvar, N. Vigdorovich, N. P. Leonova and E. V. Lysova, "The resistivity of Cu-Mn-Ge, Cu-Mn-Sn and Cu-Ge-Sn alloys", Metall. Term. Obrab. Metallov, 1984, N 2, p.61-64.

5. N. R. Bochvar, N. P. Leonova,E. V. Lysova, A. M. Sasov and V. I. Sidorenko, "Phase composition and structure of Cu-based alloys in massive and film states", Nov. Material. dlay Microelectroniki, Sbornik Statei, Kiev, AN Uk. USSR, 1984, p.39-43.

182

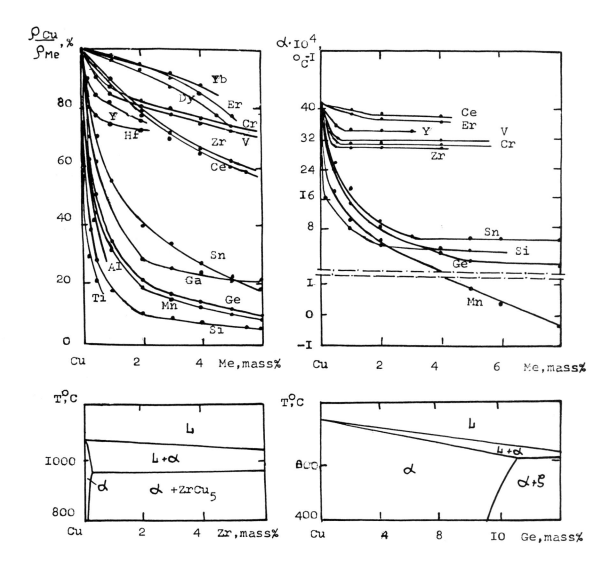

Fig. 1. The influence of composition on $\dfrac{\rho CuS}{\rho Me}$ and α and the Cu-rich corner of Cu–Cr and Cu–Ge phase diagrams.

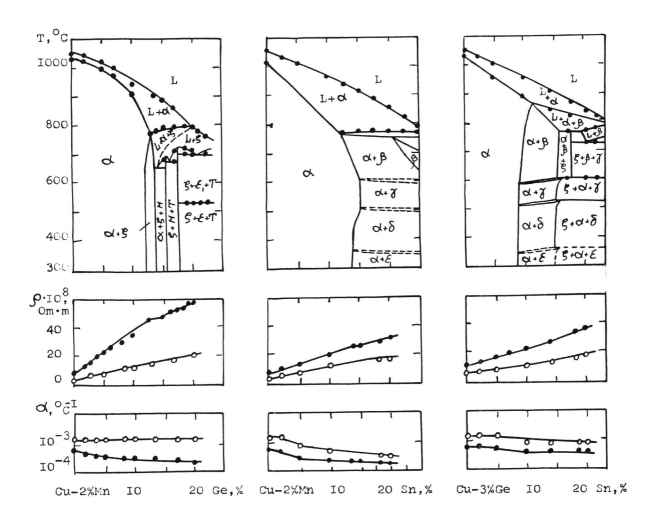

Fig. 2. Vertical sections for the Cu–Mn–Ge, Cu–Mn–Sn, Cu–Ge–Sn systems and the influence of composition on ρ and α in vacuum evaporated alloys (• - alloys, o - films).

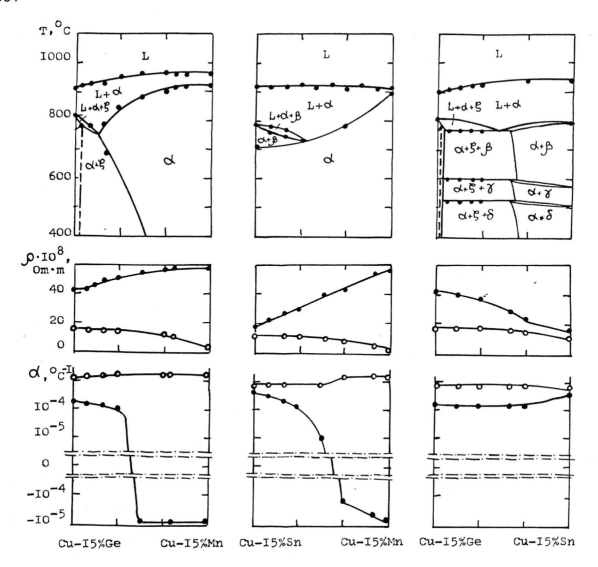

Fig. 3. Vertical sections for the Cu–Mn–Ge, Cu–Mn–Sn, Cu–Ge–Sn systems and the influence of composition on ρ and α in vacuum evaporated alloys (• - alloys, o - films).

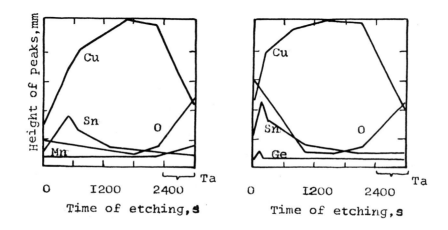

Fig. 4. Distribution of elements in Cu–5%Sn–2%Mn and Cu–5%Sn–3%Ge alloy films.

24

ROLE OF MISCIBILITY GAPS IN MAGNETIC ALLOYS: ALNICO, Fe-Cr-Co AND Co-Cr

K. Ishida and T. Nishizawa

Department of Materials Science, Faculty of Engineering,
Tohoku University, Sendai 980, Japan

ABSTRACT

The role of miscibility gaps in the microstructural design of Alnico, Fe-Cr-Co and Co-Cr magnetic alloys is presented. Miscibility gaps are affected by the presence of chemical and magnetic ordering, resulting in asymmetric immiscibility. In the Alni and Alnico alloys the miscibility gap islands of the bcc phase are formed along the order (B2) - disorder (A2) transition temperatures, while in the system Fe-Cr the miscibility gap is expanded by the addition of Co and shifted to the Fe-rich side along the Curie temperature. It is shown that the origin of high coercive force in these alloys is the result of such changes in the shape of the miscibility gaps, and that this finding can be used to advantage in magnetic alloy design. It is further pointed out that the compositional modulation in the Co-Cr thin films, developed as perpendicular magnetic recording media, is also due to the magnetically-induced miscibility gap of paramagnetic and ferromagnetic cph (εCo) phases in this system. Suggestions are put forward for further developments in magnetic alloys design using phase diagram information.

1. INTRODUCTION

The development of high coercive magnets in the precipitation alloys began in 1931 with the discovery of their occurrence by Köster[1] in the Fe-Co-W and Fe-Co-Mo systems and by Mishima[2] in the Fe-Ni-Al system. Later, Honda et al.[3] found that additions of Co and Ti to Fe-Ni-Al alloys improved the magnetic properties significantly. Since then, many precipitation alloys have been produced. Of these, a significant advancement during the last decade has been the development of ductile hard magnets of Fe-Cr-Co alloys by Kaneko et al.[4]. One of the important chracteristics of these materials is that their high coercivity is a result of magnetic hardening due to spinodal decomposition, which leads to a microstructure composed of very fine precipitates of one phase embedded in the other.

In the present paper, the importance of phase diagram information in the design of Alni, Alnico and Fe-Cr-Co magnet alloys is reviewed with particular emphasis on the role of miscibility gaps in the control of microstructure in these alloys. Also the special microstructural features associated with the perpendicular magnetic recording alloys of the system Co-Cr, developed by Iwasaki et al.[5] are explained on the basis of the presence of a miscibility gap in this system. Some suggestions are put forward for further developments in the field of magnetic alloys using phase diagram information.

2. USE OF PHASE DIAGRAMS IN THE DESIGN OF ALNI AND ALNICO MAGNETS

2.1. Miscibility gap

It is well established that the microstructures of Alni(Fe-Ni-Al base alloys) and Alnico (Fe-Ni-Al-Co base alloys) are composed of two bcc phases α_1 and α_2; α_1 being the Fe-rich disordered A2 structure and α_2 being the NiAl-rich ordered B2 structure. The miscibility gap in the Fe-Ni-Al system was first detected by Kiuchi[7] and Bradley and Taylor[8,9] and later redetermined by Bradley[10]. Figure 1 shows the isothermal (at 750K) and vertical sections taken from Bradley[10].Two interesting features regarding the miscibility gap are immediately evident. One is the presence of a "closed miscibility gap" or "miscibility gap island" in the bcc phase region of the ternary Fe-Ni-Al system, which is absent in the bcc phases of any of the component binaries Fe-Ni, Ni-Al or Fe-Al; the other is the abnormal shape of the vertical section of the miscibility gap, with a horn extending towards the solidus curve. The first feature, namely, the occurrence of the "closed miscibility gap" has been analysed thermodynamically by Meijering[11]. The second feature, namely the peculiar shape of the miscibility gap was first analyzed and explained by Hardy[12] who took into account the effect of order-disorder reaction in the bcc phase. A more advanced and quantitative treatment of the effect of atomic and magnetic orderings on the miscibility gap, was developed by Nishizawa et al.[13,14]. In that treatment it was shown, how ordering when taken into account, could explain both

186

the raised temperature of the summit of the miscibility gap and also the developement of a ridge along the order-disorder transition and Curie temperature lines in both the isothermal and vertical sections.

Figure 2 shows the partial isothermal sections between 850 and 1000°C for Fe-Ni-Al and Fe-Ni-Al-Co systems determined by Hao et al.[15]. Since the composition difference of Co between disordered α_1 and ordered α_2 phases is less than 2 at.%, the quaternary diagrams of the Fe-Ni-Al-Co system are constructed on the sections of equiconcentration for Co. Figure 3 shows the vertical section diagrams of the Fe-Ni-Al-Co system. It can be seen that the miscibility gap between α_1 and α_2 phases is narrowed and shifted to the Fe-rich corner by the addition of Co, and also that the order-disorder transition line is shifted to the Fe-rich side. Ti, which is one of the important alloying elements in Alnico magnets has the same effect of shifting the summit of the miscibility gap to the Fe-rich side[16]. From these figures it is clear that the miscibility gap develops a ridge along the order-disorder transition line, not only in the vertical, but also in the isothermal sections as predicted by thermodynamic calculations[12,14]. This particular feature of the phase diagram can be used to advantage in obtaining microstructures with desired magnetic properties. Good magnetic properties in the precipitation alloys are obtained under the following conditions[6,17]:

(i) the ferromagnetic α_1 phase is present as isolated particles embedded in a weakly magnetic or nonferromagnetic matrix.

(ii) these particles are highly ferromagnetic and uniaxially aligned, and

(iii) the volume fraction of this phase is in the range 0.4-0.6.

The first and third factors are strongly influenced by the composition of alloys and the shape of the miscibility gap; hence the importance of phase diagram information in the design through microstructural selection.
For instance, the addition of Co or Ti to Alni shifts the summit of the miscibility gap to the Fe-rich side. The consequence of such a shift is the increased volume percent of the ferromagnetic phase and decreased ferromagnetism of the matrix phase[16].

2. 2. Design of ductile magnetic alloys
Another example of the use of phase diagram information is in the area of producing ductile magnets. It is known that the Alni and Alnico magnet alloys suffer from the considerable disadvantage of being very hard and brittle rendering post-machining and grinding operations very difficult and expensive. It is therefore desirable to look for ways of obtaining ductile magnets.

Utilising phase diagram information, the present authors have recently been able to identify a composition close to the conventional Alni alloy that could be fabricated by hot forging and hot rolling[18]. Fig.4(a) shows the appearance of Fe-20 at.%Ni-15 at.%Al alloy sheet after hot rolling and the as-cast microstructure is shown in Fig.4(b). It can be seen from Fig. 4(b) that a very small amount of fcc γ phase significantly improves the hot workability by being present as a ductile film on the grain boundaries of ordered α_2-phase thus eliminating the grain boundary weakness. The same principle can be extended to the alloys of the Alnico system. The procedure would then be (i) to identify a composition from the phase diagram that would contain a small amount of γ-phase at the hot-working temperature, (ii) to choose an appropriate solution treatment temperature (this temperature may be influenced by the shape of the γ-loop) to remove the γ-phase completely and subsequently, (iii) to age at a temperature in the miscibilty gap to obtain the desired microstructure. Although precise control of alloy composition, hot-working condition and heat treatment are required, there is still the significant possibility of producing ductile Alni and Alnico magnets.

3. ROLE OF THE MISCIBILITY GAP IN THE DESIGN OF Fe-Cr-Co MAGNETS
Fe-Cr-Co alloys were known to be of potential use for permanent magnet applications because of their good ductility and excellent magnetic properties[4,17]. The fact, that addition of cobalt to the Fe-Cr binary alloys resulted in an expanded miscibility gap, was taken advantage of designing better magnetic alloys in the ternary system. The basic concepts of alloy design as related to the phase diagram have been discussed in detail by Inden[19]. A brief outline of the use of phase diagram information in designing magnetic alloys in this system is presented below.

Even though σ phase is the most stable phase to precipitate at conventional magnetic ageing temperatures used in Fe-Cr-Co alloys, Kaneko et al.[20] could detect the presence of a metastable miscibility gap in this system. However, they did not determine its precise shape. From a theoretical thermodynamic analysis, Nishizawa et al.[13] have predicted that the Fe-Cr-Co system should exhibit a metastable miscibility gap with a peculiar shape shown in Fig. 5 that is characterised by the development of a ridge along the Curie temperature which becomes progressively prominent with increasing additions of cobalt. Minowa et al.[21] have experimentally confirmed the existence of the miscibility gap ridges as shown in Fig. 6.

Homma et al.[22] have shown how suitable heat treatments for improving the magnetic properties of the alloys could be formulated by taking advantage of the presence of the "ridge" regions in the miscibility gap of these alloys. Their idea for development of better magnetic properties in Fe-Cr-Co alloys by thermomagnetic treatment (TMT) are based on the expected desirable characteristics of the "ridge" alloys which are as follows. In these alloys, (i) the small difference in the composition of the two phases formed within the ridge region would lead to a decrease in the elastic and interfacial energies and (ii) the Curie temperature would be located very close to the miscibility gap.

These two characteristics would enable a significant increase in the efficiency of the applied magnetic field during TMT of these alloys. Applying this idea they have been successful in identifying thermomagnetic treatment schedules in the ridge alloys, which are effective in aligning and elongating the FeCo-rich particles parallel to the applied magnetic field direction. According to them the optimum TMT temperature is located at $15 \pm 5°C$ below the miscibility gap as shown in Fig.7. The above example is a typical case of how precise phase diagram information has led to a practical development.

4. MISCIBILITY GAPS IN OTHER MAGNETIC ALLOYS

Besides the Alnico and Fe-Cr-Co alloys there also exist other groups of permanent magnet alloys whose phase diagrams show miscibility gaps and spinodal decomposition. Therefore, it may be worthwhile comparing the phase diagrams of these systems, if only, to emphasize the special role of the miscibility gap in all of these alloys.

4. 1. Fe-Mo-Co and Fe-W-Co magnets

These magnets are known as Köster's alloys[1]. Stable microstructures in these alloys show precipitates of intermetallic compounds in a ferritic matrix. However, there exists a metastable phase diagram showing a miscibility gap between Fe-rich bcc α and Mo- and W-rich bcc α' phases as shown in Fig.8(a) and 8(α')[16,23]. In fact, during ageing it has been verified that a spinodal decomposition takes place as required by the presence of the metastable miscibility gap[24]. Since the matrix phase in this case is the ferromagnetic α phase, the coercivity is due to the pinning of domain walls by the fine precipitates of the non-magnetic Felean α' phase.

4. 2. Cu-Ni-Co and Cu-Ni-Fe Alloys

The precipitation of phases in these systems is controlled by the presence of stable miscibility gaps in the Cu-Co and Cu-Fe binaries which extend into the ternaries as shown in Fig. 8(b). It has been verified, that spinodal decomposition occurs in Cu-Ni-Co and Cu-Ni-Fe magnets during ageing, and high coercivity is achieved because of the impeding action of the matrix on the rotation of the magnetic particles, resulting in large shape anisotropy. The composition of these alloys lies in the Cu-rich side as illustrated in Fig.8(b) and 8(b'), to enable the precipitation of the ferromagnetic γ_f phase in the Cu-rich γ' matrix. It is difficult in these alloys to obtain a proper volume fraction of the ferromagnetic phase because the miscibility gap is nearly symmetrical.

4. 3. Alnico and Fe-Cr-Co alloys

In section 2 the role of miscibility gaps in the choice of alloys and heat treatment schedules in these systems has already been discussed. Fig. 9 again shows the shape of the miscibility gaps in the pseudo-binary and ternary sections of these alloy systems. The miscibility gap of Alnico magnet is formed along the order-disorder transition line, while that of Fe-Cr-Co alloy is expanded along the Curie temperature. The use of the phase diagram information in alloy design for achieving the high coercivity in these systems has already been explained in section 2.

5. MISCIBILITY GAP IN THIN FILM Co-Cr ALLOYS USED IN PERPENDICULAR
MAGNETIC RECORDING

Since the advent of Co-Cr alloy thin films as the first group of alloy materials to be used as "perpendicular magnetic recording" media[5], it has been shown that the films of such alloys with compositions ranging from Co-15 to 30 at.% Cr have a compositionally modulated structure. Although numerous investigations have been carried out to clarify the origin of the modulated microstructure, no satisfactory explanation has been presented so far. The authors have recently suggested in a review on the Co-Cr phase diagram that the observed compositional inhomogeneity can be attributed to the "magnetically-induced miscibility gap" in the cph (εCo) phase[25]. Figure 10 shows the equilibrium Co-Cr phase diagram based on the thermodynamic calculations of Hasebe et al.[26].

188

This predicted diagram shows, that there exist complicated ferromagnetic-paramagnetic phase equilibria involving the presence of miscibility gaps in the Co-rich region. They arise solely because of the inclusion of magnetic interactions in the calculations. These miscibility gaps appear along the Curie temperature, not only in the fcc (αCo) phase but also in the cph (εCo) phase. The paramagnetic (αCo) decomposes into ferromagnetic (αCo) and paramagnetic (εCo) by eutectoid reaction below about 1000 K, and the miscibility gap between ferromagnetic and paramagnetic (εCo) phases is formed below about 940 K. It is quite reasonable to postulate therefore, that the fine memory units that are responsible for the magnetic behaviour in these alloy films, must be the Co-rich cph columnar crystals with [0001] axis as their growth direction, embedded in a Cr-rich paramagnetic matrix. This has been experimentally shown to be the actual case by Maeda and Takahashi[27] recently who have directly observed the existence of a modulated microstructure within sputtered Co-22 at.% Cr film grains using high-spatial resolution X-ray microanalysis. Figure 11 shows the STEM image of the film and the corresponding X-ray line analysis profile. The mean Cr content in Cr-rich region is 28.5 at.% Cr, while that in Co-rich region ranges from 15.5 to 18.6 at.% Cr. Since the critical composition of transition from ferromagnetic to paramagnetic states is considered to be 23 - 25 at.% Cr at lower temperatures, the modulated structure is composed of the ferromagnetic Co-rich and paramagnetic Cr-rich phases. The recent study of Mössbauer spectroscopy by Parker et al.[28] also confirms this finding.

The above explanation for the behaviour of Co-Cr films based on their special microstructure can now be used as a basis for search for other cobalt based alloys that are potential perpendicular recording media. The requirements for a suitable Co-X binary system, for example would be that:

(i) the alloying element X be a cph (εCo) phase stabilizer and that (ii) it decreases the Curie temperature of (εCo) phase drastically, thereby promoting the formation of a magnetically-induced miscibility gap. From a consideration of several of Co-X binaries, it appears that the Co-Si and Co-Ge systems could show the special characteristics required of (εCo) phase. As shown in Fig.12[29], both the systems Co-Si and Co-Ge can be expected to develop microstructures similar to those of the Co-Cr system. It would seem that the Co-Ge system is rather preferable because the (εCo) phase is formed in a wider range of compositions.

6. SUMMARY

The phase diagrams of Alnico, Fe-Cr-Co hard magnets and Co-Cr thin film alloys for perpendicular magnetic recording, are briefly reviewed and discussed, with particular emphasis on the role of miscibility gaps. Magnetic and chemical ordering strongly affect the size and shape of miscibility gaps in these alloys. It is shown, that the miscibility gap plays a significant role, not only in the understandings of the magnetic properties but also in the choice of suitable heat treatments to achieve the best microstructures for optimum magnetic properties. Suggestions are made for further developments of magnetic alloys using phase diagram information.

ACKNOWLEDGEMENTS

One of the authors (K.I) would like to thank Prof. A.P.Miodownik of the University of Surrey, for providing the opportunity and facilities for preparing this manuscript during his stay at the University of Surrey. The authors would also like to thank Dr. L.Chandrasekaran of the University of Surrey for help in preparation of this manuscript. Thanks are also due to Dr. Y.Maeda at NTT basic Research Laboratories and Prof. M.Okada at Tohoku University for providing us the recent information on this work.

REFERENCES

1. W. Köster, Arch. Eisenhüttenw., 6(1932) 17.
2. T. Mishima, Ohm, 19(1932) 353.
3. K. Honda, H. Masumoto and Y. Shirakawa, Sci. Rep. Tohoku Imp.univ. 23(1934) 365.
4. H. Kaneko, M. Homma and K. Nakamura, A.I.P.Conf. Proc., No.5 (1971) 1088.
5. S. Iwasaki and Y. Nakamura, IEEE Trans. Magn., Mag-13(1977) 1272.
6. K.J. de Vos, Magnetism and Metallurgy, A.E. Berkowitz and E. Kneller, eds., Academic Press, vol.1, (1969) 473.
7. S. Kiuchi, Rept. of Aeronautical Res. Inst. Tokyo Imp. Univ., 12 (1937) 179; 13 (1938) 555.
8. A. J. Bradley and A. Taylor, Nature, 140 (1938) 1012.
9. A.J. Bradley and A.Taylor, Proc. Roy. Soc., A, 166 (1938) 353.
10. A.J. Bradley, J. Iron Steel Inst., 163 (1949) 19; 168 (1951) 233; 171 (1952) 41.
11. J.L. Meijering, Phil.Res.Rep., 5 (1950) 333; 6 (1951) 183.

12. H.K. Hardy, Acta Metall., 1 (1953) 210.
13. T. Nishizawa, M. Hasebe and M. Ko, Acta Metall., 27 (1979) 817.
14. T. Nishizawa, S.M. Hao, M. Hasebe and K.Ishida, Acta Metall., 31 (1983) 1403.
15. S.M. Hao, T. Takayama, K. Ishida and T. Nishizawa, Metall. Trans., 15A (1984) 1819.
16. S.M.Hao, K.Ishida and T.Nishizawa, Metall. Trans., 16A (1985) 179.
17. M. Okada and M. Homma, Recent Magnetics for Electronics, Japan Annual Reviews in Electronics, Computer & Telecommunications, Ohshima and Northholland, ed. by Y.Sakurai, vol.15 (1984) 231.
18. K. Ishida, R. Kainuma, N. Ueno and T. Nishizawa, To be published in Metall. Trans.
19. G. Inden, User Applications of Phase Diagrams, ASM, ed. by L.Kaufman, (1986) 25.
20. H. Kaneko, M. Homma, K. Nakamura, M. Okada and G. Thomas, IEEE Trans. Magn., Mag-13 (1977) 1325.
21. T. Minowa, M. Okada and M. Homma, IEEE Trans. Magn., Mag-16 (1980) 529.
22. M. Homma, M. Okada, T. Minowa and E.Horikoshi, IEEE Trans. Magn., Mag 17(1981) 3473.
23. M.Ko and T.Nishizawa, J. Jpn. Inst. Met., 43 (1979) 126.
24. T. Kozakai and T. Miyazaki, Trans.Jpn.Inst.Met., 24 (1983) 633.
25. K. Ishida and T. Nishizawa, Bull. Alloy Phase Diagrams, in press.
26. M. Hasebe, K. Oikawa and T. Nishizawa, J. Jpn. Inst. Met., 46 (1982) 577.
27. Y. Maeda and M. Takahashi, Jpn. J. Appl. Phys., 28 (1989) L248.
28. F.T. Parker, H. Oesterreicher and E. Fullerton, J. Appl. Phys., 66 (1989) 5988.
29. H. Enoki, K. Ishida and T. Nishizawa, J. Less-Common Met., in press.

190

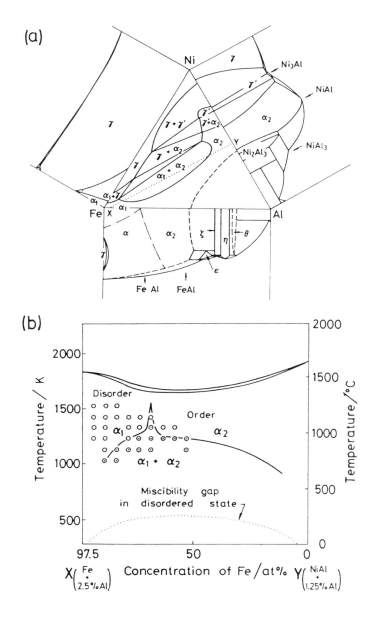

Fig. 1. Phase diagram of Ni-rich portions in the Fe-Ni-Al system after Bradley [10]. (a) Isothermal (750°C) and (b) vertical section diagrams.

191

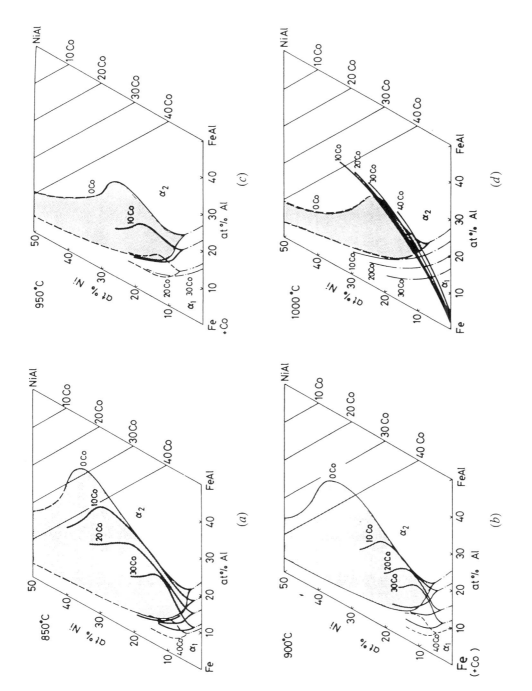

Fig. 2. Isothermal section diagrams in the Fe-Ni-Al and Fe-Ni-Al-Co systems [15]. (a) 850°C (b) 900°C (c) 950°C and (d) 1000°C.

192

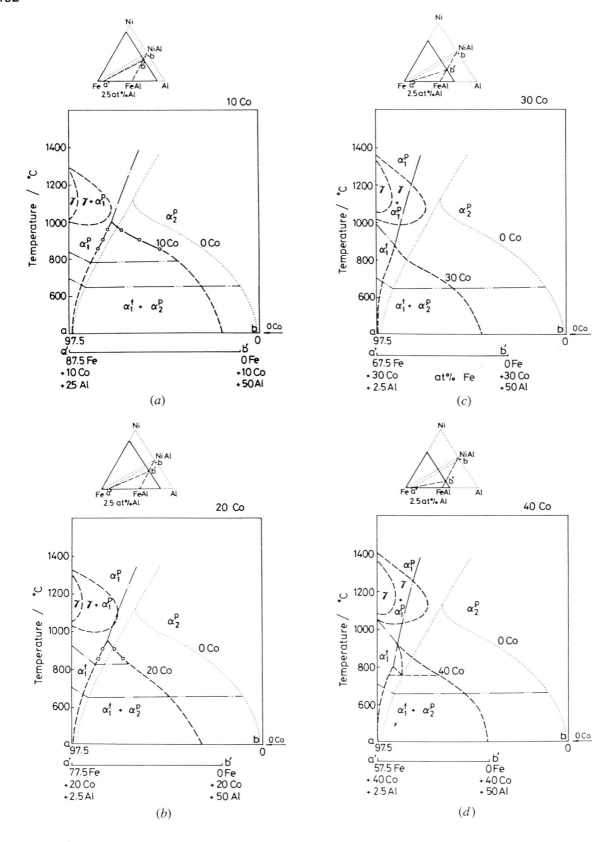

Fig. 3. Vertical section diagram in the Fe-Ni-Al-Co system [15].

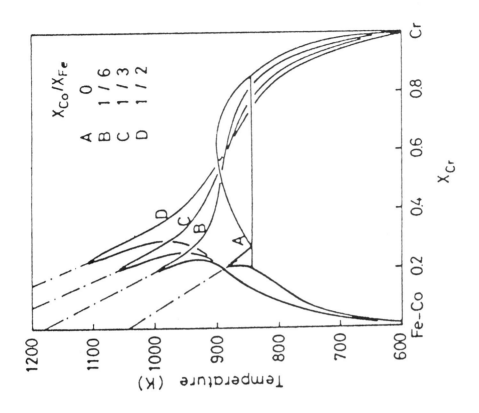

Fig. 5. Calculated miscibility gap of the α phase in the Fe-Cr-Co system [13].

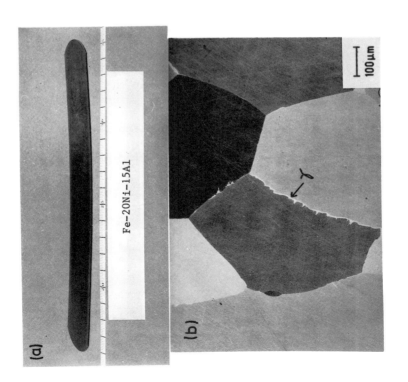

Fig. 4. Ductile Alni (Fe-20 at.% Ni-15 at.% Al) alloy [18]. (a) Appearance of specimen after hot rolling and (b) as-cast microstructure.

194

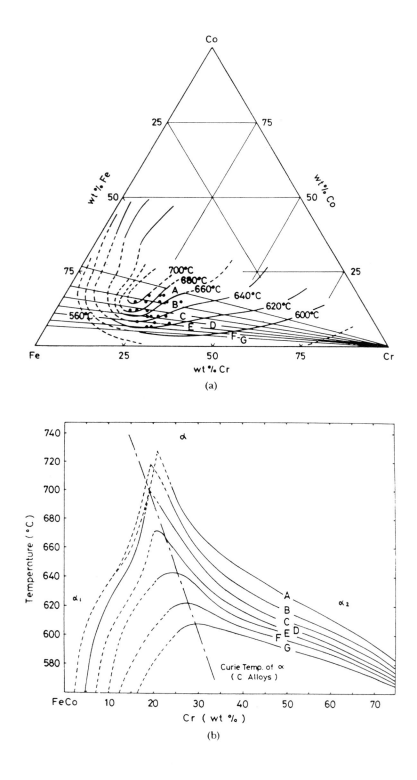

Fig. 6. Observed miscibility gap of the α phase in the Fe-Cr-Co system [21]. (a) Isothermal and (b) vertical sections.

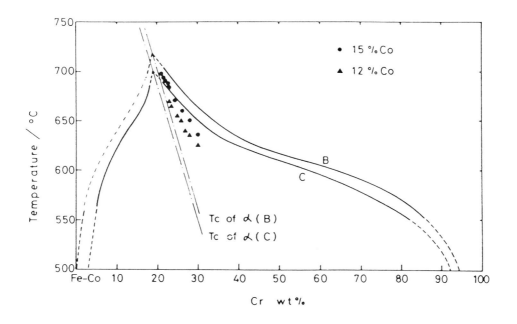

Fig. 7. Schematic plot of optimum TMT temperature of Fe-Cr-(12-15) wt% Co alloys [22].

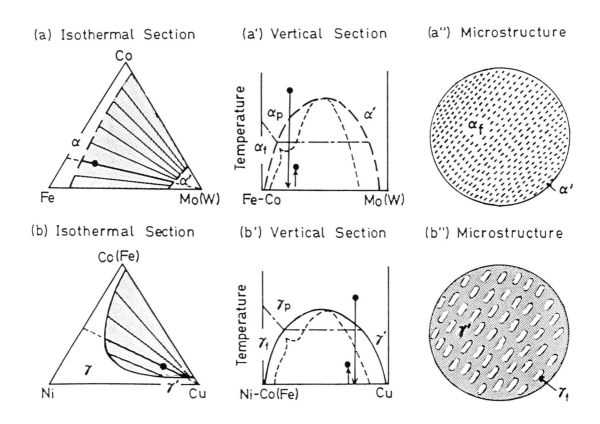

Fig. 8. Illustration of phase diagrams and microstructure of Fe-Co-Mo(W) and Cu-Ni-Co(Fe) magnets [16].

196

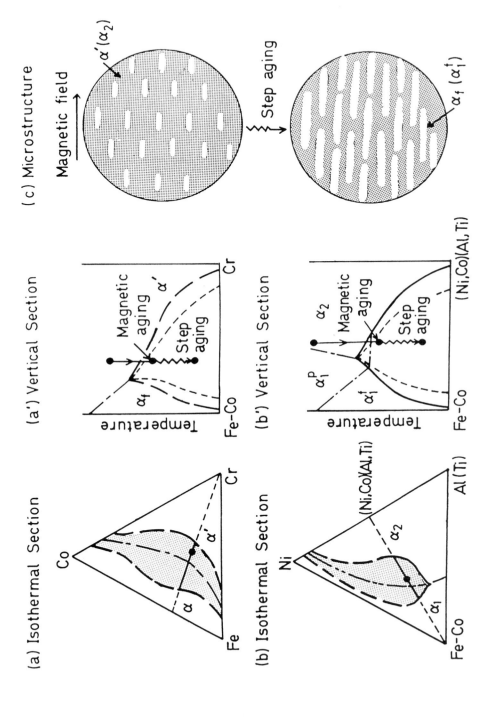

Fig. 9. Illustration of phase diagrams and microstructure of Fe-Cr-Co and Alnico magnets [16].

197

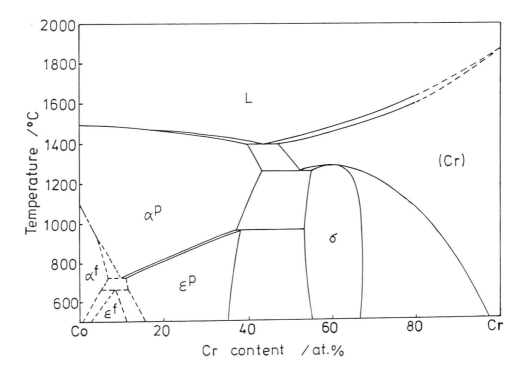

Fig. 10. Co-Cr phase diagram [25].

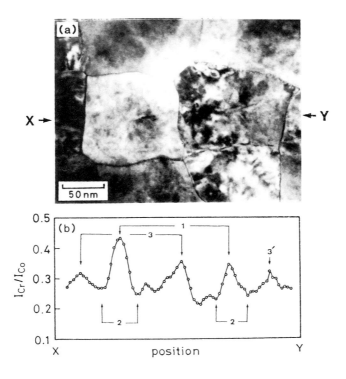

Fig. 11. Microstructure of Co-22 at.% Cr alloy [27]. (a) STEM image and (b) the corresponding X-ray line analysis profile.

198

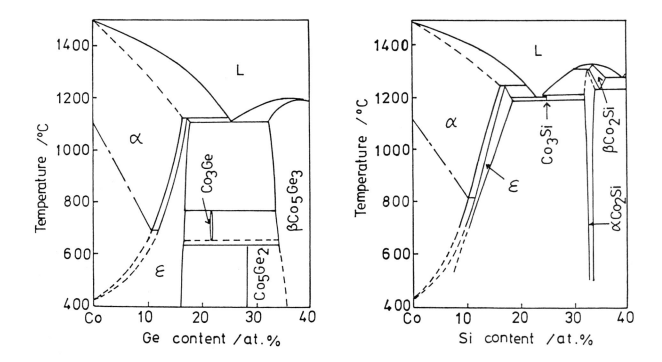

Fig. 12. Co-Ge and Co-Si phase diagrams [29].

25
PHASE EQUILIBRIA AS GUIDELINES IN PROCESS METALLURGY

D. Neuschütz

Lehrstuhl für Theoretische Hüttenkunde und Metallurgie der Kernbrennstoffe, Rheinisch-Westfälische Technische Hochschule Aachen D-5100 Aachen, Germany

1. INTRODUCTION

Thermochemistry has proved to be a useful tool for process metallurgists for many years. Increasing availability of data, of software and of computers makes thermochemical calculations quicker and more reliable than ever. The following examples are thought to demonstrate the value of phase equilibria assessments in discussing existing and novel metallurgical processes. The examples are intentionally taken from a broad variety of fields, namely hydrometallurgy, steelmaking, and powder metallurgy.

2. COPPER ELECTROWINNING

Oxidic copper ores are preferably processed via the hydrometallurgical route. Table 1 gives the sequence of process stages and typical operating data of an electrowinning plant (Example: Cerro Verde, Peru). The question is to calculate, as precisely as possible, the energy efficiency of this electrowinning plant. The electrode and cell reactions and the corresponding Nernst equations are :

$$\text{electrode } \alpha : 2e\,(\alpha) + Cu^{++} = Cu^{O}$$
$$\text{electrode } \beta : H_2O = 2\,H^+ + 1/2\ O_2 + 2e(\beta)$$
$$\text{overall} : 2e\,(\alpha) + Cu^{++} + H_2O = Cu^{O} + 2H^+ + 1/2\ O_2 + 2e(\beta)$$

$$E(T) = E^{O}(T) + \frac{RT}{2F} \ln \frac{a^2_{H^+}\cdot p O_2^{1/2}}{a_{Cu^{++}}}$$

The determination of the cell EMF E at operating conditions requires values for $E^{O}(333K)$ and the activities of the respective species at those conditions.

From standard thermodynamic data for 298 K, E(333K) can be obtained with good accuracy using the approximation

$$E(333K) = \frac{1}{2F}\left[\Delta G^{O}(298) - \Delta S^{O}(298)(333\text{-}398)\right]$$

The ionic activities are given as

$a_{H^+} = m_{H^+}\cdot \gamma_{H^+}$

$a_{Cu^{++}} = m_{Cu^{++}}\cdot \gamma_{Cu^{++}}$

Assuming complete dissociation of $CuSO_4$ and H_2SO_4 in the electrolytic solution, we have

$m_{H^+} = 2m_{H_2SO_4}$,

$m_{Cu^{++}} = m_{CuSO_4}$.

Since individual ionic activity coefficients have no significance, mean ionic activity coefficients

$\gamma^{\nu}_{\pm} = \gamma^{\nu_+}_+ \cdot \gamma^{\nu_-}_-$ with $\nu = \nu_+ + \nu_-$

are introduced:

$H_2SO_4 = 2H^+ + SO_4^-$

$\gamma^3_{\pm}(H_2SO_4) = \gamma^2_{H^+}\cdot \gamma SO_4^-$

$CuSO_4 = Cu^{++} + SO_4^-$

$$\gamma_{\pm}^2 (CuSO_4) = \gamma_{Cu^{++}} \cdot \gamma SO_4^-$$

Thus,

$$\frac{a_{H^+}^2}{a_{Cu^{++}}} = \frac{m_{H^+}^2}{m_{Cu^{++}}} \frac{\gamma_{H^+}^2}{\gamma_{Cu^{++}}} = \frac{(2\, m_{H_2SO_4})^2}{m_{CuSO_4}} \cdot \frac{\delta_{\pm}^3 (H_2SO_4)}{\gamma_{\pm}^2 (CuSO_4)}$$

With $P_{O_2} = 1$ bar and thermodynamic data [1 - 4] listed in Table 2, the EMF of the electrowinning cell under the operating conditions is

E(333K) = 0.9156 V.

While this cell voltage corresponds to a theoretical energy consumption of 772 kWh/t Cu, the operational data [5] (Table 1) of 2.0 V cell voltage and 85 % current efficiency yield 1984 kWh/t Cu. Thus, the energy efficiency η of that electrowinning plant is 38.9 %.

This calculation requires a knowledge of the activity coefficients as a function of temperature and molality in highly concentrated solutions, as they are rarely available. It may therefore be of interest to compare the result with simpler approaches:

(a) Cell temperature 298 K, activities = unity
$E = E^O(298K) = 0.8893$ V $\qquad \eta = 37.8$ %

(b) Cell temperature 333 K, activities = unity
$E = E^O(333K) = 0.8597$ V $\qquad \eta = 36.5$ %

(c) Cell temperature 333 K, acitivities = correct figures
$E = 0.9156$ V $\qquad \eta = 38.9$ % (correct value)
The simplest approach (a) deviates by not more than 3 % from the correct result.

3. ZINC ELECTROLYSIS

About 75 % of the world's primary zinc production makes use of the hydrometallurgical route, where sulphide zinc concentrates are roasted and leached, the zinc sulphate solutions purified and electrolyzed so that metallic zinc is deposited at the cathode and oxygen evolved at the insoluble anode. Table 3 summarizes the process steps and typical operating conditions for the electrolytic zinc recovery [6].

The electrical energy consumption of the electrowinning process amounts to about 3300 kWh/t Zn, thus representing a major cost factor of primary zinc production.

A novel concept [7] has therefore been suggested and developed to the pilot scale, replacing the conventional anode by a reversible hydrogen electrode, Fig. 1. By means of equilibrium considerations, a comparison is carried out between the novel concept and the conventional zinc electrowinning cell.

I: Conventional cell
$$\text{Electrode } \propto : 2e\,(\propto) + Zn^{++} = Zn^O$$
$$\text{Electrode } \beta : H_2O = 2\, H^+ + 1/2\, O_2 + 2e(\beta)$$
$$\text{Overall reaction } 2e\,(\propto) + Zn^{++} + H_2O = Zn^O + 2H^+ + 1/2\, O_2 + 2e(\beta)$$

II: Novel cell
$$\text{Electrode } \propto : 2e\,(\propto) + Zn^{++} = Zn^O$$
$$\text{Electrode } \beta : H_2 = 2\, H^+ + 2e(\beta)$$
$$\text{Overall reaction: } 2e\,(\propto) + Zn^{++} + H_2 = Zn^O + 2H^+ + 2e(\beta)$$

For calculating the equilibrium cell voltages, we make use of the results obtained for the copper electrowinning cell, where the standard EMF at unit activities deviated only by 3 % from the correct EMF at operating temperature and activities. Since the concentrations of sulphuric acid and metal sulphate are similar in both cases, it seems reasonable to use the standard EMF's for the comparison of the two electrowinning cells.

The standard equilibrium cell voltages are

$E_{298}^O\,(I) = 1.99$ V,

$E_{298}^O\,(II) = 0.763$ V.

These values can directly be taken from the E-p_H-diagram by Pourbaix [8] for zinc, Fig. 2, since all acitivities are taken as unity.

From operational results for both the conventional and the (pilot scale) novel cell, the actual cell voltages and energy consumptions are known [6, 7]. Combining the theoretical cell voltages and theoretical energy demand with these figures, we find a current efficiency of 82 % for both cells, while the energy efficiencies are 49 % for the conventional and 42 % for the novel cell. Table 4 summarizes the theoretical, observed and calculated values for both zinc electrowinning cells. It should be remembered that the energy consumed by ohmic resistance of the electrolyte does not depend on the EMF of the cell. Therefore, the overall energy efficiency decreases with decreasing EMF for equal current densities and temperatures.

The savings in electric energy are about 1800 kWh/t Zn, when the conventional anode is replaced by a reversible hydrogen electrode. On the other hand, the novel cell requires special anodes and consumes at least the stoichiometric amount of hydrogen per unit amount of zinc. The profitability of this concept will finally depend upon the price relation of electric energy versus hydrogen gas.

4. STAINLESS STEELMAKING

There are not many examples in metallurgy, where the process technology is such a straightforward consequence of chemical thermodynamics as in the case of stainless steelmaking [9].

Discovered in 1912, stainless steel has found a steadily increasing demand with some 8 million tonnes of annual production in the Western world today and long-term growth-rate of 3.5 % per year.

The classical austenitic stainless steel contains about 18 % Cr and 8 % Ni and less than 0.05 % C to avoid intergranular corrosion.

The problem existing with the making of high-chromium low-carbon steels is that chromium and carbon have a similar affinity to oxygen. Since chromium is commonly traded as charge chrome containing about 7 % carbon, a refining step is required to oxidize the carbon without slagging the chromium.

The competing equilibria are thus

$$2 \underline{C} + O_2 = 2 CO$$

and $\frac{4}{3} \underline{Cr} + O_2 = \frac{2}{3} (Cr_2O_3)$

For the austenitic steel mentioned above, these equilibria have been calculated [10] taking into account the activity coefficients of carbon and of chromium in liquid steel, while the activity of Cr_2O_3 in the slag, the CO partial pressure and the temperature were taken as variables. In Fig. 3, the result is presented in the form of O_2 partial pressure versus reciprocal temperature with CO partial pressure and Cr_2O_3 activity as the parameters. Each of the reactions is represented in the figure by a set of three parallel lines. As an example, take the C/CO line for p_{CO} = 1 bar and the Cr/Cr_2O_3 line for a $_{Cr_2O_3}$ = 0.1. At the conditions of their intersection, i.e. 10^{-8} bar O_2 and 2000°C, 18 % Cr co-exist in the liquid steel with 0.05 % C. At lower temperatures, Cr is preferentially oxidized, carbon at higher temperatures. At p = 1 bar, the production of an 18-8 stainless steel with 0.05 % C without noticeable Cr losses requires about 2000°C operating temperature.

In fact, the early manufacturing method, which was applied from 1920 until 1965, was a one-stage electric arc furnace process, where the CO partial pressure was close to 1 bar. Temperatures had to be extremely high, decarburization was minimized by using expensive low-carbon ferrochrome, and final carbon contents of the steels were around 0.25 %.

It was not before the sixties that the carbon monoxide partial pressure was discovered as a technically feasible parameter in stainless steelmaking. As seen from Fig. 3, the decrease of p_{CO} to 0.1 bar at a $_{Cr_2O_3}$ = 0.1 lowers the equilibrium operating temperature from 2000°C to 1700°C, and whenever p_{CO} = 0.01 bar, the temperature must only be around 1500°C to reach low carbon contents without Cr oxidation. Since that time stainless steels have been made in two-stage processes, a melt-down stage in an electric arc furnace to produce an alloy with about 1.5 % C, and a refining stage in a separate vessel, where p_{CO} can be lowered in the course of carbon removal to values as low as 3 to 5 mbar. The reduction of the CO partial pressure is achieved either by decreasing the total pressure (Vacuum Oxygen Decarburization = VOD) or by dilution of the gas phase with inert gas, mostly with argon, at a total pressure of 1 bar (Argon Oxygen Decarburization = AOD).

The VOD process introduced in 1965 was made technically possible through the development of sufficiently large vacuum pumps and vacuum equipment. The AOD process that started operation in 1968 was the by-product of the large-scale air liquefaction required for oxygen steelmaking that replaced the air-based Basic Bessemer Process in the late fifties: Argon became available at reasonable prices, proving to be an ideal inert diluent in stainless steelmaking .

At present, about 75 % of all stainless steel is made by AOD or related technology, while 25 % is produced via VOD. The latter is said to make lower final C-contents possible, but needs somewhat longer process times. Meanwhile, a number of process variations have been developed from these principles [9].

In the thermodynamic discussion, we assumed the Cr_2O_3 activity in the slag to be 0.1. To keep chromium losses low, the final slag should contain only very little chromium oxide, i.e. definitely less than its saturation value, where a_{Cr2O3} would be unity.

Geiseler et al. [11] gave a thorough analysis of the changes in slag composition that take place during metal decarburization in the AOD process, Fig. 4. The slag coming from the melt-down stage contains only a little chromium oxide. At the beginning of oxygen refining, lime is added thereby increasing the slag basicity drastically. In the later part of carbon removal, chromium tends to be oxidized together with carbon, which may lead to Cr_2O_3 saturation of the slag, until in the final stage, Cr_2O_3 is again reduced from the slag by means of FeSi additions to the converter. Further additions of lime for good final desulphurisation increases the slag basicity again. The final slag shows only small solubility for Cr_2O_3. This time-dependent behaviour of the slag is further complicated by the fact that, under reducing conditions, chromium monoxide CrO is formed [12] which most probably has somewhat higher solubility in basic slags than Cr_2O_3.

Thus, the assumption of $a_{Cr2O3} = 0.1$ seems to be justified in order to describe an equilibrium system with low chromium losses. Practical experience shows that neither VOD nor AOD – although the thermodynamics are favorable – keep chromium oxidation low enough to operate without the addition of FeSi or CaSi.

5. HOT ISOSTATIC PRESSING

While geologists consider total pressure one of their major variables, metallurgists tend to neglect its influence on chemical equilibria since in most cases their processes operate at or below atmospheric pressure.

There are, however, a number of processes with phase transitions taking place under elevated pressure, such as in squeeze forging or in hot isostatic pressing.

As an example, the pressure dependence of the eutectic temperature of the reaction $\gamma' + \gamma$ = liquid in nickel-based superalloys is discussed. This temperature is of interest in the HIP process of superalloy powder. Since during hipping the γ'-phase is supposed to dissolve completely in the γ-phase, the highest possible temperature is chosen for hipping. On the other hand, the formation of a liquid phase is to be avoided, because on solidification micro-porosity may form which is harmful to the mechanical properties. Compared to a densification at or around atmospheric pressure, the HIP process is operated at about 2000 bar. As the subsequent calculation will show, the eutectic temperature is about 10 K higher at that pressure, thus allowing the operating temperature to be increased correspondingly.

The calculation of this pressure dependence of the eutectic reaction can only be carried out approximately. First, the complex alloy is substituted by the better known two-component system Ni-Al [13], Fig. 5. The eutectic reaction then is γ (AlNi$_3$) + γ (Ni) = Liquid.

Second, a number of assumptions has to be made regarding the thermochemical data and relevant compositions.

Generally, the pressure-temperature relationship for a three-phase reaction $\alpha + \beta = \gamma$ in a two-component system with mole fraction 1 - x for component I and x for component II is represented by

$$\frac{dT}{dP} = \frac{\left(x^\gamma - x^\beta\right) V^\alpha + \left(x^\alpha - x^\gamma\right) V^\beta + \left(x^\beta - x^\alpha\right) V^\delta}{\left(x^\gamma - x^\beta\right) S^\alpha + \left(x^\alpha - x^\gamma\right) S^\beta + \left(x^\beta - x^\alpha\right) S^\delta}$$

with V and S the integral molar volumes and entropies of the respective phases. This equation may be abbreviated to

$$\frac{dT}{dp} = \frac{\Delta V^{\alpha + \beta = \gamma}}{\Delta S^{\alpha + \beta = \gamma}} \text{ for eutectic reactions, } \frac{dT}{dp} = \frac{\Delta V^{\alpha = \beta + \gamma}}{\Delta S^{\alpha = \beta + \gamma}} \text{ for peritectic reactions.}$$

In the case of a one-component, two-phase equilibrium, e.g. the melting reaction of a pure substance, the relationship reduces to the well-known Clausius-Clapeyron equation

$$\frac{dT}{dp} = \frac{\Delta V^{\alpha = \beta}}{\Delta S^{\alpha = \beta}}$$

The data required to apply these equations to the eutectic reaction $\gamma+\gamma'$ = liquid in the Ni-Al system are summarized in Table 5. The majority of values are based on assumptions, interpolations or comparisons with similar metals [10]. Actually, there is a striking lack of precise data with respect to pressure dependence calculations.

The results are for the melting point of Ni

$$Ni(s) = Ni(li)$$
$$dT/dp = 2.87\cdot10^{-8} \text{ K/Pa}$$

and for the eutectic reaction

$$\gamma' \text{ (AlNi}_3) + \gamma \text{ (Ni)} = \text{Liquid}$$
$$dT/dp = 4.8\cdot10^{-8} \text{ K/Pa}$$

Thus, for a HIP pressure of 2000 bar = 0,2 GPa, the melting temperature of pure nickel is raised by 5.7 K, and the eutectic temperature of the γ' / γ reaction by 9.6 K. Consequently, under hipping conditions, the maximum operating temperature may be chosen to be about 10 K higher than the annealing temperature at atmospheric pressure.

6. CONCLUSION

Phase equilibria calculations have become a common tool in describing metallurgical processes either for process control or development. Even where precise thermochemical data are lacking, reasonable assumptions can be made to give a useful evaluation of process features in materials technology.

ACKNOWLEDGEMENTS

The calculations and assessments related to chapter 5 were carried out by Dr. Klaus Hack of GTT, Aachen, whose support is gratefully acknowledged.

REFERENCES

1. I. Barin, O. Knacke and O. Kubaschewski: Thermochemical properties of inorganic substances, Springer-Verlag, Berlin / Verlag Stahleisen, Düsseldorf, 1973 and 1977.
2. O. Knacke, R. Schlim: Erzmetall 37 (1984) 544-548.
3. H. Majima and Y. Awakura: Met. Trans. B, Vol. 17B, 621-627.
4. M. Jaskula and J. Hotlos: The mean thermodynamic activity coefficient of $CuSO_4$ in the ternary system $CuSO_4$-H_2SO_4-H_2O at 60°C, Hydrometallurgy, accepted for publication.
5. H. Hilbrans and P. Paschen: Erzmetall 34 (1981) 639-644.
6. F. Pawlek: Metallhüttenkunde, W. de Gruyter Berlin, 1983.
7. A. v. Röpenack: Future Changes in the Physico-Chemistry of Zinc Electrowinning; Proc. Lead-Zinc '90, Ed. T. S. Mackey et. al, The Minerals, Metals and Materials Soc., 1990.
8. M. Pourbaix: Atlas d'Equilibres Electrochimiques à 25°C, Paris, 1963, pp. 405-413.
9. H.-U. Lindenberg, K.-H. Schubert and Z. Zörcher: Stahl u. Eisen 107 (1987) 1197-1204.
10. Databank THERDAS, Lehrstuhl f. Theor. Hüttenkunde, RWTH Aachen.
11. J. Geiseler, K. Grade and P. Valentin: Stahl u. Eisen 103 (1983) 1013-1017.
12. K. Morita and N. Sano: A view of mechanism of reduction of chromium oxides in the molten slag, Tetsu-to-Hagane Vol. 74 (1988) 2361-3.
13. M. Hansen and K. Anderko: Constitution of Binary Alloys, McGraw Hill, New York, 1958, p.119.

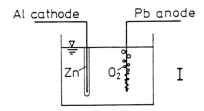

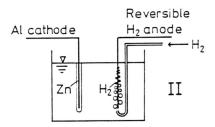

Fig. 1. Conventional and novel cells for zinc electrowinning.

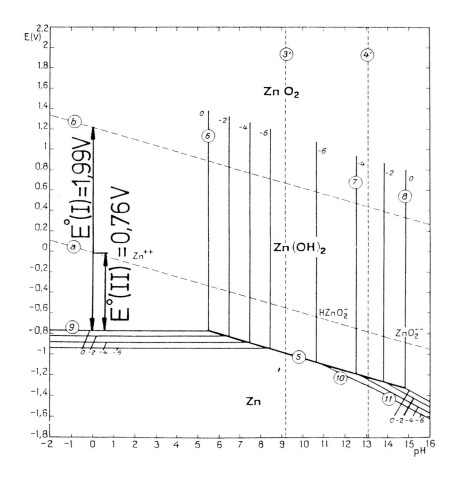

Fig. 2. E-pH-diagram for zinc acc. to M. Pourbaix, containing standard EMF's for conventional (I) and novel (II) electrowinning cells.

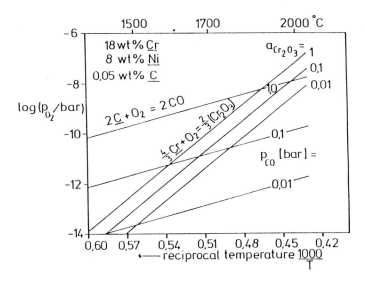

Fig. 3. Equilibrium relations for C/CO and Cr/Cr$_2$O$_3$ in stainless steelmaking.

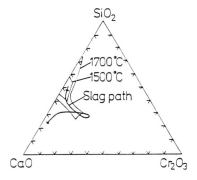

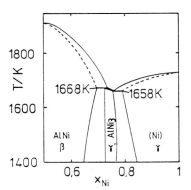

Fig. 4. Slag path during stainless steel refining in AOD process.

Fig. 5. Ni-Al phase diagram   acc. to Hansen and Anderko.

Table 1 Operating data of copper electrowinning

Processing_of oxidic_copper ores:

Leaching - solvent extraction - electrowinning - casting.

Operating data_of electrowinning:

Cathodes	Cu
Anodes	Pb + 0.5 % Ca
Electrolyte	0.6 m $CuSO_4$
	1.7 m H_2SO_4
Temperature	60 °C
Cell voltage	2.0 V
Current efficiency	85 %

Table 2 Thermodynamic data to calculate EMF of copper electrowinning cell

$$Cu^{++} + H_2O = Cu^O + 2 H^+ + 1/2 O_2$$

$\Delta G^O_{298} = 171\ 610$ J/mol   (Ref. 1,2)

$\Delta S^O_{298} = 162.88$ J/mol K   (Ref. 1,2)

Mean activity coefficients for 1.7 m H_2SO_4 + 0.6 m $CuSO_4$:

$\gamma_\pm (H_2SO_4) = 0.12$ (Ref. 3)

$\gamma_\pm (CuSO_4) = 0.026$ (Ref. 4)

Table 3 Operating data of zinc electrolysis

Hydrometallurgical_processing of_sulfidic zinc_ores:

Roasting - leaching - leach purification - zinc electrolysis - casting.

Average operating conditions_for_the_zinc electrolysis

Cathodes   Al sheets
Anodes   Pb + 0.1 % Ca
Electrolyte   0.7 m H_2SO_4 + 2.0 m $ZnSO_4$
Temperature   35 ... 40 °C

Table 4 Comparison of theoretical and operational data for the conventional and the novel zinc electrowinning cell

	Conventional cell	Novel cell
Theor. cell voltage E^0_{298} at standard conditions, V	1.99 V	0.763 V
Theor. energy demand, kWh/t Zn	1630	625
Operational cell voltage, V	3.3	1.5
Operational energy consumption, kWh/t Zn	3300	1500
Current efficiency, %	82	82
Energy efficiency, %	49	42

208

Table 5 Data used to calculate the pressure–temperature relationships in Ni–Al system

(a) Pure nickel: $\Delta V = 2.93 \cdot 10^{-7}$ m^3/mol

$\Delta S = 10.21$ J/mol K

(b) Eutectic reaction γ'(AlNi$_3$) + (Ni) = liquid (1658 K)

	γ'	γ	liquid
X_{Ni}	0.740	0.798	0.750
V m^3/mol	$7.87 \cdot 10^{-6}$	$8.24 \cdot 10^{-6}$	$8.56 \cdot 10^{-6}$
S *) J/mol K	2.168	4.184	15.209

*) solid Al and Ni as reference states

26

APPLICATION OF THE CALCULATION OF PHASE EQUILIBRIA TO THE PYROMETALLURGICAL EXTRACTION FROM SULPHIDE ORES

J. R. TAYLOR AND A. T. DINSDALE*

Johnson Matthey Technology Centre, Sonning Common, Reading RG4 9NH, UK
* Division of Materials Metrology, National Physical Laboratory, Teddington, Middlesex TW11 0LW, UK

ABSTRACT

In order to achieve a better understanding of the pyrometallurgical treatment of sulphide ores, experimental work on the equilibration and simulated converting of Cu-Fe-Ni-S mattes with Fe-O-SiO$_2$ slags was undertaken. In addition. a thermodynamic model was developed which allowed calculation of the equilibrium relationships between the phases at various stages of the process. The development of the multicomponent models for the liquid matte phase, the solid alloy and solid sulphide phases from the basic binary phase equilibria and thermodynamic data is described. Experimental results are presented and compared with the model predictions. Results for the matte/slag/gas equilibration experiments under controlled partial pressures of SO$_2$ are in excellent agreement with those calculated from the model. Experimental results for matte sulphur contents during converting and Cu and Ni slag contents are presented and interpreted with the help of the model. Two methods of calculating the sequence of events during matte solidification are considered.

1. INTRODUCTION

Sulphide ores are a major source of copper, nickel and precious metals. It is common to extract these metals by blowing air into the molten sulphide (matte), with a slag cover, to oxidise sulphur into the gas phase and iron into the slag phase. On cooling. the matte separates into solid sulphides and a metal phase from which the precious metals can be extracted by aqueous chemistry techniques. The sulphur content of the matte at the end of the converting stage is critical in establishing the desired phase formation during the cooling stage.

In order to achieve a better understanding of the process an experimental programme was undertaken which involved equilibration and simulated converting of matte/slag systems. In addition, a major effort was put into the development of a thermodynamic model to allow calculation of the equilibrium relationships between the phases involved in the process. This provided an independent check on the experimental results, gives a quantitative assessment of the effect of variables which may be difficult to investigate experimentally and, once the relationship between model and process has been established, provides a useful tool for assessing the likely effect of any process changes.

Mathematical expressions were required to describe the thermodynamic behaviour of all the phases involved including the gas, matte, slag, solid metallic and solid sulphide phases. The starting point for modelling the condensed phases was the representation of the equilibrium phase relationships for all the component binary systems. The final model allows calculation of the phase relationships in the multicomponent matte/slag/gas system.

The work described in the first half of the paper concerns the development of the data and models for the solid and liquid phases for the Cu-Ni-Fe-S system. A more detailed account is given elsewhere (84DIN). Some applications of the data from this assessment have been presented previously (88DIN/HODa. 88DIN/HODb). The data for the gas phase were taken from the SGTE data base (86DIN, 87ANS/SUN) while the data for the slag phase were taken from a published assessment for the CaO-Fe$_x$O-SiO$_2$ system (80GOE/KEL).

In the second half of the paper experimental results are presented and compared with the model calculations. The use of the model in analysing and interpreting the experimental results is emphasized.

2. DEVELOPMENT OF MODEL

Representation of Data for the Liquid Phase

The major problem associated with this work was the representation of the thermodynamic data in terms of composition for the different phases. The liquid phase is particularly difficult to represent because in all metal-sulphur systems its thermodynamic properties show marked changes at particular compositions. These compositions differ from system to system (eg at $x_s = 0.5$ for the Fe-S system and $x_s = 0.33$ for the Cu-S system) and it has been proposed that this phenomenon was due to some sort of ordering. Conventional power series expressions such as the Redlich-Kister expression are not capable of representing the data.

Much research has already been carried out into methods for representing the data for these sulphide liquids. In particular it is worth emphasising the work of Kellogg, Larrain and their colleagues (79LAR/KEL, 80LEE/LAR, 79LAR), and Chang and his colleagues (79CHA/SHA, 82CHU/CHA, 85CHU/HSI, 87HSI/CHA). Both groups have used the associated solution model for the liquid phase.

For this work an extended two sublattice model which has been advocated by Hillert and colleagues (75HIL/STA, 76HIL/STA, 81FER/HIL), was adopted. This model was chosen because it seems in principle physically more realistic than the associated solution model for melts of substances that as crystals have extended rather than molecular bonding. Because of this it should offer superior prospects for the extrapolation of the binary thermodynamic data for the liquid phase into the ternary and quaternary systems without the necessity of introducing an excessive number of ternary or higher order parameters.

According to the sublattice model the first sublattice would contain metal ions and positively charged vacancies and the second sublattice sulphur ions and negatively charged vacancies. There is however no suggestion that these charges are formally present. The magnitude of the charge on a vacancy would be equivalent to the average charges on the ions on the other sublattice. Copper is assumed to be a univalent ion while iron and sulphur are divalent. Therefore for the Fe-S system the metal sublattice would consist of Fe^{2+} ions and vacancies with a charge of +2 while on the sulphur sublattice there would be S^{2-} ions and vacancies with a charge of -2. This would lead to an equal number of sites on the two sublattices. However, for the Cu-S liquid phase the metal sublattice would contain Cu^+ ions and vacancies with a charge of +2 while the sulphur sublattice would contain S^{2-} ions and vacancies with a charge of -1. Here the ratio of sites between the two sublattices is not fixed but is defined by a charge balance between the various ions and charged vacancies.

For the ternary Cu-Fe-S liquid phase the metal sublattice would consist of Fe^{2+} and Cu^+ ions and vacancies with a +2 charge. The sulphur sublattice would consist of S^{-2} ions with one type of vacancy having a charge which would be non-integral, varying according to the relative amounts of copper and iron in the phase.

The general multicomponent model for metal-sulphide liquids is complicated and the interested reader is recommended to consult reference 84DIN for further details.

Representation of data for the other phases

The thermodynamic data for the bcc, fcc and L12 phases of the Cu-Fe-Ni system were represented by the widely used Redlich-Kister power series for the excess Gibbs energy of formation. It was assumed that the solubility of sulphur in these phases could be neglected.

For the purpose of this work where the metal-rich part of the phase diagram is of interest, and in order to simplify the ternary and quaternary phase diagram calculations. sulphur solubility was also neglected in the binary phases FeS (pyrrhotite), Cu_2S (digenite) and NiS even though it is known to be appreciable. A study of ternary experimental information showed that none of the several phases of higher sulphur content than digenite and pyrrhotite has any effect on the phase equilibria in the metal rich part of the system. From experimental data it appeared that FeS and NiS are completely miscible and for convenience this whole phase has been labelled pyrrhotite for these calculations. In the literature the phase is often referred to as the monosulphide solid solution. Pyrrhotite and digenite (based about Cu_2S and sometimes labelled as bornite) are not completely miscible although both phases appear to intrude into the ternary systems in the direction of the other and this was assumed for these calculations. The model used to represent the data for these phases was a simplification of the model used to represent the data for the liquid phase but with the added assumption that no vacancies were present. Data were required for the hypothetical forms of FeS and NiS in the digenite phase and Cu_2S in the pyrrhotite phase and these data were derived from ternary thermodynamic and phase diagram information.

The data for the beta phase based about the composition Ni_3S_2 were modelled, for simplicity using a Redlich-Kister expression. This is almost certainly inappropriate and some sort of sublattice description should in principle be used. Unfortunately the phase is apparently unquenchable which makes the exact choice of model difficult. The situation is further complicated. Certain evidence (78LIN/HU, 76RAU) suggests that in the Ni-S binary system the phase region consists of two distinct phases separated by a narrow two-phase region. Both Fe and Cu are appreciably soluble in this phase which also shows a wide range of sulphur solubility The use of the Redlich-Kister substitutional model, although aesthetically inappropriate allowed the simple extension of the data into the multicomponent system.

The remaining solution phase of interest in the metal rich part of the Cu-Fe-Ni-S system is the pentlandite phase

$(Fe,Ni)_9S_8$. It is believed that the solubility of Cu and S in this phase is small and could therefore be neglected. This phase was modelled as a conventional substitutional solution between the Fe and Ni atoms on the metal sublattice with a second sublattice occupied solely by sulphur atoms. The ratio of sites between the sublattices is 9:8.

The remaining phases of interest in this system are the low temperature binary stoichiometric phases chalcocite (Cu_2S), millerite (NiS), heazlewoodite (Ni_3S_2) and Ni_7S_6.

Assessment or Validation of data for the Cu-Fe-Ni system
The data used for the base metal systems were taken from an assessment of Chart (82CHA/GOH). The phase diagrams and thermodynamic data for the component binary systems are fairly well established and the datasets used were essentially modifications of existing critically assessed data. The adopted thermodynamic data for the Cu-Fe system were taken from the assessment of Kubaschewski et al. (77KUB/SMI). The phase diagram calculated using these data is shown in Fig. 1.

By combining the binary datasets and introducing ternary parameters for the fcc phase Chart et al. were able to calculate the phase diagram for the Cu-Fe-Ni system between the temperatures 673 K and 1673 K in good agreement with available experimental data. Figure 2 shows the phase diagram calculated for 873 K. The data provide a reliable base for calculations of equilibria in the Cu-Fe-Ni-S system. More recently the Cu-Fe-Ni system has been assessed by Jansson (87JAN).

Provision of data for the metal-sulphur binary systems
The three metal sulphur binary systems have a number of basic features in common including a low mutual solubility between sulphur and the metallic elements in the solid state and the formation of a very stable congruently melting compound. The compounds also exhibit non-stoichiometry to sulphur-rich compositions. The liquid phase shows a pronounced tendency to immiscibility on either side of the composition corresponding to this compound phase and sharp changes in the thermodynamic data at about this composition. These data imply ordering of the liquid phase as confirmed by electrical conductivity measurements.

Fernández Guillermet et al. (81FER/HIL) had used the two sublattice model to represent the data for the liquid Fe-S system and this assessment was used for this work.

Data were derived for the Cu-S system to be consistent with the experimental phase diagram and thermodynamic data. Most attention was focused on the metal rich part of the system (82DIN/CHA).

The Ni-S system is rather more complicated than the Fe-S or Cu-S systems. The metal rich part of the phase diagram is dominated by a deep eutectic between essentially pure nickel and the so-called beta phase, based around the composition Ni_3S_2, which transforms on cooling to the low temperature stoichiometric phase heazlewoodite (Ni_3S_2). The other high temperature phase in the metal-rich part of the system is the monosulphide $Ni_{1-x}S$ phase isostructural with pyrrhotite FeS. A set of data were derived to represent the experimental thermodynamic data and also the phase diagram. The calculated phase diagram is shown in Fig. 3.

Calculation of phase equilibria in the Cu-Fe-S system
According to Craig and Scott (74CRA/SCO) "more time and effort has been expended in the determination of phase equilibria and minerological relationships among the Cu-Fe sulphides than any other ternary sulphide system". They go on to say that despite this many aspects of the phase diagram are unknown. This is hardly surprising in view of the large number of ternary phases present in the system. Fortunately as far as the modelling of pyrometallurgical extraction is concerned most of these ternary phases are stable only at low temperatures and are in any case sulphur rich of the composition line joining Cu_2S and FeS and can therefore be ignored.

The experimental work on the phase diagram and thermodynamic data of the Cu-Fe-S system has been reviewed by Craig and Scott (74CRA/SCO) and by Chang et al. (79CHA/NEU, 76CHA/LEE). A comparison was made between the experimental thermodynamic data and those calculated simply from the assessed data for the binary systems, ie with no ternary parameters included, and very good agreement was found.

Both Cu_2S and FeS intrude into the ternary system to a considerable degree. Data were developed for the solubility of Cu in FeS and Fe in Cu_2S and good agreement was obtained between calculation and a phase diagram for the pseudo-binary section as reviewed by Chang et al. (79CHA/NEU, 76CHA/LEE). The metal-rich part of the phase diagram calculated for 1273 K is shown in Fig. 4 and is in excellent agreement with the diagram provided by Chang et al.

Calculation of phase equilibria in the Cu-Ni-S system
The phase diagram of the Cu-Ni-S system has also been the subject of much study and this work has also been reviewed by Chang et al. (79CHA/NEU). The high temperature phase diagram is dominated by three separate homogeneous liquid regions as in the Cu-S binary system. The miscibility gap between the matte phase and the metal-rich liquid phase has been studied by a number of groups of which, according to Chang et al., the data of Schlitt et al. (73SCH/CRA) and Lee et al. (75LEE, 80LEE/LAR) are the most reliable. A number of investigations

of the phase diagram have been made for lower temperatures (62KUL/YUN, 69KUL/YUN, 40KOS/MUL, 63MOH/KUL, 74MOH/KUL, 74CHI/GUL, 14FRI, 14HAY, 24STA/FAI).

The liquid phase of the ternary Cu-Ni-S system has been modelled successfully by Larrain and Lee (80LAR/LEE) and Chuang and Chang (82CHU/CHA). However neither group attempted to provide data for the solid phases in the ternary system. For the Cu-Fe-S system as described above it was found that the two sublattice model was able to represent the thermodynamic data for the liquid phase without the need for any ternary interaction data. For the Cu-Ni-S system, however, it was found that ternary parameters were necessary in order to represent both the experimental thermodynamic data and the miscibility gap in Cu-Ni-S system emanating from the Cu-S binary system. Acceptable agreement was obtained although it was found difficult to reproduce the miscibility gap exactly. Figure 5 shows the calculated partial pressures of sulphur expressed as $\log(P_{s2})/2$ for 1473 K and 1673 K for the pseudo-binary cut through the ternary system corresponding to a constant $x_{Ni}:x_{Cu}$ ratio of 3. Superimposed on these curves are the experimental data of Lee et al. (75LEE, 80LEE/LAR). The agreement is very good for these Ni rich compositions. For Cu rich compositions the agreement is not quite so good and this is reflected by small differences between the experimental and calculated phase boundary of the miscibility gap which is very sensitive to subtle changes in the thermodynamic data.

Data were developed to model the intrusion of Cu_2S and NiS into the ternary system to agree with known experimental data and to model the appreciable solubility of Cu in the beta phase based upon the approximate composition Ni_3S_2. Figure 6 shows the calculated diagram for 873 K. A ternary eutectic is predicted from the calculations and this is in excellent agreement with experimental information (79CHA/NEU).

Calculation of phase equilibria in the Fe-Ni-S system

The Fe-Ni-S system is of great general interest since phases based in this system are found in many ores. The experimental data for the Fe-Ni-S system has been reviewed by Hsieh et al. (82HSI/CHA) and Craig and Scott (74CRA/SCO).

There have been few experimental studies to derive thermodynamic data for the Ni-Fe-S system. Naldrett and Craig (68NAL/CRA) and Kao (80KAO) have studied the partial pressure of sulphur above the monosulphide solid solution (pyrrhotite). Kao, who also studied the two-phase region between pyrrhotite and the beta phase, paid particular attention to the phase relationships and the representation of the thermodynamic data. Scott et al. (72SCO/NAL, 74SCO/NAL) investigated the activity of FeS in the pyrrhotite phase at 1203 K and found that NiS and FeS mix ideally for a given mole fraction of sulphur. Studies have been carried out on dilute solutions of sulphur in the liquid phase (66BAR, 60ALC/CHE, 55COR/CHI, 59ZEL/PAY, 59SPE/JAC). Concentrated liquids have been studied by Vaisburd et al. (69VAI/REM) who measured the activities of Fe in concentrated liquids at 1573 K while Byerley and Takebe (72BYE/TAK), Meyer (72MEY) and Bale (80BAL) have studied the partial pressures of sulphur above the liquid phase at high temperatures.

As was found for the Cu-Ni-S system, it was necessary to introduce ternary parameters to obtain close agreement between predicted and experimental thermodynamic and phase diagram data.

Data for the pyrrhotite phase were developed to be consistent with the measured ideal mixing of FeS and NiS (72SCO/NAL, 74SCO/NAL) at 1203 K and the report that a miscibility gap forms at temperatures below 573 K (73CRA, 73MIS/FLEa, 73MIS/FLEb, 74MIS/FLE). Data were also developed for the beta phase which appears to be stabilised by solution of Fe.

The other phase of interest in this system is the pentlandite phase $(Fe,Ni)_9S_8$ which is stable below 883 K. Data were derived for this phase to be consistent with this observation and to reproduce the extent of stability of this phase at 673 K shown by Craig et al. (68CRA/NAL). Figure 7 shows the region of stability of pentlandite and its predicted phase relationships for 673 K.

Calculations in the Cu-Fe-Ni-S system

In the quaternary system experimental data are available for the partial pressure of sulphur above the matte as a function of sulphur concentration for certain relative proportions of Cu, Fe and Ni (75LEE, 80BAL). Calculations were made using the data derived for the binary and ternary systems to compare the experimental and predicted thermodynamic data for the liquid phase. The agreement was found to be excellent without the need to introduce quaternary parameters.

3. EXPERIMENTAL

Matte/Slag/Gas Equilibration

Approximately 15g in total of prefused Ni_3S_2, Cu_2S and FeS with a Ni:Cu weight ratio of 2:1, held in a silica crucible, were heated in a vertical tube furnace to the reaction temperature, typically 1540 K. Finely-divided silica (0.5g) was added to the initial mix to aid the formation of a slag in situ during equilibration. The partial pressure of SO_2 over the matte was controlled by introducing gas mixtures of known N_2/SO_2 ratios at 40-100 ml/min into the furnace via a silica tube held just above the surface of the melt. because of deterioration of the crucible, runs were not extended beyond 6 hours. To speed the attainment of equilibrium 1% O_2 was added to the gas stream and the gas bubbled through the melt in the middle stages of the run. At the end of the run the whole crucible was quenched by raising it into the cold top of the furnace. It was noted that independent of the initial iron level the final matte iron content was always below about 5 wt%. The slag formed as a 'collar' between the matte and crucible wall rather than covering the matte completely.

A hydrogen reduction method was used to analyse the matte sulphur. Other elements in the matte and slag were analysed by atomic absorption spectrophotometry.

Converting Trials
Simulated converting runs were carried out on 300g of prefused matte with 300g of prefused slag held in clay/graphite crucibles in an open muffle furnace. 800 ml/min of oxygen was blown into the matte and the matte and slag sampled by suction or dip sampling at intervals throughout the blow. Temperatures (and oxygen partial pressures) were measured from a ZrO_2-Y_2O_3 oxygen probe held in the slag phase. SO_2 partial pressures were estimated by continuously sampling the atmosphere 50mm above the melt and passing the gas through an infra-red gas analyser. A temperature of 1543 K was aimed for throughout the blow.

4. APPLICATION OF MODEL

Matte/Slag/Gas Equilibration
The results for these runs are presented in Fig. 8 in terms of the matte sulphur content for a given iron content. The values calculated from the model for the same conditions, assuming 30 wt% SiO_2 in the slag, are shown for comparison. The agreement is excellent for the three sets of results with 5% SO_2 while the experimental sulphur contents at 50% SO_2 are slightly lower than predicted. Where differences between the model and experiment occur the approach adopted has been to assume that these are due to a real effect which requires further explanation. In the present case it is known that high P_{so_2} atmospheres produce high sulphur mattes with high equilibrium values of P_{s_2}. This may lead to direct loss of sulphur as S_2 gas. A non-equilibrium state could be produced in which removal of iron through the reaction

$$Fe_{matte} + SO_2 = FeO_{slag} + S_{matte} \qquad (1)$$

is relatively slow, depending on reaction with the matte phase, and lags behind the more rapid evaporation of sulphur. From the model it may be calculated that the sulphur pressure would be 6 times higher in a matte in equilibrium with 50% SO_2 at 1533 K compared with 5% SO_2 at 1573 K. This mechanism would also explain the experimental difficulty of producing an equilibrated matte phase with a high iron content as noted previously.

From the experimental results and the model it can be shown that for a given matte iron content the sulphur content increases with increasing P_{so_2} and silica content but decreases with increasing temperature.

Converting Trials
The results from the simulated converting operation are shown in Fig. 9 with a semi-quantitative indication of the P_{so_2} value at the time of taking the sample. The line drawn shows the matte composition for 5% SO_2 suggested by the results. At the beginning of the blow when the iron content of the matte is high the main reaction is oxidation of iron to iron oxides at very low values of P_{so_2}. Under these conditions it may be shown both experimentally and through the model (88DIN/HODa) that addition of oxygen results in an increase in matte sulphur level. As the iron level is reduced P_{so_2} increases and as in the case of the equilibration runs, for a given matte iron level high sulphur contents are associated with high P_{so_2} levels. However, the absolute values of the sulphur contents are consistently lower by about 1 wt% when compared with the equilibration experiments or the model. One contributory factor to this may be the presence of a temperature difference between the slag phase, in which the temperature is measured, and the matte phase, in which the exothermic oxidation of Fe and S occurs, so that the effective temperature of converting is higher than that recorded and used in the calculations. The mechanism for sulphur removal during blowing is

$$S_{matte} + O_2 = SO_2 \qquad (2)$$

while for iron removal from the matte the suggested reaction

$$Fe_{matte} + \frac{1}{2}O_2 = FeO_{slag} \qquad (3)$$

Reaction 3 depends on reaction across the matte/slag interface and is likely to be slower than Reaction 2, leading to a non-equilibrium state where Fe removal lags behind S removal giving the observed lower S contents. As an equivalent explanation consider the arrival of oxygen gas in the bulk of the matte well away from the slag interface. A local equilibrium might be set up between Fe, S and O whereby the products were SO_2 and pure iron oxides rather than iron oxide dissolved in the slag. Dissolution in the slag would come at a later stage but after removal of S from the system. Calculations under these conditions have been carried out using the model assuming very low SiO_2 slag contents (2wt%) and 5% SO_2. Calculations were also carried out for 30 wt% SiO_2 for comparison. The results are included in Fig. 9 and show a difference of about 1 wt% S for the two SiO_2 contents which is consistent with the experimental values. Formation of iron oxide within the bulk of the matte and out of contact with the slag may account for the non-equilibrium build-up of magnetite observed in commercial copper converting practice.

Cu and Ni in Slag

The copper and nickel contents of the slags for the equilibration and simulated converting runs are shown in Figs. 10 and 11 respectively. The line drawn through the equilibration results in Fig. 10 has been superimposed onto the simulated converting results in Fig. 11. Surprisingly there is very little difference in the Cu and Ni slag contents from the two types of experiment. It might be expected that the slags from the converting runs would contain more entrapped matte and hence higher Ni and Cu contents. The controlling P_{SO_2} is also indicated on the diagram.

Within the scatter of the experimental results it appears that the Cu and Ni slag contents are independent of P_{SO_2} for mattes of the same Fe content. Using the model it is possible to investigate this more fully. Activities of Cu and Ni in the matte and P_{SO_2} values were found from the model at .05 and .5 P_{SO_2}, 1533 K and 30 wt% SiO_2 in the slag. The results are given in Table 1. If it can be assumed that the presence of sulphur in the slag has no effect then the corresponding wt% Cu and Ni in the slag may be calculated from the relationships developed for sulphur-free $Fe-O-SiO_2$ slags (75TAY/JEFa, 75TAY/JEFb)

$$\text{wt\% Cu in slag} = 35.a_{Cu}.P_{O_2}^{\frac{1}{4}}.\, e^{(7362/T - 2.6389)} \qquad (5)$$

$$\text{wt\% Ni in slag} = 65.3 a_{Ni} P_{O_2}^{\frac{1}{4}}.\, e^{(22969/T - 8.062)} \qquad (6)$$

The results are presented in Fig. 12. To a first approximation Cu and Ni levels at widely different values of P_{SO_2} are similar. However it can be seen that increasing P_{SO_2} decreases the slag Cu content but increases the slag Ni content. This difference in behaviour may be explained in terms of the different valencies of Cu and Ni. For Cu,

$$\text{\% Cu in slag} \propto P_{O_2}^{\frac{1}{4}}$$

whilst for Ni,

$$\text{\% Ni in slag} \propto P_{O_2}^{\frac{1}{4}}$$

Increasing P_{SO_2} increases both P_{S_2} and P_{SO_2} for a given matte iron level. Increasing P_{S_2} increases the matte sulphur level and decreases the activity of both Cu and Ni (see Table 1). In the case of Ni, the increase in P_{S_2} is more than sufficient to offset the decrease in nickel activity whereas in the case of Cu. for which P_{O_2} appears to the power $^1/_4$, the decrease in Cu activity has a greater effect than the increase in P_{O_2}. The model also suggests that the slag content of both Cu and Ni increases with increasing temperature.

By comparing Figs. 10 and 11 with Fig. 12 it can be seen that the experimental Cu and Ni slag contents are higher than those calculated assuming a sulphur-free slag. It is not clear to what extent this additional Cu and Ni content is due to entrapment of matte particles or a true enhanced solubility brought about by the presence of sulphur in the slag (72SEH/IMR, 74NAG). In either case the additional Cu and Ni should be present as sulphides.

It has been suggested that sulphides may be dissolved preferentially from a slag by treatment with a bromine/methanol mixture (69SPI/THE). Nine slags from the converting trials were treated in this way. The results, plotted in terms of Cu and Ni slag content before and after treatment, are shown in Fig. 13 and compared with the calculated 'oxidic' content at 5% SO_2. The agreement is very good and hence consistent with the view that there is an oxidic and sulphidic contribution to the Cu and Ni slag content which may be identified in this way. In support of there being a true sulphide solubility it was noted that the Ni/Cu ratio in the mattes in contact with the slag samples varied from 1.6 to 2.2, with a mean value of 1.8, whilst the ratio of Ni to Cu removed by the bromine treatment was much lower at 1.02 ± 0.38 for the nine samples and 1.02 ± 0.11 for seven of the nine samples. The relatively low scatter of the experimental results in Figs. 10 and 11 and the generally close agreement between the Cu and Ni levels in the equilibration and simulated converting experiments also suggests that sulphide solubility (as opposed to matte entrappment) is an important factor in determining the overall Cu and Ni slag content and that an extension of the slag model to incorporate S would be required to account for the total slag Cu and Ni loss.

Solidification of Mattes

Control of the matte sulphur level during converting and phase formation during the subsequent slow-cooling of the matte are important factors in the process for the recovery of platinum group metals. Representation of the solidification of the matte was approached in two ways. In the simpler method full equilibrium was calculated for a given overall matte composition at a series of temperatures. Alternatively a 'sequential' cooling method was used whereby liquid at equilibrium at one temperature was 'removed' from any accompanying solid phases and a new equilibrium calculated for the liquid at a lower temperature, this process being repeated at a series of temperatures.

The results for a particular matte (6 wt% Fe, 1:1 Cu:Ni ratio, 20.8 wt% S) are given in Fig. 14 in terms of the percent of original liquid left as cooling proceeds. Results are presented for both equilibrium cooling and sequential cooling using different temperature intervals. In the present case it appears that cooling in step sizes of 5° or below gives very similar results. The main difference in the two sets of calculations is that more liquid is retained to lower temperatures in the sequential cooling calculation. In both cases an alloy phase is formed accompanied by sulphide phases based on Cu_2S and Ni_3S_2. However in the sequential cooling calculation a pentlandite phase is produced with the liquid at low temperatures which is not found in the equilibrium calculation results. Limited experimental evidence suggests that the sequential cooling calculations more accurately represent the final matte structure and phases.

ACKNOWLEDGEMENTS

This work was funded in part by Matthey Rustenburg Refiners. Discussions with MTDS staff at NPL, particularly T. I. Barry and T. G. Chart, and the programming work of S. M. Hodson, are gratefully acknowledged.

REFERENCES

14FRI K. Friedrich, Metall u. Erz, 1914, 11, 160.

14HAY C. R. Hayward, Trans. AIME, 1914, 48, 141.

24STA/FAI A. Stansfield, W. V. Faith, Trans. R. Soc. Can., 1924, 18, 325.

40KOS/MUL W. Koster, W. Mulfinger, Z. Elektrochem., 1940, 46, 135.

55COR/CHI J. A. Cordier, J. Chipman, Trans. AIME. 1955, 203, 905-907.

59SPE/JAC R. Speiser, A. J. Jacobs, J. W. Spretnak, Trans. Met. Soc. AIME, 1959, 215, 185.

59ZEL/PAY G. R. Zellars, S. L. Payne, J. P. Morris, R. L. Kipp, Trans. Met. Soc. AIME, 1959, 215, 181.

60ALC/CHE C. B. Alcock, L. L. Cheng, J. Iron Steel Inst., 1960, 195, 169-73.

62KUL/YUN G. Kullerud, R. A. Yund, J. Petrol., 1962, 3, 126-75.

63MOH/KUL G. H. Moh, G. Kullerud, Carnegie Inst. Washington Year Book, 1963, 62, 189.

64MOH/KUL G. H. Moh, G. Kullerud, Carnegie Inst. Washington Year Book, 1964, 63, 209.

66BAR M. R. Baren, PhD Dissertation. University of Pennsylvania 1966.

68CRA/NAL J. R. Craig, A. J. Naldrett, G. Kullerud, Carnegie Inst. Washington Year Book, 1968, 66. 440.

68NAL/CRA A. J. Naldrett, J. R. Craig, Carnegie Inst. Washington Year Book, 1968, 66, 436.

69KUL/YUN G. Kullerud, R. A. Yund, G. H. Moh, "Magmatic Ore Deposits" ed. H. D. B. Wilson, Economic Geology Publishing Co.,Monograph, 1969, 4, 323

69SPI/TEM P. Spira, N. J. Themelis, J Metals, April 1969,35-42.

69VAI/REM S. E. Vaisburd, T. F. Remen, A. B. Sheinin, Zhur. Fiz. Khim., (Russ. J. Phys. Chem.) 1969, 43, 1780.

72BYE/TAK J. J. Byerley, N. Takebe, Metall. Trans., 1972, 3. 559-64.

72MEY G. A. Meyer, D. Eng. Sci. Thesis, Columbia University, New York, 1972. "A Thermodynamic study of the Systems Ni-S and Ni-Fe-S".

72SCO/NAL J. R. Scott, A. J. Naldrett, E. Gasparrini, Econ. Geol., 1972, 67, 1010.

72SEH/IMR F. Sehnalek, I. Imris, Advances in Extractive Metallurgy and Refining, ed. M. J. Jones, Inst. Min. Metal., London 1972, 39-62.

73CRA J. R. Craig, Am. J. Sci.. 1973, 273A, 496.

73MIS/FLEa K. C. Misra, M. E. Fleet, Econ. Geol., 1973, 68, 518.

73MIS/FLEb K. C. Misra, M. E. Fleet, Mat. Res. Bull. 1973, 8, 669-678.

73SCH/CRA W. J. Schlitt, R. H. Craig, K. J. Richards, Metall. Trans., 1973, 4, 1994.

74CHI/GUL D. M. Chizhikov, Z. F. Gulyanitskaya, N. V. Belyanina, L. I. Blokhina, Russ. Metall., 1974, No. 3, 42.

74CRA/SCO J. R. Craig, S. D. Scott, "Sulphide Mineralogy" ed. P. H. Ribbe (Washington DC: Mineral. Soc. Amer.), 1974, CS-1.

74MIS/FLE K. C. Misra, M. E. Fleet, Econ. Geol., 1974, 69, 391.

74NAG M. Nagamori, Metall. Trans., 1974, 5, 531-538.

74SCO/NAL S. D. Scott, A. J. Naldrett, E Gasparrini, Int. Mineral. Assoc. Ninth Gen. Meet., Berlin 1974, "Regular solution model in the Fe_{1-x}-Ni_{1-x}-S (mss) solid solution".

75HIL/STA M. Hillert, L. -I. Staffansson, Metall. Trans., 1975, 6B, 37.

75LEE S. L. Lee, D. Eng. Sci. Thesis, 1975, Columbia University, New York.

75TAY/JEFa J. R. Taylor, J. H. E Jeffes, Trans IMM, 1975, 84, C18-24.

75TAY/JEFb J. R. Taylor, J. H. E. Jeffes, Trans IMM, 1975, 84, C136-148.

76CHA/LEE Y. A. Chang, Y. E. Lee, J. P. Neumann, "Extractive Metallurgy of Copper" (ed. J. C. Yannopoulos and J. C. Agarwal) Vol. 1, Chapter 2, The Metallurgical Society AIME, New York, 1976.

76HIL/STA M. Hillert, L. -I. Staffansson, Metall. Trans., 1976, 7B, 203.

76RAU H. Rau, J. Phys. Chem. Solids, 1976, 37, 929-930.

77KUB/SMI O. Kubaschewski, J. F. Smith, D. M. Bailey, Z. Metallkd., 1977, 68, 495.

78LIN/HU R. Y. Lin, D. C. Hu, Y. A. Chang, Metall. Trans. B., 1978, 9B, 531-538.

79CHA/NEU Y. A. Chang, J. P. Neumann, U. V. Choudary, "Phase diagrams and Thermodynamic Properties of ternary Copper-Sulfur-Metal systems", Incra monograph VII, The Metallurgy of Copper, 1979.

79CHA/SHA Y. A. Chang, R. C. Sharma, Proc. Conf. "Calculation of Phase Diagrams and Thermochemistry of Alloy Phases", ed. Y. A. Chang, J. F. Smith, 145-174.

79LAR J. M. Larrain, CALPHAD, 1979, 3, 139-157.

79LAR/KEL J. M. Larrain, H. H. Kellogg, Proc. Conf. "Calculation of Phase Diagrams and Thermochemistry of Alloy Phases", ed. Y. A. Chang, J. F. Smith, 130-144.

80BAL C. W. Bale, Private Communication, 1980.

80GOE/KEL R. P. Goel, H. H. Kellogg, J. M. Larrain, Metall. Trans., 1980, 11B, 107.

80KAO M. Kao, M.S. Thesis, University of Wisconsin-Milwaukee, Milwaukee, WI, USA, 1980. "A Thermodynamic Study of the Fe-Ni-S System at 973 K".

80LAR/LEE J. M. Larrain, S. L. Lee, Can. Met. Quarterly, 1980, 19, 183.

80LEE/LAR S. L. Lee, J. M. Larrain, H. H. Kellogg, Metall. Trans. B., 1980, 11B, 251.

80SHA/CHA R. C. Sharma, Y. A. Chang, Metall. Trans., 1980, 11B, 139-146.

81FER/HIL A. Fernández Guillermet, M. Hillert, B. Jansson, B. Sundman, Metall. Trans., 1981, 12B, 745.

82CHA/GOH T. G. Chart, D. D. Gohil, Z. shu. Xing, NPL Report DMA (A) 54, August 1982, "Calculated Phase Equilibria for the Cu-Fe-Ni System".

82CHU/CHA Y-Y. Chuang, Y. A. Chang, Metall. Trans. B., 1982, 13B, 379.

82DIN/CHA A. T. Dinsdale, T. G. Chart, T. I. Barry, J. R. Taylor, High Temp.-High Press., 1982, 14, 633-640.

82HSI/CHA K. C. Hsieh, Y. A. Chang, T. Zhong, Bull. Alloy Phase Diagrams, 1982, 3, 165.

84DIN A. T. Dinsdale, Thesis "The Generation and Application of Metallurgical Thermodynamic Data", Brunel University, 1984.

85CHU/HSI Y. -Y. Chuang, K. -C. Hsieh, Y. A. Chang, Metall. Trans, 1985, 16B, 277.

86DIN A. T. Dinsdale, NPL News, Summer 1986, p26-7, "Scientific Group Thermodata Europe".

87ANS/SUN I. Ansara, B. Sundman, CODATA Report, "Computer Handling and Dissemination of Data", 1987, 154-8.

87JAN A. Jansson, Report Royal Institute of Technology, TRITA-MAC-0340, 1987.

87HSI/CHA K. -C. Hsieh, Y. A. Chang, Can. Metall. Quart., 1987, 26, 311.

88DIN/HODa A. T. Dinsdale, S. M. Hodson, T. I. Barry, J. R. Taylor, Proc. Conf. 27th Annual Conference of Metallurgists, CIM, Montreal 1988, "Computations using MTDATA of Metal-Matte-Slag-Gas Equilibria".

88DIN/HODb A. T. Dinsdale, S. M. Hodson, J. R. Taylor, Proc. Conf. 3rd International Conference on Molten Slags and Fluxes, 27-9 June 1988, University of Strathclyde, Glasgow. p246-53, Institute of Metals.

Table 1 Activities and compositions for a matte phase (Ni:Cu wt ratio 2:1) in equilibrium with a Fe-O-SiO$_2$ slag (30 wt% SiO$_2$) at fixed pSO$_2$ and 1533K

pSO$_2$ = 0.05

wt% Fe	wt% S	aCu	aNi	pO$_2$	pS$_2$
2.0	21.9	0.153	0.309	1.13E–8	1.43E–4
4.0	23.7	0.100	0.241	6.86E–9	3.92E–4
6.0	24.8	0.075	0.205	5.17E–9	6.91E–4

pSO$_2$ = 0.50

wt% Fe	wt% S	aCu	aNi	pO$_2$	pS$_2$
2.0	24.5	0.070	0.200	4.30E–8	9.96E–4
4.0	25.8	0.044	0.154	2.69E–8	2.54E–3
6.0	26.7	0.032	0.131	2.06E–8	4.33E–3

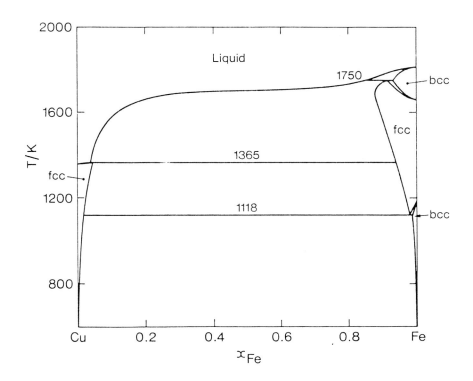

Fig. 1. Calculated Cu-Fe phase diagram.

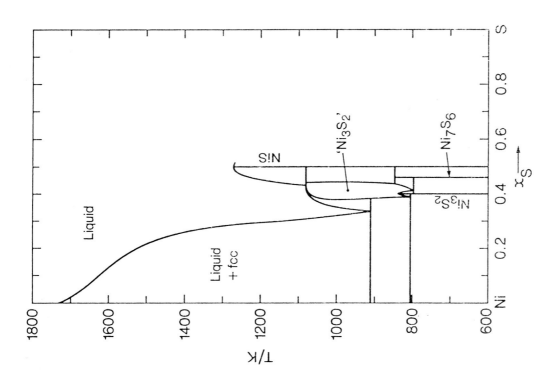

Fig. 3. Calculated Ni-S phase diagram.

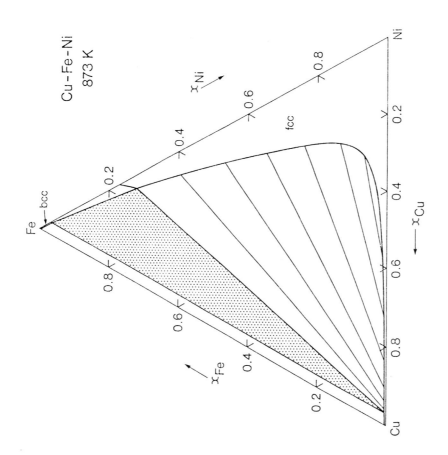

Fig. 2. Calculated isothermal section of the Cu-Fe-Ni system at 873 K.

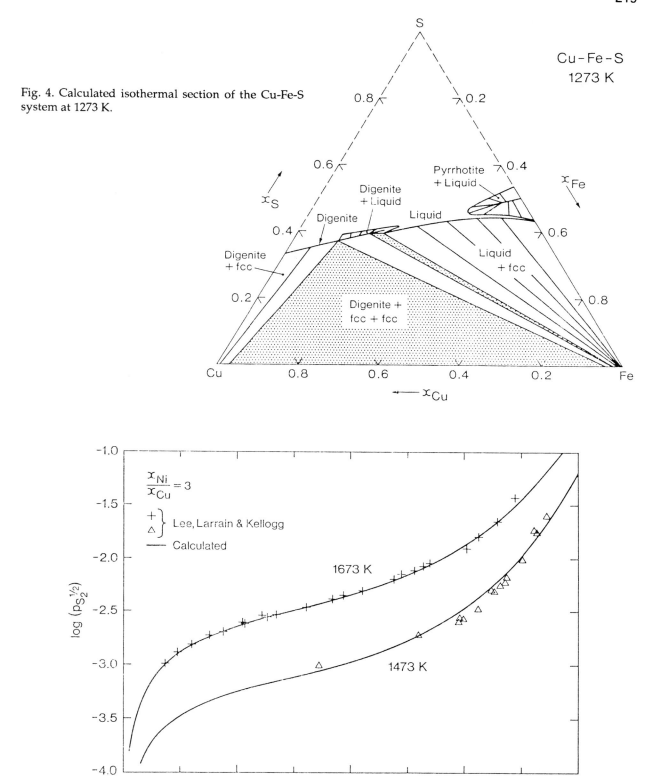

Fig. 4. Calculated isothermal section of the Cu-Fe-S system at 1273 K.

Fig. 5. Comparison between calculated and experimental partial pressures of sulphur (atm.) in the Cu-Ni-S system.

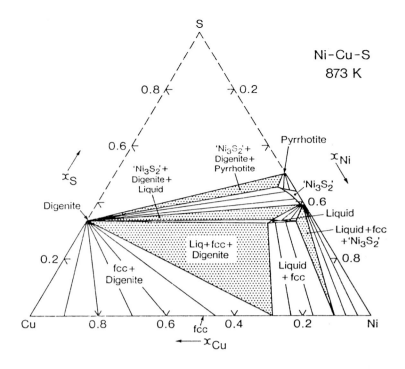

Fig. 6. Calculated isothermal section of the Ni-Cu-S system at 873 K.

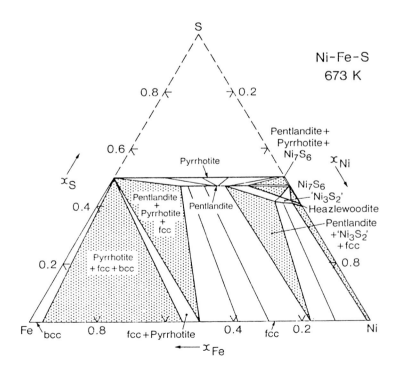

Fig. 7. Calculated isothermal section of the Ni-Fe-S system at 673 K.

221

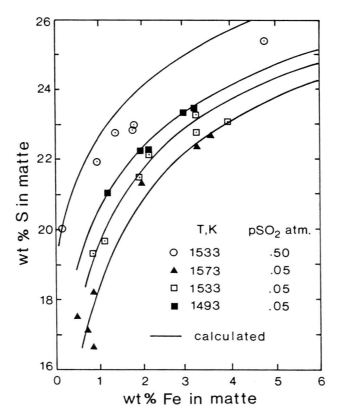

Fig. 8. Calculated and experimental matte compositions (Ni:Cu wt ratio 2:1) at different temperatures and SO$_2$ partial pressures.

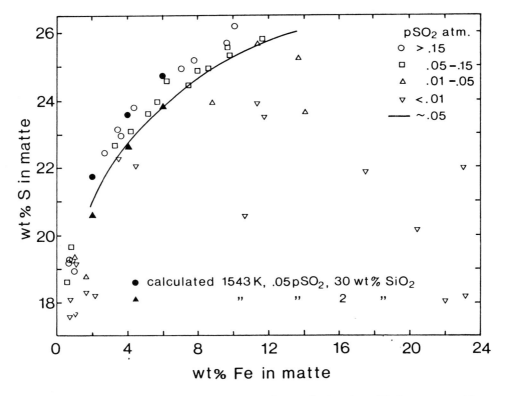

Fig. 9. Effect of pSO$_2$ on matte composition during converting. Some calculated equilibrium compositions are shown for comparison.

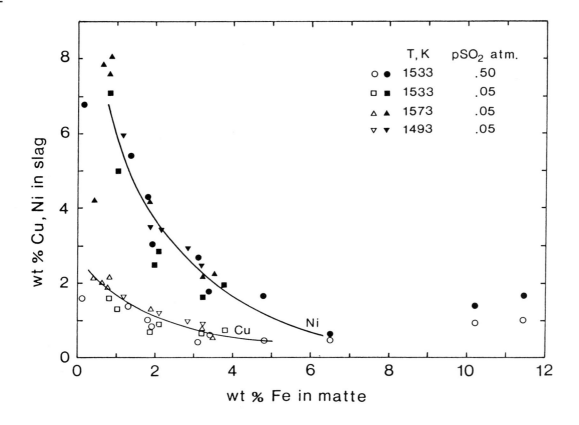

Fig. 10. Cu and Ni contents of slag from equilibration experiments.

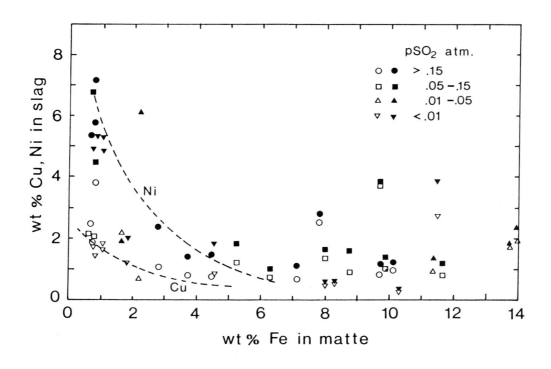

Fig. 11. Cu and Ni slag contents during simulated converting operation.

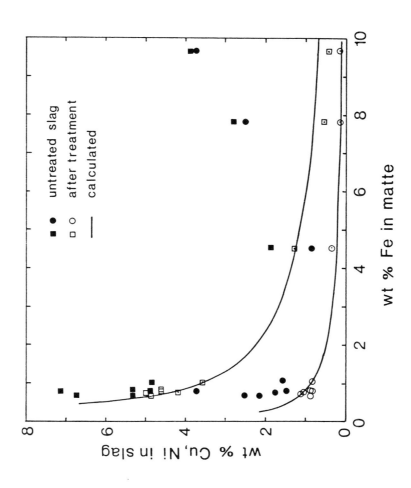

Fig. 13. Cu and Ni contents of slag from simulated converting operation before and after treatment with bromine/methanol.

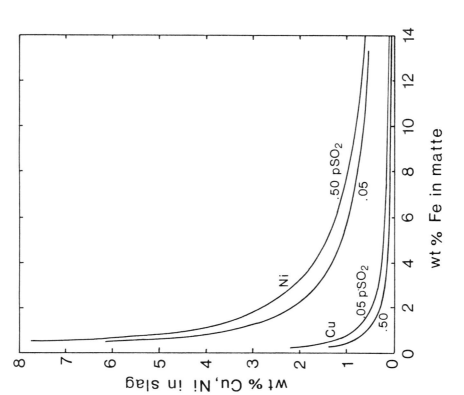

Fig. 12. Calculated effect of pSO_2 on the Cu and Ni contents of Fe-O-SiO_2 slag at 1533 K.

224

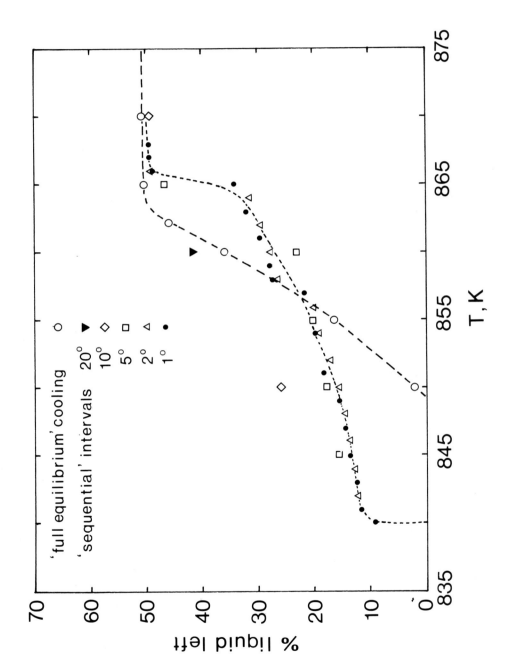

Fig. 14. Percent of original liquid remaining during matte solidification using two different methods of calculation.

27
THE OXY-CHLORINATION OF CHROMITE ORE

F. R. SALE AND Z. KOSUMCU

Manchester Materials Science Centre, University of Manchester and UMIST, Grosvenor Street,
Manchester, M1 7HS, UK

ABSTRACT

A study of the oxy-chlorination of an Australian chromite ore is reported. The ore contained an exsolution spinel phase within a chromite matrix, which contained aluminium and magnesium in addition to chromium and iron. The exsolution phase contained relatively more iron and magnesium than the matrix chromite. Gibbs free energy data for the formation of chromyl chloride (chromium dioxy dichloride, CrO_2Cl_2) have been used to predict the relationship between temperature and partial pressure of chromyl chloride generated at various total pressures by the oxy-chlorination of chromic oxide. Predominance area diagrams of the relevant metal-chlorine-oxygen systems have been calculated and used to predict the conditions required for the production of chromyl chloride and to predict possible reactions of the other metallic elements in the ore. Pure Cr_2O_3 and the chromite ore have been subjected to oxy-chlorination at temperatures in the range 973 K to 1200 K in a fluidised bed reactor.

It was found necessary to use temperatures of the order of 1200 K, as well as high flow rates of chlorine, to produce CrO_2Cl_2 in accordance with thermochemical predictions. The largest yield of oxychloride was obtained using a partial pressure ratio, $P_{Cl2}:P_{O2}$ of 4 : 1. The oxy-chlorination process occurred in a topochemical manner in which a $MgO.Al_2O_3$ spinel was left as a porous solid residue. The overall reaction occurred in two concomitant steps which took place at different rates. Iron was first chlorinated to give $FeCl_3$ which left a deposit of Cr_2O_3 on the newly formed $MgO.Al_2O_3$ spinel. In the second process the Cr_2O_3 was oxychlorinated to yield CrO_2Cl_2.

1. INTRODUCTION

Chromite, the solid solution spinel of general formula $(Mg, Fe)^{2+} (Cr, Al, Fe)_2^{3+} O_4$ is the major commercial source of chromium metal. This spinel also occurs extensively as the refractory material chrome-magnesite. The gradual depletion of high grade chromite and recent price increases and supply difficulties has necessitated the processing of low grade ore and the recycling and extraction of chromium from secondary materials. One obvious secondary source is "used" refractories of the chrome-magnesite family. Chromite ores tend to be used in the "as-mined" state or after subjection to simple physical concentration processes. The chromium-iron ratio in chromite ores cannot be improved significantly by physical methods alone since the iron is contained within the spinel lattice. As a result, chemical means of destroying the spinel structure are necessary for chromium enrichment.

The chlorination of low-grade chromite ores is one method of chemical beneficiation. The first studies date back to the 1920s. Most previous investigations have been performed under reducing conditions either to chlorinate the iron oxides selectively or to achieve total chlorination, which would be followed by fractional distillation to achieve separation of the metal chlorides. In the early work, chlorination was attempted using mixtures of CO and Cl_2,[1] H_2 and HCl[2] or following selective reduction of the iron oxide content with H_2 and C[3]. Doerner[4] was the first worker to quantify data on the formation of the various chlorides from chromite. Of the more recent studies of the chlorination of different ores the work of Pokorny[5], using either HCl or Cl_2, with or without the presence of carbon and the detailed study of Magidson[6], who considered the use of Cl_2 in the presence of carbon and reported effects of time and temperature on the kinetics of extraction and on the comparative chlorination and sublimation rates of the products, are valuable in establishing the background to the present study. Similarly, much work on Indian chromites by Athawale and Altekar[7] and Khundar and Talukdar[8], and work on Egyptian chromites by Hussein and El-Barawi[9][10] have established the role of chlorination, in reducing or neutral conditions, as a means of upgrading ores or extracting a number of metals by chloride volatilisation. The concomitant "in-situ" generation of HCl, by the decomposition of hydrated calcium chloride, and hydrochlorination of Cr and Fe from Australian chromites has been reported by Maude and Sale[11], Whilst this process simplified the supply of the chloridising agent by reducing the process to a chloridising roast it did not give products which were easily separable. As such, in common with other processes using reducing conditions, further distillation was required to achieve a clean separation of the products.

The chlorination of chromite under oxidising conditions offers an extra flexibility in that the production of oxy-chlorides becomes feasible. Emmeneggar and Petermann[12] and Schäfer and Odenbach[13] have considered the vapour transport of Cr_2O_3 under oxy-chloridising conditions and Peshev[14] has reported the preparation of single crystals of Cr_2O_3 by vapour transport under oxy-chloridising conditions. Saeki[15] reported that chromium dioxy-dichloride was formed during the chlorination of pure Cr_2O_3, in disagreement with the claims of Morozov and Fefelova[16] that the oxy-chloride was not stable under the conditions used for vapour transport or oxy-chlorination. Belton and Sano[17] studied the volatilisation of chromic oxide in oxygen-chlorine-argon mixtures, over the temperature range 900-1250 K and determined, using a transpiration method, the Gibbs free energy for the reaction between chromic oxide, chlorine and oxygen to give chromyl chloride vapour. The close consistency of the results was taken to suggest only a minor, if any, interference from other gaseous species.

From this background on the chlorination upgrading of chromite and on the possibilities of the vapour transport of Cr_2O_3 using oxygen-chlorine mixtures, a logical development seemed to be the use of oxy-chlorination for the selective volatilisation processing of chromite ores.

2. EXPERIMENTAL PROCEDURES

2.1 Characterisation of chromite ore

The density of the ore was determined to be 4.44 g cm^{-3} using a density bottle technique. Particle size analysis was carried out using the usual "cone and quartering" method for sieve analysis. Optical microscopy was performed by mounting the ore with a dispersion of Al_2O_3 particles in both "scandiplast" and "araldite" resin. The dispersion of alumina was necessary to increase the hardness of the resin matrix so that excessive rounding of the chromite grains did not occur during polishing. Reflected light microscopy was found to be most suitable for examination of the ore because of its opacity to transmitted light. An oil immersion objective was used to resolve the fine exsolution precipitates that were present in the ore as shown in Fig. 1. Analytical TEM and SEM and microprobe analysis were used to determine the composition and structure of both the matrix and the exsolution phase, which were shown to be spinels of the average composition given in Table 1. A small amount of impurity, in the form of light coloured grains, was also present with the chromite. This was identified to be a pyroxene which approached enstatite ($MgSiO_3$) in composition. X-ray diffraction analysis was used to confirm the phases present in the as-received ore and to identify the phases present in samples retrieved during the course of oxy-chlorination.

2.2 Chlorination Kinetics

All chlorination experiments were carried out in a silica, laboratory scale, fluidised bed reactor. The reactor was approximately 3 cm diameter and 35 cm height and was lowered into a furnace of constant hot zone of the order of 10 cm. This allowed the fluidised charge, some 3 cm in height, to be kept at a constant temperature and gave sufficient distance above the fluidised charge to prevent any "carry-over" of ore particles.

It also gave a reasonable temperature gradient within the reactor (from 1200 K in the hot zone to 400 K at the top of reactor) to allow some condensation of chlorides to occur. The off-take gases were taken from the reactor to a condenser which was cooled using crushed solid CO_2 to ensure collection of any chromyl chloride. In later experiments this "dry" collection of chromyl chloride was abandoned and carbon tetrachloride was used as a collection solvent for the oxy-chloride. The chlorine gas was supplied, with a 99.99% purity by BOC along with the oxygen and argon which were of chemical grade, high purity quality. Monel needle valves and mixing assemblies were used to reduce corrosion problems. All gases were dried using anhydrous $CaCl_2$ and supplied at rates of up to 1 l min^{-1} to the reactors. Preliminary experiments were carried out using "analar" grade Cr_2O_3, which contained less than 0.01 wt% chloride and less than 0.2 wt% sulphate.

2.3 Characterisation of reaction products

Atomic absorption spectroscopy was used to analyse solutions made up by the dissolution of the condensed chlorides and oxy-chlorides. To minimise interference effects a nitrous oxide-acetylene flame was used. Standards were analysed frequently to determine the extent of this interference which occurs between Fe and Cr, particularly when air-acetylene flames are used. X-ray diffraction analysis was carried out on the solid condensates and also on the residues left behind in the fluidised bed reactors. A dry box was found to be necessary to prevent oxidation/hydrolysis of the chlorides. The solid residues were also characterised by scanning and optical metallography and electron probe microanalysis.

3. THERMOCHEMICAL PREDICTIONS

3.1 Formation of CrO_2Cl_2

As indicated previously Sano and Belton[17] have determined the equilibrium constant and standard Gibbs free energy change for the formation of CrO_2Cl_2 by the oxy-chlorination of Cr_2O_3 :-

$$Cr_2O_3(s) + \tfrac{1}{2}O_2(g) + 2Cl_2(g) \rightarrow 2CrO_2Cl_2(g)$$

$$\log K_p = -4750/T + 0.86$$

and $\Delta G^\theta = 90.900\ (\pm\ 1200) - 16.46\ (\pm\ 1.2)T \quad J$

The extrapolated standard Gibbs energy of reaction remains positive up to 5522 K. The equilibrium constant for the reaction may be given as:

$$K_p = p^2_{CrO_2Cl_2} / p^2_{Cl_2} \cdot p_{O_2}{}^{\tfrac{1}{2}}$$

assuming the chromic oxide to have unit activity. According to the stoichiometry of the reaction if x is the degree of conversion of Cr_2O_3 at equilibrium, then (1-x) moles Cr_2O_3 remain in the solid phase and $\tfrac{1}{2}$(1-x) moles of oxygen, 2(1-x) moles of chlorine and 2x moles of CrO_2Cl_2 exist in the gas phase giving a total of $\tfrac{1}{2}$(5-x) moles of gas at equilibrium. Accordingly, the partial pressures of the gaseous species may be expressed in terms of the degree of conversion and the total pressure in the system e.g.

$$P_{O_2} = \frac{(1\text{-}x)P}{(5\text{-}x)} \ , P_{Cl_2} = \frac{4(1\text{-}x)P}{(5\text{-}x)}$$

and $PCrO_2Cl_2 = \dfrac{4x\,P}{(5\text{-}x)}$ such that the equilibrium constant can be given by:

$$K_p{}^2 = x^4(5\text{-}x) / (1\text{-}x)^5\,P$$

The polynomial that results from this expression can be solved for several absolute values of K_p (temperature) and total pressure P in the system.

A computer program, which made use of the NAG subroutine CO2AEF for the solution of the fifth degree polynomial, was written to allow the calculation of the equilibrium partial pressures of CrO_2Cl_2 for various temperatures and total pressures. These data are shown in Fig. 2, which indicates that at 1200 K and 1 atmosphere total pressure, a value of approximately 11 mm Hg may be expected for the partial pressure of CrO_2Cl_2, which is high enough for appreciable chlorination - volatilisation. This value is in good agreement with the figure derived by Sano and Belton[17] who assumed that P_{Cl_2}: P_{O_2} equalled 4:1 and that the total pressure of 1 atmosphere was approximated by P_{Cl_2} plus P_{O_2} (i.e. $P_{CrO2Cl2}$ was ignored).

4.2 Predominance Area Diagrams

Figure 3 shows the diagram for the Cr-O-Cl system at 1200 K, where the condensed phases are represented by complete lines and regions of stability of gaseous phases are shown by dashed lines. The phase boundaries between CrO_2Cl_2 and Cr_2O_3, and between CrO_2Cl_2 and $CrCl_4$, were calculated for a pressure of 11 mm Hg for CrO_2Cl_2. For a chlorine partial pressure of just less than 1 atmosphere the area between the vertical dotted lines on Fig. 3 represents the region where Cr_2O_3 will not undergo chlorination. At this partial pressure of chlorine, chlorination is only possible at values of $\log P_{O_2}$ lower than -7.5, with the formation of $CrCl_3$ (or $CrCl_4$) or at values of $\log P_{O_2}$ greater than approximately -1, where CrO_2Cl_2 may be formed at a partial pressure of 11 mm Hg. At low chlorine partial pressures under reducing conditions it can be seen that the formation of the undesirable liquid $CrCl_2$ is favoured. At extremely low $\log P_{O_2}$ and $\log P_{Cl_2}$ values the stable solid phase is chromium metal. However, this is unlikely to be achieved in practice because carbon would most likely be used to produce the low P_{O_2} and it would react to give one of the chromium carbides.

Figures 4, 5, 6 and 7 are predominance area diagrams for the Fe-O-Cl, Mg-O-Cl, Al-O-Cl and Si-O-Cl systems respectively at 1200 K. These diagrams were all calculated assuming unit activity for the relevant oxides. It was realised that these activities would not apply to the chromite however it was felt that a useful indication of reaction possibilities could be obtained from the diagrams. It is quite evident that by carrying out chlorination under oxidising conditions the likelihood of producing chlorides of the contaminants Al and Mg (present in the spinel) and Mg and Si (present in the enstatite based phase) is reduced. The Si-O-Cl system also allows prediction of possibilities of reaction of the reactor itself under the oxy-chlorination conditions. The dashed lines for all the gaseous species in Figs. 4 to 7 are calculated for 1 atmosphere partial pressure of the volatile compound. It is interesting to note that Fig. 4 indicates that Fe_2O_3 is the least likely of the iron oxides to be chlorinated when P_{O_2} values exist under which the oxides would normally be stable. At approximately 1 atmosphere P_{Cl_2} it appears that $\log P_{O_2}$ must be less than approximately -4 for chlorination to yield $FeCl_3$. However, Fig. 4 was calculated for P_{FeCl_3} = 1 atm. and volatilisation of $FeCl_3$ may occur at pressures far lower than unity. The effect of lowering the pressure of $FeCl_3$ is to swing the $FeCl_3$-Fe_2O_3 boundary in a clockwise direction. No contribution was expected from Fe_2Cl_6 or FeOCl since the former decomposes to $FeCl_3$ at 1050 K and latter is stable only to 810 K.

The Mg-O-Cl system shown in Fig. 5 indicates that the chlorination of MgO to $MgCl_2$ is barely possible under oxidising conditions with values of P_{Cl2} close to unity. If chlorination of MgO is required it is clear that reducing conditions are required, particularly if P_{Cl2} values lower than unity are considered. The Al-O-Cl system (Fig. 6) indicates that the chlorination of Al_2O_3 will be feasible with a log P_{O2} value of -6.6 and lower for a 1 atmosphere partial pressure of chlorine. Al_2Cl_6 is stable up to 1150 K. At 1200 K approximately 2% Al_2Cl_6 is in equilibrium with $AlCl_3$. SiO_2, present in the enstatite contaminant and also as the construction material for the reactor, is relatively stable under oxy-chlorination conditions. Chlorination is only possible at 1200 K when the oxygen partial pressure is 10^{-9} or lower (for 1 atmosphere pressure of product $SiCl_4$). However, as for Fe, the volatile chloride boundary on the diagrams may be shifted by consideration of lower partial pressures of the chloride vapour species.

All predominance area diagrams were calculated using the tabulated data of Barin and Knacke,[18] except that for CrO_2Cl_2 which was taken from Sano and Belton.[17]

4.3 Volatilities of relevant chlorides

During any chlorination or oxy-chlorination process the volatilities of the possible products are very important because both the reaction kinetics and recovery of products are enhanced with increasing volatility. The volatilities are also important in the consideration of subsequent separation processes if a clean separation is not to be obtained during the initial chlorination or oxy-chlorination reaction. Figure 8 represents graphically the saturated equilibrium vapour pressures as a function of temperature of the chlorides and oxy-chloride likely to be stable at the temperatures of reaction. The values of the vapour pressures were taken from Barin and Knacke[18], except that for CrO_2Cl_2 which was taken from Sano and Belton[17].

Magidson[6] has shown that at 1173 K both solid $CrCl_3$ and liquid $MgCl_2$ were formed during the chlorination of chromite in the presence of carbon and reported that the chlorination process was more rapid than the vaporisation of the resultant chlorides. Saeki[15] reported a similar effect during the reduction-chlorination of Cr_2O_3. In other studies, particularly where low partial pressures of chlorine have been used, liquid $FeCl_2$ or $CrCl_2$ have been reported as the products of reaction. These observations can all be explained by reference to Fig. 8 on the basis of the relatively low volatilities of the chlorides in question. The dimers Al_2Cl_6 and Fe_2Cl_6 are more volatile than the respective monomers and so dimer formation enhances the volatility of these species. However, it is clear from Fig. 8 that CrO_2Cl_2 is the most volatile of the likely products, with a boiling point of 389 K, such that it may be readily separated from any other chlorides that may form.

5. RESULTS AND DISCUSSION

To demonstrate the applicability of the thermochemical predictions to the oxy-chlorination process a summary of the results of the chlorination experiments will be presented. A fuller presentation of the experimental data and its interpretation in terms of the mechanism of oxy-chlorination and the rate controlling steps involved in the process are beyond the scope of this present paper.

5.1 Oxy-chlorination of pure Cr_2O_3

Qualitative experiments were carried out at temperatures of 973, 1073 and 1173 K using P_{Cl2} : P_{O2} ratios of 1:3, 1:1, 3:1 and 4:1 as well as using pure Cl_2. The results are presented in Table (2), where it can be seen that temperatures of the order of 1200 K and a P_{Cl2} : P_{O2} ratio of 4:1 gave the most rapid rate of reaction. The product, chromyl chloride, was observed as a reddish vapour and the condensed phase was a cherry-red clear liquid which decomposed and became darker in colour on exposure to light. The product, collected in the dark, was analysed to contain 33.7 wt% Cr and 45.5 wt% Cl which agrees extremely well with the theoretical composition of 33.57 wt% Cr, 45.77 wt% Cl and 20.66 wt% O, thus indicating that the product of oxy-chlorination of Cr_2O_3 at approximately 1200 K with a 4:1, Cl_2:O_2 ratio was CrO_2Cl_2 as predicted previously. Under these conditions no evidence was found for any of the other chlorides of chromium in the reaction product. These observations thus supported the predictions of Sano and Belton[17] and the experimental observation of Saeki[15] on the chlorination of Cr_2O_3 in the absence of carbon, but without oxygen additions to the vapour phase. It is interesting to note that no silicon was found in any of the condensed products which indicated that the prediction of the likely stability of SiO_2 as a constructional material for the reactor under the oxy-chlorination conditions was correct.

5.2 Oxy-chlorination of chromite

5.2.1 Product characterisation

After the qualitative observations of the oxy-chlorination of Cr_2O_3 had been made, the characterised chromite ore was studied. During the initial stages of the work the products of oxy-chlorination were collected carefully for identification and characterisation. In addition to chromyl chloride, which was, as in the case of the oxy-chlorination Cr_2O_3, collected in the solid CO_2 cooled condenser, a brownish red chlorination product was observed on the reactor walls where the temperatures were below approximately 573 K. This product was unstable when exposed to air and changed colour to orange, yellow and finally brown in a relatively short period of time. X-ray diffraction analysis of the product after exposure to air revealed FeOCl. The earlier thermodynamic predictions had shown that FeOCl was not stable at the reaction temperatures although its stability had been shown to increase rapidly with decreasing temperature. Initially, it was postulated that the FeOCl may have been produced by a gas phase reaction in the cooler regions of the reactor. However, when the solid reaction product was collected under argon and prepared for X-ray diffraction analysis in an argon dry box it was shown to be $FeCl_3$ (with traces of $FeCl_2$). Hence, it was apparent that the FeOCl was produced by reaction at room temperature with the atmosphere.

The chromyl chloride was collected and analysed to give the data shown in Table (3). It can be seen that there is no detectable contamination of the CrO_2Cl_2 by silicon, aluminium or magnesium, and that the iron carry-over in the simple fluidised bed system was extremely small. A mass balance indicates that the iron contamination seems to have occurred as a result of $FeCl_3$ carry-over.

5.2.2 Reaction Residue

X-ray diffraction data for the chromite ore and for the reaction residue obtained at various points in time are shown diagramatically in Fig. 9. It is clear that the original chromite spinel is changed as reaction proceeds to yield the $MgO.Al_2O_3$ spinel. An interesting observation is that in the intermediate stages of reaction Cr_2O_3 can be detected as part of the solid residue. Eventually the Cr_2O_3 is removed from the product. SEM micrographs of the end product, a partially reacted particle and an original ore particle are shown in Fig. 10 to 12. It is evident that as the reaction proceeds new platelets of spinel ($MgO.Al_2O_3$) are formed as a solid reaction residue as a topochemical reaction occurs. Cr_2O_3 is left behind as a surface layer on the platelets of $MgO.Al_2O_3$ in the partially-reacted material. An added observation is that the new platelets seem to grow on the plates of the exsolution phase. Microprobe analyses have been carried out across the diameters of polished, partially reacted particles. In all cases an increase in Al and Mg is seen in the region of the platelets of reaction residue with much reduced levels of Fe and Cr. A sharp interface was detected in all cases between the unreacted core and the reaction residue.

5.2.3 Kinetics of oxy-chlorination

Figures 13 and 14 show the relationship between time and wt% oxide chlorinated (which has been calculated from amounts of collected chlorides) for reaction temperatures of 1000, 1100 and 1200 K for chromium and iron respectively. The yields of greater than 25% of the Cr_2O_3 contained in the chromite are larger than those determined by Pokorny[5] (10%) and those of Saeki[15] (10-12%) during the chlorination of pure Cr_2O_3. The values of 40-45% determined for the conversion of iron-oxide are lower than those of Pokorny[5]. The differences in oxide contents of the ores, and differences in particle size, make a direct comparison difficult. However, it is clear that deliberate oxy-chlorination has helped the recovery of chromium.

Analysis of the residue from the reactor showed that significantly larger amounts of chromium were removed from the ore than were collected in the condenser. Examination of a grey-white deposit, which was detected immediately above the fluidised bed, showed it to be Cr_2O_3. Hence, it is apparent that some chromyl chloride had decomposed to yield Cr_2O_3, chlorine and oxygen immediately after reaction, in a manner similar to that shown by Schäfer and Odenbach[13] and Emmenegger and Petermann[12] in the growth of Cr_2O_3 from chromyl chloride vapour. It is apparent that to increase the yield of chromium from the chromite the elevated temperature zone of the reactor needs extension away from the fluidised particles and a rather more sophisticated collection system needs incorporation into the reactor.

6. CONCLUSIONS

(1) The use of predominance area diagrams has helped predict the conditions necessary for the oxy-chlorination of chromite to give chromyl chloride.

(2) Predominance area diagrams have also helped to predict the behaviour of other components in the chromite and the materials used for reactor construction during oxy-chlorination.

(3) During oxy-chlorination at 1200 K, iron is volatilised as $FeCl_3$ whilst the chromium is volatilised as CrO_2Cl_2. Aluminium and magnesium are not volatilised from the spinel during the oxy-chlorination process. The end result is that as chromium and iron are removed from the chromite a porous, platelike solid residue of $MgO.Al_2O_3$ is left behind. Similarly, silicon, which was present in the silica reactor and the enstatite impurity in the ore, was not volatilised during the oxy-chlorination process.

230

(4) The production of $FeCl_3$ and CrO_2Cl_2 occur concomitantly as the reaction proceeds in a topochemical manner. $FeCl_3$ is produced more rapidly in the overlapping reactions which leaves some Cr_2O_3 for subsequent reaction and volatilisation as CrO_2Cl_2.

REFERENCES

1. E. P. Venable and D. H. Jackson (1920), J. Elisha Mitchell Sci. Soc., 35, p.87.
2. W. H. Dyson and L. Atchison (1920), Brit. Patent 176 729 and 179 201.
3. V. S. Yatlow and A. V. Popova (1933), J. Appl. Chem. USSR., 6, p.1049.
4. H. A. Doerner (1930), U.S. Bur. Min. Bull., R.I. 3049 and 2999.
5. E. A. Pokorny (1957), "Extraction and Refining of the Rarer Metals - a symposium", London, Inst. Min. Metall., p.34.
6. I. A. Magidson (1971), Zhurn. Priklad. Khim., 34, p.953.
7. A. S. Athawale and V. A. Altekar (1969), Trans. Ind. Inst. Metals, June, p.29.
8. M. H. Khundar and M. I. Talukdar (1963), Pakistan J. Sci. Ind. Res., 6, p.218 .
9. M. K. Hussein and K. El-Barawi (1971), Trans. Inst. Min. Metall., C80, p.C7.
10. Idem, (1974) ibid, C83, p.C154.
11. C. R. Maude and F. R. Sale (1977), ibid, C86, p.C82.
12. J. Emmeneggar and H. Petermann (1968), J. Crystal Growth, 2, p.34.
13. H. Schäfer and H. Odenbach (1966), Z. Anorg. Allg. Chem., 346, p.127 .
14. V. Peshev (1973), Mat. Res. Bull., 8, p.1011.
15. Y. Saeki, R. Matsuzaki and H. Morita (1971), Kogyo Kagaku Zasshi, 74, p.344.
16. I. S. Morozov and G. F. Fefelova (1971), J. Appl. Chem. USSR, 44, p.1174.
17. N. Sano and G. R. Belton (1974), Metall. Trans., 58, p.2151.
18. I. Barin and O. Knacke (1973), "Thermochemical Properties of Inorganic Substances", Springer Verlag.

Fig. 1. Optical micrograph of "as-received" chromite.

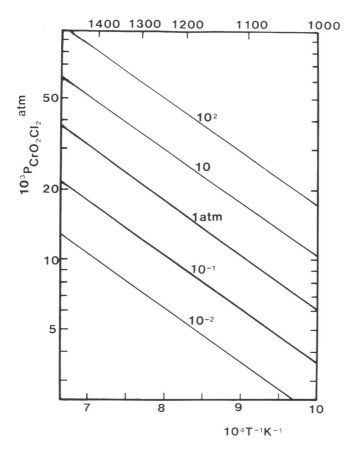

Fig. 2. Relationships between partial pressure of CrO_2Cl_2 and reciprocal absolute temperature for various total pressures.

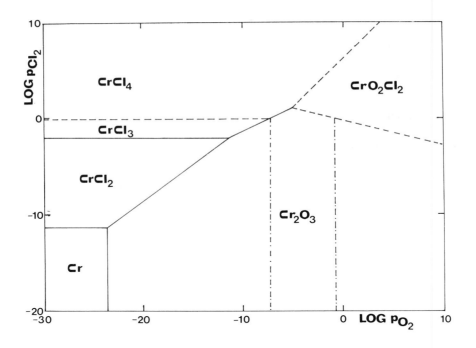

Fig. 3. Predominance area diagram for Cr-O-Cl system at 1200 K.

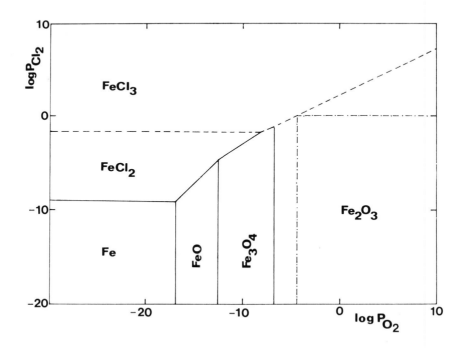

Fig. 4. Predominance area diagram for Fe-O-Cl system at 1200 K.

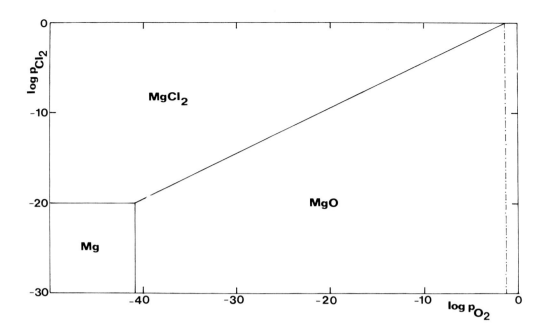

Fig. 5. Predominance area diagram for Mg-O-Cl system at 1200 K.

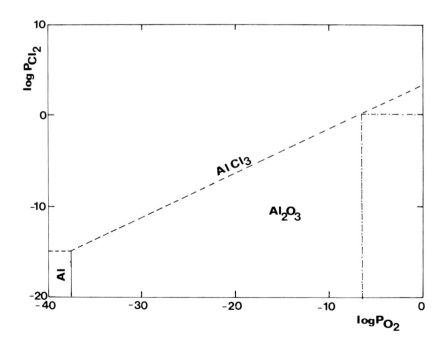

Fig. 6. Predominance area diagram for Al-O-Cl system at 1200 K.

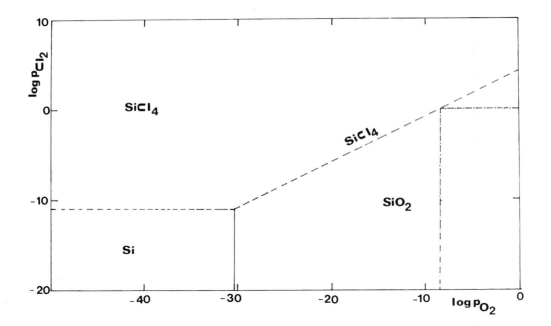

Fig. 7. Predominance area diagram for Si-O-Cl system at 1200 K.

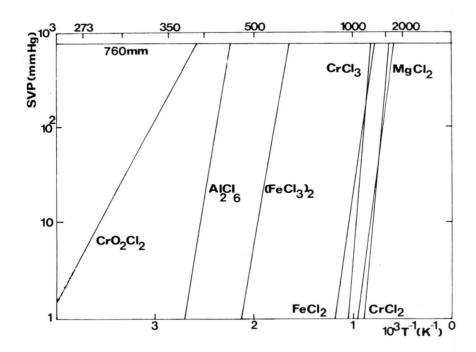

Fig. 8. Relationships between saturated equilibrium vapour pressures of the likely chloride products and temperature.

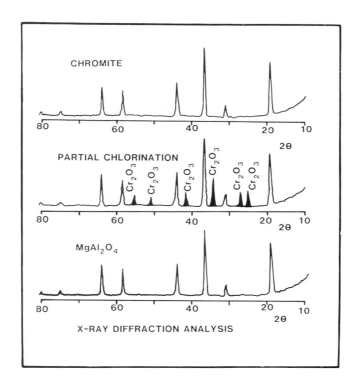

Fig. 9. X-ray diffraction spectra for chromite ore, products of partial oxy-chlorination and product at end of oxy-chlorination.

Fig. 10. Scanning electron micrograph of "as received" chromite grains.

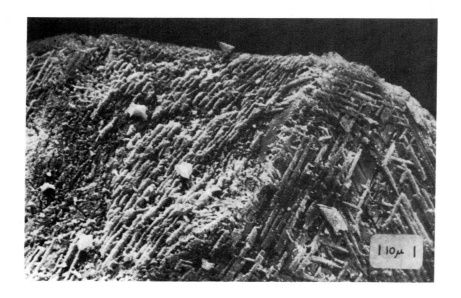

Fig. 11. Scanning electron micrograph of a partially reacted chromite grain.

Fig. 12. Scanning electron micrograph of the residue left after reaction.

237

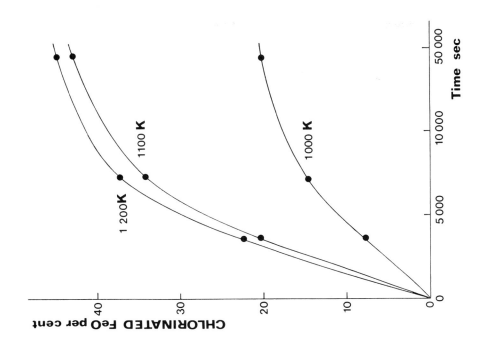

Fig. 14. Relationship between time and *wt%* of FeO chlorinated from chromite for different reaction temperatures.

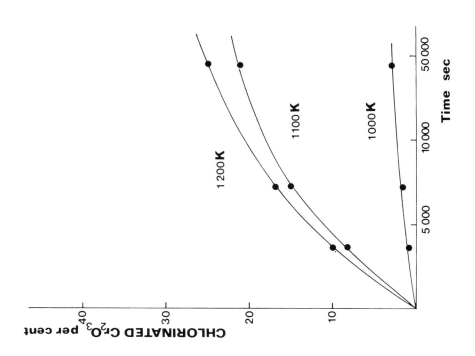

Fig. 13. Relationship between time and wt% of Cr_2O_3 chlorinated from chromite for different reaction temperatures.

238

Table 1 Electron microprobe analysis of chromite grains

Element	wt%	Mole%	Oxide wt%
Matrix Chromite			
Cr	37.39	10.04	54.65
Fe	11.57	2.89	14.88
Mg	9.02	5.18	14.96
Al	7.59	3.93	14.34
			————
			98.83
Exsolution Phase			
Cr	30.69	7.98	44.75
Fe	19.63	4.76	25.25
Mg	10.58	5.89	17.55
Al	7.13	3.58	13.48
			————
			101.13

Table 2 Quantitative observations on the oxy-chlorination of Cr_2O_3

Temp. K	$Cl_2:O_2$ ratio	Visual indication of reaction
973	1:3	None
973	1:1	None
973	3:1	None
973	4:1	None
973	100% Cl_2	None
1073	1:3	None
1073	1:1	None
1073	3:1	Slow
1073	4:1	Medium
1073	100% Cl_2	Slow
1173	1:3	None
1173	1:1	None
1173	3:1	Slow
1173	4:1	Fast
1173	100% Cl_2	Medium

Table 3 Analysis of product from CO_2-cooled condenser after oxy-chlorination of chromite at 1200 K, $P_{Cl2}: P_{O2}$ = 4, time of reaction 90 min

Element	Analysis wt%	Ideal CrO_2Cl_2 wt%
Cr	33.62	33.57
Fe	0.14	–
Cl	46.01	45.77
O	20.03	20.66
Si	Not detected	–
Al	Not detected	–
Mg	Not detected	–

28

THE USE OF PHASE DIAGRAMS IN THE DEVELOPMENT OF ADVANCED MAGNESIUM EXTRACTION PROCESSES

A. M. CAMERON

Manchester Materials Science Centre, University of Manchester and UMIST, Grosvenor Street, Manchester M1 7HS, UK

ABSTRACT

Magnesium metal is currently produced by electrolysis of anhydrous magnesium chloride or by pyrometallurgical reduction of magnesium oxide.

The recovery and manufacture of anhydrous $MgCl_2$ from seawater or brine is problematic and accounts for 50% of the production cost of electrolytic magnesium. In spite of this, electrolysis accounts for some 80% of world production levels due in part to the scale of operations required for economic operation of an electrolytic plant.

By contrast feedstocks suitable for the pyrometallurgical processes, such as dolomite and magnesite, are abundantly available. When taken in conjection with the lower capital expenditure and high space time yields available from a high temperature process, this observation suggests that there is considerable scope for a more widespread adoption of a pyrometallurgical technique.

This paper describes how information derived from appropriate phase diagrams can be used to help circumvent some of the difficulties associated with the traditional pyrometallurgical technologies.

INTRODUCTION

The bases of the early pyrometallurgical technologies are readily understood by considering the relative thermodynamic stabilities of the commonly occurring oxides. This can most readily be accomplished by means of an Ellingham diagram which clearly reveals both the high thermodynamic stability of magnesium oxide, and the limited choice of suitable reductants.

Carbon and silicon have both been employed as primary reductants and aluminium has found some limited application as a secondary or supplementary reductant but its use is restricted on economic grounds. Two main process routes have therefore been employed at industrial scale. These are; Carbothermic, based upon the reaction

$$MgO_{(s)} + C_{(s)} = Mg_{(g)} + CO_{(g)} \qquad (1)$$

and Silicothermic

$$2MgO_{(s)} + Si_{(s)} = 2Mg_{(g)} + SiO_{2(s)} \qquad (2)$$

In their simplest forms both processes are subject to severe technical constraints. The carbothermic process demands a high purity seawater magnesia since at the high temperatures involved coreduction of impurities leads to contamination of the product metal. Since silica is inherently less stable than magnesia at industrially acceptable operating temperatures, vacuum furnaces are required to render the silicothermic route feasible. These technical limitations have been overcome by utilising the thermodynamic information derivable from appropriate phase diagrams. The development of modern magnesium production processes also highlights the value of thermodynamic modelling of complex multi-component oxide systems.

CARBOTHERMIC TECHNOLOGIES

Hansgirg, [1],[2], is generally credited with the development of the earliest carbothermic process. The operation at Permanente, [3], was based on the use of an electric arc furnace containing a bed of metallurgical coke maintained at 2000°C. Briquettes of magnesia and coke were charged to the furnace and the resulting mixture of gaseous magnesium and carbon monoxide was passed to a quench unit into which a stream of cold natural gas was injected. This quench technique was incapable of producing the heat and mass transfer rates required to prevent the reverse reaction taking place. Since the selectivity of the reduction reaction was also poor, the product consisted of a mixture of fine magnesia, carbon dust and an impure pyrophoric magnesium dust. This material was an explosion hazard and required sublimation in order to recover a pure metal.

In recent years there has been renewed interest in the development of a carbothermic technology,[4]. Research efforts were concentrated on improvements to quench efficiency by, for example, replacing gas injection by liquid metal spraying, however considerable effort was also expended on enhancing the selectivity of the reduction reaction. This was accomplished by replacing the traditional hot coke bed with a molten slag system which acts as both a heat transfer medium and as a chemical 'sink' for feedstock impurities.

SELECTION AND THERMODYNAMICS OF A CARBOTHERMIC SLAG SYSTEM

In essence there are two essential features of a suitable slag system. For production of a pure metal, magnesia activities should be as high as possible. The slag should also be selected in such a way as to minimise the activities of the remaining components, since this will promote dissolution of feedstock impurities and help avoid unwanted side reactions. It is particularly important to prevent reactions which lead to the formation of volatile suboxides as these will have a detrimental influence on both quench efficiencies and product purity. Silica will gradually accumulate in the slag due to its presence in natural magnesites and the ash content of the reductant. To maximise the cycle time prior to the onset of SiO formation the process slag was selected to be an aluminate based system. Since calcium oxide is the major component of calcined magnesite the main slag constituents are therefore CaO, MgO and Al_2O_3.

At a total pressure of 1 atmosphere the equilibrium reaction temperature for the carbothermic reduction of magnesia is 1760°C. High magnesia activities can therefore be ensured by the selection of slag compositions from within the periclase primary recrystallisation field of the CaO-MgO-Al_2O_3 system, provided the compositions lie on or close to the 1750°C (2023K) liquidus line.

Consideration of the appropriate phase diagram, Fig. 1, reveals that the maintenance of a constant MgO content in the range 18-20 wt% will result in liquidus temperatures of 1700-1800°C over a wide range of CaO:Al_2O_3 ratios. Several other issues had to be resolved in order to define the range of slag compositions suited to the production of a pure metallic vapour, these were;

(i) The degree of magnesia undersaturation which could be tolerated,
(ii) Determination of the optimum CaO:Al_2O_3 ratio,
(iii) How much silica could be allowed to accumulate in the slag before it begins to react with the carbon?

In order to answer these questions, a simple thermodynamic model was developed which was based upon the following observations:

(a) Magnesia saturation levels are adequately described by assuming ideal mixing of supercooled liquid MgO and binary slags of various CaO:Al_2O_3 ratios, Fig. 2.
(b) Up to a concentration of at least 10 wt% MgO, substitution of CaO by MgO has little effect on silica activities in the quaternary CaO-MgO-Al_2O_3-SiO_2 system, [5],[6]. CaO and MgO can therefore be treated as a single component allowing activities of species other than magnesia to be estimated from measured data pertaining to the CaO-Al_2O_3-SiO_2 ternary system.

Activities obtained from the model, which has been described in greater detail elsewhere, [4], were used as input to a Gibbs energy minimisation routine enabling prediction of magnesium vapour purities as a function of slag composition. Predicted gas compositions had Mg contents in excess of 99% over a wide range of slag compositions and this was experimentally confirmed, Table 1. Pilot scale trials confirmed the feasibility of the process and showed that the slag performs well as a 'sink' for feedstock impurities. Development of an efficient quench system remains as an obstacle to the industrial implementation of the process.

SILICOTHERMIC PROCESSES

A key advantage of the silicothermic processes is the fact that there is no oxidising species associated with the gaseous reaction product. Complex gas quench technology can therefore be replaced by a relatively simple condensor.

The Pidgeon process was the earliest such technique and various modifications of the original technology still find industrial application. This is somewhat surprising given the numerous shortcomings of the process.

In essence briquettes of calcined dolomite and ferrosilicon are charged to tubular retorts which are heated to a temperature of about 1200°C. By evacuating each retort to a pressure of 0.2 torr the reaction

$$2\,CaO.MgO_{(s)} + (\underline{Si})_{(s)} = 2\,CaO.SiO_2(s) + 2Mg_{(g)}$$

is promoted. The magnesium is collected as 'crowns' on a water cooled section of the retort which is external to the gas fired furnace. The combination of external heating of the retorts and solid state nature of the reaction results in low space-time yields and the Pidgeon process has consequently been superseded by the Magnetherm process.

In the Magnetherm process, bauxite is used to flux the solid dicalcium silicate reaction product. The slag produced has a nominal composition of 58 wt% CaO, 25%SiO$_2$, 11% Al$_2$O$_3$ and 6% MgO and has a liquidus temperature of approximately 1850°C which compares with 2130°C for the congruently melting compound 2CaO.SiO$_2$. Examination of the solidus surface of the quaternary reveals that final solidification does not occur until a temperature of 1300°C has been reached. No problems have been reported with respect to the tapping of the slag at the normal operating temperature of 1550°C.

It is of interest to consider the solidification path of a Magnetherm slag in a little more detail since this has a fundamental bearing on the thermodynamics of the process.

In spite of the addition of the alumina, the normal Magnetherm slag composition remains in the dicalcium silicate phase field. As dicalcium silicate is precipitated on cooling, the remaining liquid becomes progressively richer in both MgO and Al$_2$O$_3$. It is easily shown that the liquid phase compositions will change in the manner shown in Table 2. By plotting these compositions on appropriate sections of the phase diagram, Christini, [7], has shown that liquid composition reaches the di-calcium silicate/periclase phase boundary at a temperature of 1550°C. Provided the process is operated at, or close to this temperature, magnesia activities may be taken as unity. Since the equilibrium constant of the reaction may be expressed as,

$$K = \frac{a_{SiO_2} \cdot P^2_{Mg}}{a_{Si} \cdot a_{MgO}}$$

and since the slag composition is essentially invariant, it is clear that the magnesium partial pressure is a function of the silicon content of the spent reductant. In practice, the silicon content of the residue is in the vicinity of 20 wt% for an operating temperature and pressure of 1550°C and 0.05-0.1 atmospheres respectively.

Although the Magnetherm process has superseded the Pidgeon technology it is still subject to some of the severest limitations of the early technology. In particular, the vacuum requirement ensures that the process can only be operated in a batch mode. Leakage of air into the condensor causes excessive product loss due to both oxide and nitride formation. A means of eliminating the need for vacuum has long been recognised as a valuable, but elusive, improvement of the technology.

It is obvious that there is a large increase in entropy associated with the silicothermic reduction reaction. Consequently, the thermodynamics of the process should be favoured by increasing the reaction temperature. Unfortunately, unless a different slag system is employed, any advantage accruing will be offset by the fact that the Magnetherm slag rapidly becomes undersaturated with respect to MgO at temperatures higher than 1550°C.

Mintek, [8], have overlooked this fact in attempting to employ a D.C. arc plasma reactor to attain atmospheric operation. Calculations clearly show that P$_{Mg}$ can only exceed the pressure of their plasma gas if unacceptably high silicon contents are accepted in the residual ferrosilicon. Even so, the temperatures employed would have to be significantly higher than those reported in their description of the process, Fig. 3. In effect, the Mintek process is using the plasma as a means of 'sweeping' low pressure magnesium gas from an otherwise conventional Magnetherm reactor.

It is probable that the Magnetherm and carbothermic technologies both suffer a heat-transfer limitation, [4],[7]. In both cases the highly endothermic reduction reactions take place at the surface of the slag bath. It was the recognition of this heat transfer limitation that first caused the potential of D.C. plasma arcs to magnesium processing to be explored [9]. Initial trials revealed the possibility of attaining high localised surface slag temperatures which open the way to genuine atmospheric operation of the silicothermic process. To take full advantage of this observation, suitable slag compositions must be identified. The key requirements of such slags are now reasonably well defined;

- the slag should remain at, or close to magnesia saturation throughout the process,
- low silica activities should be maintained,
- the slag should be effective in passivating other oxides which may be introduced as impurities and could potentially react to contaminate the gaseous product.

High magnesia activities can be ensured in a manner analogous to that employed in the carbothermic process previously described. In this case, however, the use of a silicon reductant requires that a means be found of describing the liquidus isotherms of the periclase region in terms of quaternary slag compositions. This is less problematic than it may appear and it can be shown that the liquidus isotherms are adequately described in terms of the excess base parameters of the slags. Excess base is related to the slag composition by the expression

$$E.\ B. = n\ MgO + n\ CaO - 2/3\ nAl_2O_3 - n\ SiO_2$$

where n is the number of moles in 100g equivalent of slag.

This relationship can be utilised to delineate suitable slag compositions. All that is required is that the desired reaction temperature be identified. As pointed out earlier, this may then be related to an appropriate magnesia content via the CaO-MgO-Al$_2$O$_3$ ternary. If this magnesia content is held constant then the CaO:Al$_2$O$_3$ ratio needed to maintain a near constant liquidus temperature becomes a function only of silica content. An example based on a reaction temperature in the vicinity of 1900°C is given in Table 3.

In addition to a high activity of MgO the other essential characteristic is that the slag should have a low silica activity to avert the reaction;

$$(SiO_2) + \underline{Si} = 2\ SiO_{(g)}.$$

Clearly, a wide range of SiO$_2$ contents are available. However, in terms of process economics, it is advantageous to minimise the quantity of alumina flux employed. In reality, therefore, the SiO$_2$ content will be a compromise between two conflicting demands. Once again CaO and MgO can be treated as a single component and the preferred silica concentration determined on the basis of its activity in the CaO-Al$_2$O$_3$-SiO2 system. It can be shown that the process is, in fact, extremely flexible with respect to slag composition and operating temperature.

It may, at first sight, seem surprising that liquidus isotherms in a complex quaternary system are so readily related to an empirical parameter such as excess base. This would overlook the fact that excess base parameters have been successfully employed in the interpretation of complex equilibria involving slag systems of quite diverse compositions. For example, the sulphur capacity of many steelmaking slags, which is determined by the equilibrium,

$$\tfrac{1}{2}S_{2(g)} + \left(O^=\right) = \left(S^=\right) + \tfrac{1}{2}O_{2(g)}$$

has been shown to be a function of excess base, [10]. Excess base parameters are therefore an empirical measure of the concentration of 'free' oxygen ions (O$^=$) in a molten slag. This being the case, it is less perplexing that it should show some correspondence to the activity of basic (or network modifying) oxides and hence liquidus temperatures in the periclase phase field.

In fact, silicate slag structures are generally thought to comprise a three dimensional network of SiO$_4$ tetrahedra in which each oxygen is shared between two silicon atoms, (0°). Addition of a basic oxide such as MgO or CaO disrupts the network and gradually introduces increasing numbers of oxygens bound to one silicon, (O$^-$), and ultimately free oxygen ions, (O$^=$). Many statistical thermodynamic models of molten slag systems are based upon the occurrence and distribution of these three types of oxygen ion and have been shown capable of accurately predicting phase relations in multi-component oxide systems. Such models have recently been reviewed by Masson, [11]. They can provide a valuable check on predictions of component activities for slag systems such as those discussed in this paper.

CONCLUSIONS

A working knowledge of phase relations in the system(s) MgO-CaO-Al$_2$O$_3$(-SiO$_2$) has been applied to eliminate many of the shortcomings of magnesium production processes.

The excessive cycle times of the Pidgeon process are eradicated by using an alumina flux to produce a molten slag system which is doubly saturated with respect to 2CaO.SiO$_2$ and MgO. The vacuum requirement of the Magnetherm process is eliminated by using plasma heating. Phase relations can be employed to derive possible slag compositions suited to this process.

Trials of the plasma process have been conducted in a 0.6 MW, reactor [12]. These trials confirm that high purity magnesium vapour is generated at atmospheric pressure. The silicon content of the spent reductant was observed to be as low as 1.2 wt% for one version of process which compares with 20 wt% if Magnetherm technology is employed. Phase diagram studies can therefore result in substantial economic benefits when applied to extractive metallurgy.

ACKNOWLEDGEMENTS

The author wishes to express his gratitude to Billiton Research B. V., Arnhem, The Netherlands for their willingness to conduct trials of the atmospheric plasma process described. In particular thanks are due to Drs V. G. Aurich and D. L. Canham without whose assistance the trials would have been impossible.

REFERENCES

1. F. J. Hansgirg, The Iron Age, 18th Nov. 1943, p.51.

2. F. J. Hansgirg, The Iron Age, 25th Nov. 1943, p. 52.

3. T. A. Dungan, 'Production of Magnesium by the Carbothermic Process at Permanente', Trans AIMME, vol. 159, 1944, pp. 308-314.

4. A. M. Cameron, et al, 'Carbothermic Production of Magnesium' in 'Pyrometallurgy '87'. Proceedings of a symposium. Published by The Institution of Mining and Metallurgy, Sept. '87, pp. 195-223 .

5. D. A. R. Kay and J. Taylor, Trans. Farad. Soc., vol. 56, p. 1372, 1960.

6. R. H. Rein and J. Chipman, Trans. Met. Soc. A.I.M.E., Vol. 233, p. 415, 1965.

7. R. A. Christin, 'Equilibria among metal, slag and gas phases in the Magnetherm process', Light Metals, New York, 1980, pp. 981-995.

8. A. F. A. Schoukens, 'Plasma-arc process for the production of magnesium', In Extraction Metallurgy '89, Proceedings of Symposium. Published by The Inst. of Mining and Metallurgy, July 1989, pp. 204-223.

9. G. F. Warren and A. M. Cameron, 'Process for producing magnesium', European Patent Appl. No. 84201741.0, 30th Jan. '85. (Shell Int. Research).

10. S. Brown, Ph.D. Thesis, University of Strathclyde, 1977.

11. C. R. Masson, 'The Chemistry of Slags - An Overview', In Proceedings of Second Int. Symp. on Metallurgical Slags and Fluxes. Editors H. A. Fine and D. R. Gaskell, published by The Met. Soc. of AIME, Nov. 1984.

12. A. M. Cameron, 'Plasma powered magnesium production'. Journals of Metals, April 1990, pp. 46-48.

246

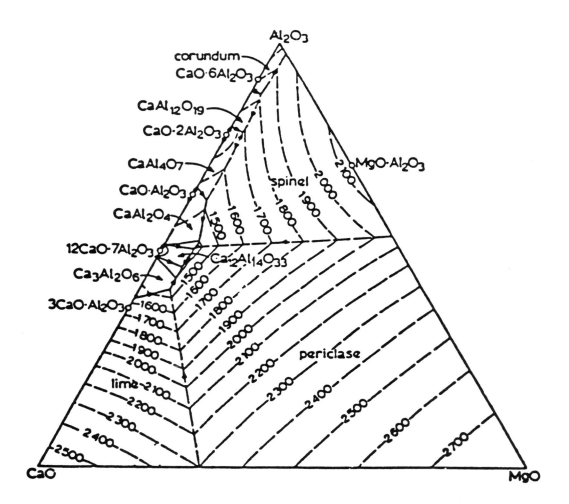

Fig. 1. Phase diagram for the CaO-MgO-Al₂O₃ system.

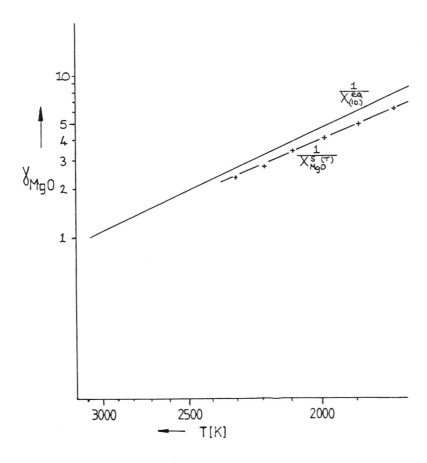

Fig. 2. Comparison of $^\gamma$MgO (assuming ideal behaviour between a CaO:Al_2O_3 binary and supercooled liquid MgO) with the value obtained from the phase diagram.

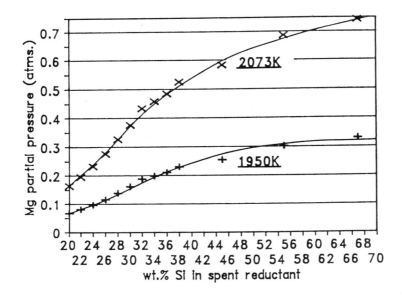

Fig. 3. Theoretical magnesium partial pressures for a modified version of the Magnetherm process as operated by Mintek.

Table 1 Carbothermic reduction from a molten slag bath. Comparison of predicted and experimentally determined gas compositions

Description	Nominal Slag Composition				Predicted Gas Composition				Experimental Composition			
	MgO	CaO	Al$_2$O$_3$	SiO$_2$	Mg	Ca	Al	Si	Mg	Ca	Al	Si
'Standard' Slag	20	35	45	-	99.44	0.55	4.7×10^{-3}	-	99.28	0.63	8.4×10^{-2}	-
High CaO/High SiO$_2$ Slag	20	35	45	15	99.40	0.21	9.8×10^{-4}	0.39	98.55	0.74	1.7×10^{-1}	0.54
High Al$_2$O$_3$/High SiO$_2$ Slag	30	20	35	15	99.45	0.04	1.8×10^{-3}	0.51	99.14	0.08	6.0×10^{-2}	0.73
High CaO/Low SiO$_2$ Slag	20	40	35	5	99.51	0.44	1.3×10^{-3}	0.04	98.85	0.96	1.5×10^{-1}	0.05
High Al$_2$O$_3$/Low SiO$_2$ Slag	20	30	45	5	99.59	0.18	2.3×10^{-2}	0.20	99.54	0.34	2.8×10^{-2}	0.10

*   All compositions in % (m/m)
*   Nominal slag compositions taken from 1800°C liquidus of periclase region

Table 2 Liquid phase compositions of Magnetherm slag as a function of temperature

Temperature °C	% Solids	Liquid Phase Composition (%)				Recrystallisation Field
		MgO	Al_2O_3	CaO	SiO_2	
1850	0	6.0	14.0	55.0	25.0	Dicalcium Silicate
1820	10	6.7	15.6	53.9	23.9	"
1770	20	7.5	17.5	52.5	22.5	"
1650	30	8.6	20.0	50.7	20.7	"
1550	40	10.0	23.3	48.3	18.3	Dicalcium Silicate/ Periclase.

Table 3 Variation of liquidus temperature with excess base parameter for slags containing 25 wt% MgO

Slag Composition (wt%)				Liquidus Temp (°C)	Excess Base
MgO	CaO	Al_2O_3	SiO_2		
25	33	42		1900	0.93
25	37	28	10	1950	0.93
25	39	21	15	1950	0.93
25	41	14	20	1950	0.93

29

CRITICAL ASSESSMENT OF TERNARY ALLOY PHASE DIAGRAMS: A WORLD WIDE JOINT EFFORT

G. EFFENBERG*, C. BÄTZNER, Q. RAN AND G. PETZOW

Max-Planck-Institut für Metallforschung, Stuttgart, Germany
*Materials Science, International Services GmbH, Stuttgart, Germany

OBJECTIVES

As a basis for application and optimization of materials it is vital to know the set of equilibrium phase configurations that are stable under different conditions of temperature, composition and other variables. Even if an applied material is not in its equilibrium state under use, for processing during production and achievement of certain properties it is necessary to know the equilibrium phase diagram.

Materials constitution has been of subject research for about one century. The extent of experimental work differs among the various combinations of elements according to their importance in applications. But in all cases the available data is scattered through the world literature and scanning these publications reveals the fact that many experiments are repeated by scientists, who were unaware of former investigations in the same area or were unable to get the respective publications.

Based on different experimental methods and technical facilities, the results published in many cases are inconsistent and interpretations of the results of any measurements can contain mistakes. Both of these could sometimes be removed, if someone with the necessary knowledge would assess the multicomponent system by considering all existing data. The quality of his work could be increased further, if he were given the possibility to discuss his conclusions with other scientists.

On the basis of such an evaluation a consistent and formal homogeneous report could be derived, which saves engineers and scientists from having to conduct a literature search, from reading, comparing and calculating different scales, and from interpretation of inconsistencies and contradictions.

The essential part in this is, that the work done in such a project such as the one presented here cannot be reduced to simply review of existing data but it is a critical assessment, an evaluation of all investigations done, clarifying or showing inconsistencies, mistakes and areas for further necessary experiments.

So the goal of the project for the critical assessment of phase diagrams is:

1. the development of an idea for the organization of the work.

2. the identification of all relevant literature on the systems concerned.

3. the foundation of a group of highly skilled authors for assessment. The objectives are system reports with the following attributes:

(a). based on all available data;

(b). consistent and pointing out mistakes and contradictions (e.g. also to thermodynamic rules);

(c). unified format in diagrams, text and tables.

4. the implementation of a operational structure and administration for the assessment, discussion and computer recording of text and diagrams.

5. the development of a formal characterization of the final reports.

6. the finding of an appropriate way for publication.

CONCEPT

The concept developed at the Max-Planck-Institut in Stuttgart is in principle as follows:

1. The evaluation is done in alphabetical order starting with Ag-X-Y, Al-X-Y and going on to As-X-Y.

2. The literature is searched on-line and by hand at least back to the year 1935 and, based on the references given in the publications found, back to the 19th century.

3. An international team of experienced material scientists has been established.

4. The organization and administration of the project is done by a small 'project core'.

5. The communication between authors, reviewers and management is done by workshops, meetings in Stuttgart, post, Telex, Fax and courier

6. The system reports are published in book form but can also be obtained individually as a single system report.

According to the idea of Hansen and his collection of binary systems, the project is doing the evaluation in alphabetical order throughout the whole periodic system, but only dealing with ternary systems where existing data include equilibria with a metal or a metallic phase.

The team of the authors (that is 3rd Al-volume) is made up of participants from well-known institutions in thirteen countries:

Baikov Institute of Metallurgy, Ac. Sc. USSR

Brunel Univ., Dept. Materials Technology,
United Kingdom

Centre National de la Recherche Scientifique
Centre d'Etudes de Chimie Métallurgique
Vitry-sur-Seine/Cedex, France

Forschungsinstitut für Edelmetalle und
Metallchemie Schwäbisch Gmünd, Germany

Institut für physikalische Chemie Univeristät Wien,
Austria

Istituto di Chimica Generale ed Inorganica
dell'Universita di Genova, Italy

Laboratoire de Chimie Minérale II
Faculté de Pharmacie, Université Paris-Sud
Chatenay-Malabry, France

L'viv State University
Kathedra of Inorganic Chemistry
L'viv, USSR

Manchester Materials Science Centre
University of Manchester and UMIST
Manchester, United Kingdom

Max-Planck-Institut für Metallforschung
Institut für Werkstoffwissenschaft
Pulvermetallurgisches Labor
Stuttgart Germany

Metallgesellschaft AG
Zentrallabor
Frankfurt am Main, Germany

Moscow State University
Chemical Faculty
Moscow, USSR

North Carolina State University
Department of Materials
Science and Engineering
Evanston, USA

Universidade Estadual de Campinas
Instituto de Fisica "Gleb Wataghin", DFESCM
Campinas-SP, Brazil

Technische Universität Clausthal
AG Elektronische Materialien
Clausthal-Zellerfeld, Germany

The authors are supplied with the literature relevant to their systems by Stuttgart. They work on their own responsibility as far as the evaluation is concerned, but of course in discussion and exchange with reviewers and technical staff. They are supplied with the 'Notes for authors' originating from the experiences during the

checked by desk editors for errors originating from the computerization and again for consistent description of the constitution of the system. These checks go on during all following steps of the formation of the system report until it finally goes to the publishing house.

The following graphic shows the whole process:

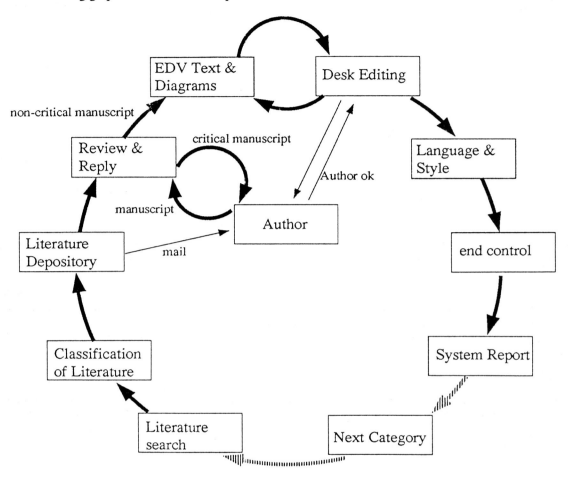

The structure of the system report described in the 'Notes for Authors' is fixed throughout all of the categories and originates from the demand to contain all necessary data to understand the constitution of a system. On the basis of the available literature the following parts can build a system-report:

Heading gives the name of the system and the full name of all authors

Introduction is a brief critical review of the published work concerning experimental methods, agreement between different authors and the amount of work done.

Binary Systems gives the sources of the binary systems used in the èvaluation. If they are not from generally known sources, the phases diagram is included in the system report as a figure.

Solid Phases gives information additional to the table 'Solid Phases' which includes data on the structure, lattice parameters and ranges of existence for the phases found in the ternary system.

Pseudobinary Systems contains information on existing or assumed pseudobinary systems, the corresponding phase diagram is included as figure.

Invariant Equilibria describes the invariant equilibria of the ternary system together with an optional table giving the concentrations of participating phases and a Scheil Reaction Scheme.

Invariant Equilibria describes the invariant equilibria of the ternary system together with an optional table giving the concentrations of participating phases and a Scheil Reaction Scheme.

Liquidus Surface is also in most cases referring to a figure and contains additional information which makes it more easy to understand the drawing and its emergence from the evaluation.

Isothermal Sections referring to included diagrams gives information on the accepted isothermal sections and on modifications made to them e.g. due to the accepted binary systems, the removal of inconsistencies between different publications or from thermodynamic rules.

Miscellaneous may contain all data the author thinks to be necessary, helpful or interesting for the understanding of the ternary system, e.g. vertical sections, magnetic or physical properties etc.

The diagrams included in a system report are generally in at.% with an additional mass % scale and vertical sections fit into horizontal sections in size.

As an example for a system report the system Al-Cd-Sb is included at the end of this document.

RESULTS

The first category Ag-X-Y was published in 1988 in two volumes containing about 480 ternary systems. The evaluation of these systems is based on 1900 relevant papers which had been identified in classifying 5000 literatures or literature abstracts.

Since 1988 two Al-Volumes containing the systems Al-Ar-O to Al-Cu-Ru have been published and the third volume will be available in December 1991. The total Al-category will contain 1080 ternary systems based on 5000 relevant papers in five volumes and will be available in the middle of 1992.

Besides the published volumes the project can offer access to about 8000 references available at the Max-Planck-Institut in Stuttgart. These concern the constitution of all ternary alloys containing Al or Ag as one component.

The group is open to any scientist or engineer, who is interested in participation or has any questions related to the subjects that project deals with. Actually the management and organization is done in cooperation between the Max-Planck-Institut für Metallforschung, Institut für Werkstoffwissenschaft, Pulvermetallurgisches Laboratorium Heisenbergstr.5, D-7000 Stuttgart 80, and the company Materials Science ⇌ International Services GmbH, c/o Dr. G. Effenberg, Heisembergstr. 5, D-7000 Stuttgart 80, Germany.

EXPERIENCES

During the eight years of work some experiences worthy of mention are as follows:

The literature search for a project like this cannot rely on computerized bibliographic services, about 30-40 % of the literature would have been undetected because of the limited periods covered by these services.

As a conservative estimate it can be said that about 50 % of the published phase diagram data exhibit serious discrepancies. They cause the critical assessment to demand a great effort and demonstrate the need for performing the evaluation.

The international character, the organizational structure and the large number of scientists cooperating in the project lead to the establishment of a platform for exchange and discussion. In addition to the routine assessment and on demand this team can provide more extensive and maybe problem oriented information.

EXAMPLE

Aluminium-Cadmium-Antimony

Rainer Schmid-Fetzer

Introduction

The ternary equilibria along a variety of composition cuts have been investigated by differential thermal analysis, microstructural analysis and microhardness measurements [85Bel]. Samples were prepared by fusion of the elemental components in evacuated silica capsules at 1077°C for 3 h with vibration and then annealed at 447°C

254

for 240 h. A final anneal at 347-247°C for 400 h caused conversion of the metastable Cd-Sb-phases to the stable system. A number of misleading interpretations are given in this paper: The eutectic liquid in the CdSb-AlSb quasibinary system is given at 1 mole % AlSb and 317°C, which would mean a depression of the CdSb-melting point of 148°C caused by 1 mole % impurity. A more realistic eutectic temperature (433°C) can be read from one of the vertical sections. Also the ternary eutectic temperature E_3 (Table 2 and Fig.4) is given at 250°C, which is very unlikely in view of its close vicinity to the Cd-melting point (321.1°C) and the almost identical position of the eutectic liquid of the accepted Cd-AlSb pseudobinary system at 307°C [85Bel]. Therefore 305°C is accepted for E_3 (Table 2). Major conflicts with basic rules of heterogeneous equilibria occur in the plotted Cd-AlSb pseudobinary system and the ternary "liquidus surface" [85Bel].

A linear variation of the energy band gap of the semiconductors CdSb and AlSb was found by [59Rad] in the range 0–40 mole % AlSb of the CdSb-AlSb section. This is inconsistent with the finding of negligible solid solubility of CdSb and AlSb [85Bel]. The influence of small Cd-additions on the electrical properties of AlSb has been studied by [56Smi].

Binary Systems
The binary systems Al-Cd, Al-Sb and Cd-Sb from [Mas] are accepted.

Solid Phases
The stable solid phases of this system are given in Table 1. A metastable Cd_3Sb_2 phase is formed easily [Mas]. A possible solid solubility of AlSb in (CdSb) remains questionable [59Rad and 85Bel].

Pseudobinary Systems
The pseudobinary Cd-AlSb system is given in Fig.1 [85Bel], the shape of the liquid miscibility gap has been corrected. The second pseudobinary system CdSb-AlSb is most probably a simple eutectic one. Figure 2 is constructed from a reinterpretation of the data given by [85Bel].

Invariant Equilibria
Data on the four-phase equilibria and the three three-phase equilibria (maximum point) in Table 2 are from [85Bel], except the liquid compositions of E_1 which were not given by [85Bel] and the temperature E_3 which had to be corrected. A reaction scheme is constructed in Fig.3.

Liquidus Surface
Based on Table 2, some additional data [85Bel] and the accepted binaries the liquidus surface has been constructed, see Fig.4. The field of primary AlSb crystallization and the liquid miscibility gap cover almost the entire ternary system. Two estimated isotherms of the AlSb-liquidus at 627°C (maximum of the L'+L"+(AlSb) monotectic) and at 600°C (estimated critical point) are also given.

REFERENCES

[56Smi] Smirous, K., "The influence of Impurities on the Properties of AlSb" (in Russian), Czechoslov. J. Phys., 6, 299-300 (1956) (Experimental, 7).
[59Rad] Radautsan, S.I. "The Ternary System Al-Cd-Sb" (in Russian), Uch. Zap. Kishnev. Univ., 39, 69-72 (1959) (Experimental, 14).
[85Bel] Belotskii, D.P., Lundich, M.S., Kotsyumakha, M.P., Makhova, M.K., Lesina, L.V. and Noval'kovskii, N.P., "The Cd-Sb-Al System", Inorg. Mater., 21, 951-954 (1985), translated from Izv. Akad. Nauk SSSR, Neorg. Mater., 21(7), 109-1096 (Equi. Diagram, Experimental, 5).

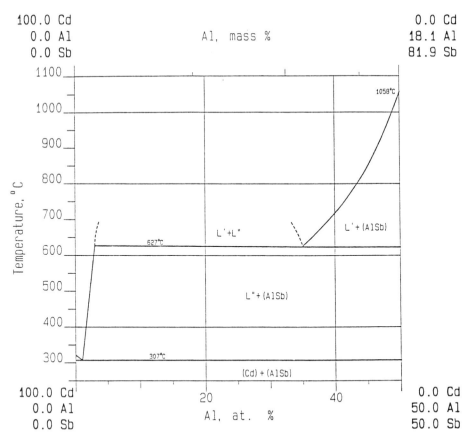

Fig.1 The pseudobinary system Cd-AlSb.

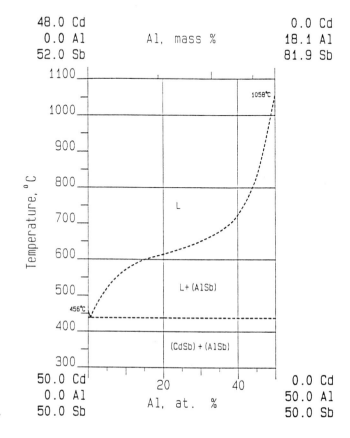

Fig.2 The pseudobinary system CdSb-AlSb.

256

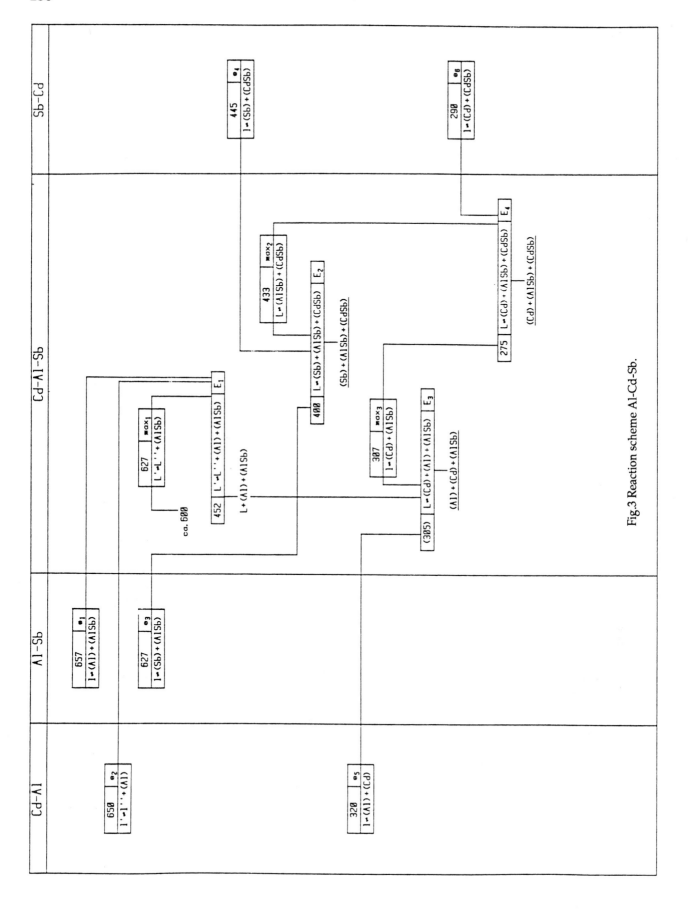

Fig.3 Reaction scheme Al-Cd-Sb.

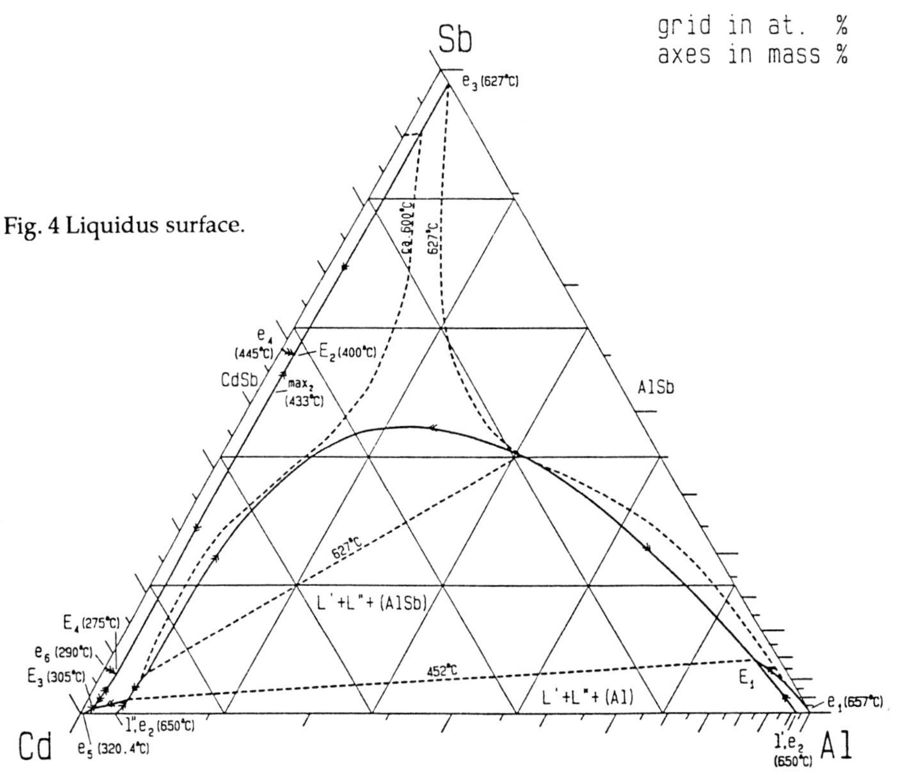

Table 2 Invariant equilibria

T(°C)	Type	Reaction	Phase	Composition (at.%) Al	Cd	Sb
627	max₁	L' ⇌ L'' + (AlSb)	L'	41	18	41
			L''	6	88	6
			(AlSb)	50	-	50
452	E₁	L' ⇌ L'' + (Al) + (AlSb)	L'	(89)*	(3)*	(8)*
			L''	(5)*	(93)*	(2)*
			(Al)	100	-	-
			(AlSb)	50	-	50
433	max₂	L ⇌ (CdSb) + (AlSb)	L	(2)	(48)	50
			(CdSb)	-	50	50
			(AlSb)	50	-	50
400	E₂	L ⇌ (CdSb) + (Sb) + (AlSb)	L	2	42	56
			(CdSb)	-	50	50
			(AlSb)	50	-	50
			(Sb)	-	-	100
307	max₃	L'' ⇌ (Cd) + (AlSb)	L''	1	98	1
			(Cd)	-	100	-
			(AlSb)	50	-	50
(305)**	E₃	L'' ⇌ (Cd) + (Al) + (AlSb)	L''	1	(98)	<1
			(Cd)	-	100	-
			(Al)	100	-	-
			(AlSb)	50	-	50
275	E₄	L'' ⇌ (Cd) + (AlSb) + (CdSb)	L''	2	92	6
			(Cd)	-	100	-
			(AlSb)	50	-	50
			(CdSb)	-	50	50

* estimated compositions
** corrected value

grid in at. %
axes in mass %

Fig. 4 Liquidus surface.

Table 1 Solid phases

Phase/ Temperature Range (°C)	Pearson Symbol/ Prototype	Lattice Parameters (pm)	Comments
(Al)	cF4 Cu	a = 404.88	[V-C]
(Cd)	hP2 Mg	a = 297.87 c = 561.66	[V-C]
(Sb)	hR2 αAs	a = 430.84 c = 1127.4	[V-C]
(AlSb)	cF8 ZnS	a = 613.55	[V-C]
(CdSb)	oP16 CdSb	a = 647.1 b = 825.3 c = 852.6	[V-C]

30
THERMODYNAMICS ASPECTS IN AMORPHIZATION: APPLICATION TO Cu–Ti AND Al–Ti

M. BARICCO AND L. BATTEZZATI*

Istituto Elettrotecnico Nazionale Galileo and Ferraris INFM-Unita' di Ricerca di Torino, Torino, Italy
*Dipartimento di Chimica Inorganica, Chimica Fisica e Chimica dei Materiali, Università di Torino, Italy

ABSTRACT

The thermodynamic factors which favour the amorphization of metallic alloys, are considered for two systems having different phase diagrams, Cu-Ti and Al-Ti.

Free enthalpy functions are calculated on the basis of a common hypothesis of ordering within the liquid phase. Its temperature dependence is expressed by means of an excess heat capacity of mixing. The glass transition temperature is evaluated, as a function of composition, using the concept of the vanishing entropy point. Metastable extensions of the liquidus and T_0 curves in the undercooled regime are obtained. In both cases a region is found for the existence of a glassy phase: it is accessible either by melt quenching or solid state reaction in Cu-Ti, but only by solid state reaction in Al-Ti.

INTRODUCTION

The formation of amorphous metallic materials either by rapid solidification or by solid state reactions, involves an interplay of thermodynamic and kinetic aspects [1]. The thermodynamic properties may be viewed as necessary conditions for amorphization to occur, if the kinetic requirements are fulfilled, i.e. if the time scale of the preparation technique meets the lifetime of a metastable phase. Therefore, a detailed thermodynamic analysis may shed light into the glass forming tendency of a given system.

The formation of metallic glasses by rapid quenching is commonly achieved for compositions in the neighbourhood of deep eutectics in the phase diagram. The presence of a deep eutectic represents an indication of stability of the liquid, relative to crystal phases, as well as of kinetic resistance to partitioning into a mixture of phases during rapid solidification. It is also widely recognized that the systems more prone to vitrify, show a negative enthalpy of mixing in the liquid state as well as a negative excess entropy of mixing, due to strong interactions among the constituents.

By means of the novel techniques of amorphization by solid state reaction [1,2], the glass forming range has been extended, in systems containing eutectics, also to compositions corresponding to intermetallic compounds, e.g. Ni-Zr, Ni-Ti. Interdiffusion at low temperature is the dominant alloying mechanism in multilayered thin films and in ball milled powders. Phase selection occurs because the time scale for the formation of the amorphous phase is shorter than that for intermetallics with complex structure, so their formation is hindered by kinetic constraints. Therefore, it is concluded that favourable conditions of stability for the glassy phase extend well beyond the eutectic compositions. To reinforce this observation, we point out that amorphization by solid state reaction has been achieved in systems which exhibit high liquidus temperatures all across the phase diagram, such as Al-Nb [3] and Al-Ti [4]. In these cases rapid solidifcation always produces crystalline phases either in a metastable or an equilibrium condition, because their nucleation cannot be avoided. In the presence of very stable, high melting phases, the liquid can hardly be quenched as a glass, even if it may share their thermodynamic stability.

In the present work we have considered two systems, which exhibit very different phase diagrams, and may be amorphized under different conditions: melt quenching and solid state reaction (Cu-Ti), solid state reaction only (Al-Ti). The free enthalpy for solution phases are computed and the conditions for amorphization are outlined.

THE CU-TI SYSTEM

Current evaluations of the thermodynamic properties of the Cu-Ti system, while giving a reasonable fit to the phase diagram, do not predict the possibility of amorphization reactions by any technique. In fact, the free enthalpy curves for the liquid, extrapolated to low temperature to represent the glassy phase, never fall below those of competing solid solutions. This implies that the locus of points of equal free energy of the liquid and solid solutions

(T$_o$ curves) stands well above the glass transition temperature, T$_g$ [5]. Under these circumstances, rapid solidification would never produce a glassy alloy, because partitionless crystallization would occur at any composition. On the other hand, amorphous Cu-Ti alloys are obtained in a wide composition range by rapid quenching, as well as solid state reactions [6].

Kinetic considerations have shown that the nucleation temperature of a competing crystal phase may occur below the glass transition, around the equiatomic composition, when very high cooling rates are employed, such as those available in laser glazing [7]. However, the amorphizing range by solid state reaction cannot be explained simply in terms of kinetic arguments [8]. In order to overcome this discrepancy, a deeper insight into the thermodynamic properties of the liquid phase seems necessary.

A peculiar feature of a metallic glass forming liquid is that its thermodynamic properties are temperature dependent, as a consequence of an increase of the degree of order on decreasing temperature. So, the enthalpy of mixing and the excess entropy of mixing become more negative as the temperature is lowered, the effect of this being more relevant in the undercooling regime. It follows that an excess heat capacity exists for the liquid state, as has been demonstrated for both metal-metal and metal-metal-loid systems [9]. Therefore, we derive a new free enthalpy curve for the liquid containing an excess term due to ordering. The existing data show that the excess heat capacity is substantial for alloy systems, and, often, is at a maximum at compositions corresponding to a compound in the solid state [10]. Pure elements have liquid heat capacities very close to those of the crystal phases at the melting point [12]. As the additive Neumann-Kopp rule is usually obeyed in crystalline alloys, the excess specific heat for liquid alloys may be approximated by the specific heat difference between liquid and crystal phases, Δ_p^{l-s}[9]. A high value of the liquid specific heat with respect to the crystal would allow the entropy of the liquid phase to approach that of the solid quickly on decreasing the temperature, according to equation (1)

$$\Delta S = \Delta S_f - \int_T^{Tf} \Delta C_p^{l-s} \, dlnT \qquad (1)$$

T$_f$ is the melting point and ΔS_f is the entropy of fusion. As a consequence, the isoentropic Kauzmann temperature T$_K$ (assumed as an ideal glass transition) would still be located at a relatively high temperature (say, 1/2 T$_f$), as well as the kinetic T$_g$ which must be reached before it on cooling.

An estimate of the effective excess specific heat has been obtained, at a few compositions close to intermetallic compounds, from the difference between the heat of fusion and the heat of crystallization of melt spun ribbons at the respective temperatures [8]. Figure 1 shows the data points to which the trend of the short range order parameter in the liquid has been superimposed as determined by Fenglai et al. [13]. It is apparent that the high values of ΔC_p^{l-s} found at glass forming compositions pertain to liquids having a substantial degree of order. Although the correlation is purely empirical, it reproduces that, already obtained, between the short range order parameter and a description of the thermodynamic properties of the liquid by means of the associate solution model [14]. In order to express ΔC_p^{l-s} as a function of composition in the following calculation, we have fitted the points with a parabola, also shown in Fig. 1. The average ΔC_p^{l-s} for pure elements has been calculated by extrapolation of the liquid data taken from the literature in undercooling regime. The assessment of the data has been described in detail in [8].

The excess specific heat is considered to drop at the ideal glass transition temperature which is calculated from a suitable entropy cycle, using various descriptions of the entropy of the liquid [10]. In a binary system an entropy balance can be established at every isoentropic temperature as follows:

$$\Delta S_{for} - X_A \Delta S_{f,A} - X_B \Delta S_{f,B} - \Delta S_{mix} = 0 \qquad (2)$$

where $\Delta S_{f,i}$ is the entropy of fusion of the constituents at T$_K$. ΔS_{mix} is the entropy of mixing in the liquid state and ΔS_{for} is the entropy of formation of the crystalline solid at the same temperature. At T$_K$, the entropy of fusion of the alloy is nil. ΔS_{for} and ΔS_{mix} may be inserted in eq. (2) either as experimental data, or from evaluation by means of suitable models. If $\Delta S_{f,A}$ and $\Delta S_{f,B}$ are expressed through eq. (1), using an effective value for ΔC_p of pure elements, the temperature T$_K$ which satisfies eq (2,) may be calculated for every alloy composition. Current literature evaluations [8,15] point to values of ΔS_{for} for Cu-Ti compounds from 0.5 to -3.35 J/molK. An average value of -1 J/molK was adopted. The entropy of mixing in the liquid state has been taken as a variable. The high temperature values already deviate from ideality: the excess value for the equiatomic composition is -2.3 J/molK at 1373 K. They are further decreased to more negative values to account for possible ordering on cooling. A decrease in ΔS_{mix} of the order of R (gas constant) for intermediate compositions leads to T$_K$ close to the experimental T$_g$'s. The trend of T$_K$ is sketched in Fig. 3.

In order to show the effect of the excess specific heat on the thermodynamic functions, the enthalpy of mixing, ΔH_{mix} calculated at 0K, is reported in Fig. 2 and compared with a calculation according to the Miedema model as adapted to glass forming systems [17]. The comparison shows that the Miedema calculation reproduces reasonably well the extrapolated enthalpy. The experimental data of [16], obtained at 1373 K and reassessed in [8], are also shown. The experimental heats of formation of the equilibrium intermetallic compounds [18] are reported in Fig. 2 as data points. Their distance from the curve representing the heat of mixing of the liquid, compare well with the experimental heats of crystallization that range from -4.6 to -7.2 kJ/mol in the composition range $0.66 > x_{Cu} > 0.33$.

The free enthalpies of the solid solution phases has been taken from a previous CALPHAD calculation [15] and used together with the new liquid free enthalpy to derive T_o curves.

At T_g, the difference in free enthalpy between liquid and equilibrium phases, will be lower than expected in the absence of a substantial specific heat difference, and the liquid phase may be even more stable than crystal phases such as solid solutions. In this case, solid state reactions of the pure elements below T_g may produce an amorphous solid if the equilibrium intermetallics have a slow kinetics of formation.

Our T_o curves (Fig. 4) for the terminal solid solutions agree with the equilibrium phase diagram at high temperature, but open up a field for amorphization in undercooling conditions. The amorphizing range is wide $0.4 > x_{Cu} > 0.8$, but is still different from the solid state reactions results: $0.28 > x_{Cu} > 0.75$ [17]. Our calculation is very much affected by the choice of the properties of pure elements: an error of 1 J/molK in the definition of the specific heat of liquid Ti would modify ΔG_{mix} at 300 K for $Ti_{60}Cu_{40}$ of about 1 kJ/mol. Correspondingly, the glass forming range would be changed of more than 10%.

Another source of uncertainty in the determination of the glass forming range, is the value of the free enthalpy of solid solutions, which become highly disordered during ball milling. A variation in specific heat of 1 J/molK can be produced by mechanical deformation of the elemental powders, as recently shown for Al-Zr [19], with a corresponding effect on the integral quantities. Moreover, in Cu-Ti, ΔH_{mix} is only moderately negative if compared to other glass-forming systems such as Ni-Zr, Ni-Ti and Al-Ti, and usual uncertainties acquire more relevance.

However, our calculation predicts a substantial glass-forming range with respect to previous ones. This is mostly due to the use of the experimental excess specific heat of mixing for the liquid phase. To demonstrate this, we have repeated the calculation of T_o for the bcc phase, using a regular solution description of the liquid at 1373 K [16]. The specific heat difference between the liquid and crystal states for pure elements was not taken into account, but the corrections due to the excess specific heat for alloys have been added to the free enthalpy curve. The results are given as thin T_o lines in Fig. 3. The high temperature branch crosses the entire diagram showing that the solid solution is more stable than the liquid. However, at low temperature, the hierarchy is reversed and a field of existence for a metastable amorphous phase is evidenced which did not appear in the original calculation [5].

THE Al–Ti SYSTEM

Amorphous phase formation at the $Ti_{75}Al_{25}$ composition has been experimentally verified by ball milling in the Ti-Al system [4], where no amorphization is possible by rapid quenching.

In order to discuss the formation of amorphous alloys, the free enthalpy of the phases which may be involved in solid state reactions has been calculated: i. e. the amorphous and terminal solid solutions, as the formation of intermetallic compounds is slower at the temperature of milling.

We have calculated T_o curves in undercooling regime, using the high temperature free enthalpy equations that originated from a CALPHAD calculation of the phase diagram by means of an optimization of Gibbs energies with respect to phase diagram and thermochemical data [20]. The extrapolation of the free enthalpy of the liquid away from the stable equilibrium range yields values that are too high to predict the formation of amorphous phases, in contrast with the experimental evidence. We have supposed a change in the thermodynamic properties of the liquid on cooling similar to that of Cu-Ti. The amorphous phase has been taken as an highly undercooled liquid with the addition of a further excess term which has been estimated by assuming an excess specific heat of mixing in the liquid state. No experimental data are available for Al-Ti concerning the glassy phase, so the excess specific heat has been taken of the same order as for Cu-Ti.

This may be justified on different grounds. A substantial ΔCp value can be linked through thermodynamic arguments to a negative volume of mixing [21]. The Al-Ti system contains atoms of the same size, so, amorphization contradicts the empirical Egami-Waseda formula, which states that the minimum concentration of a solute needed to form a glass in a given alloy is inversely proportional to the difference in atomic volume of the constituents [22]. This formula has recently been discussed, showing that, not only the difference in volume of the elements, but also the volume of mixing contribute to establish the limit for glass formation [21]. Although the only experimental evidence for a negative excess volume of mixing in Al-transition metal systems concerns Al-Ni, the volume effects on alloying computed using the Miedema model [23], are negative as for most glass forming systems. Therefore, it is likely that both a negative excess volume of mixing and a positive excess heat capacity pertain to Al-Ti.

In all known examples, the specific heat of metallic glass forming liquids increases sharply with undercooling; we have taken a simple hyperbolic dependence of it on temperature down to the glass transition. No glass transition temperature is available for Al-Ti alloys, so it has been estimated as the isoentropic temperature according to the procedure outlined above. Below T_g, the specific heat of the amorphous solid has been taken equal To that of equilibrium phases.

A different approach to the free enthalpy of the amorphous phase has been considered using the Miedema model. For calculations of the free enthalpy of mixing, the Miedema model considers the system at 0K, neglecting any entropy contribution. In the case of glass forming alloys, this turns out to be a good approximation for ΔG, because the entropy contribution is far from ideal, often close to zero and even negative. With both positions, the free enthalpy of the liquid phase becomes lower than that of competing terminal solid solutions at temperatures below 800 K, as in the case of Cu-Ti.

Figure 4 show the relative positions of the free enthalpies for solution phases at the temperature of inversion of stability between the liquid and crystal phases (Fig. 4a), and at the temperature of the milling experiments, 300 K (Fig. 4b). As can be seen, consideration of the temperature dependence of the thermodynamic properties of the liquid leads to an inversion in stability of the phases at low temperature, where its free enthalpy becomes lower than that of the solid solutions. Therefore, T_0 curves at low temperature should restrict to terminal composition ranges, leaving a region where the amorphous phase is the most stable one provided that the intermetallic compounds are left out, and is accessible by solid state reaction.

The results of the calculation show that the high temperature branches of the T_0 curves do not differ very much from those obtained when an excess specific heat in the liquid is not taken into account The curves extend over wide composition ranges at high temperature showing that crystallization of the liquid into solid solutions cannot be prevented by melt quenching (Fig. 5) [24]. As a consequence of the free enthalpy behaviour outlined in Fig. 4, the curves divide up and bend at intermediate temperature, sharply restricting the range of existence of the solid solutions in the metastable phase diagram. Within the composition range where the formation of an amorphous alloy is possible by solid state reaction, this may be more likely to occur in Ti-rich alloys, because of the higher glass transition temperature values, than in the Al-rich side of the diagram where the T_g's are estimated to be around room temperature. An analogous behaviour has been shown for the Al-Nb system the phase diagram of which bears some similarity to that of Al-Ti [3]. In this case the calculated free energy of the liquid results lower than that of crystalline solution phases at the equiatomic composition where amorphization in elemental multilayers has been achieved.

In this paper we have stressed the role of the temperature dependence of the thermodynamic properties of metallic liquid alloys with respect to vitrification. It is worth underlining the general nature of this behaviour which is shared by all glass forming materials [25]. For instance, in covalently bonded systems, such as those based on silica, the liquid shows a much higher specific heat than the solid phases, accounting for the high glass transition temperatures (typically $2/3\ T_f$). First measurements point to an effect of the mixing properties also in vitrification of the newly developed class of fluorozirconate glasses [26].

ACKNOWLEDGEMENTS

Work performed within the "Progetto Nazionale 40% MURST: Leghe e composti intermetallici". Useful discussions with G. Cocco, L. Schiffini, I. Soletta (Univ. Sassari), S. Enzo (Univ. Venezia), A. L. Greer (Univ. Cambridge) and N. Cowlam (Univ. Sheffield) are acknowledged.

REFERENCES

1. W. L. Johnson, Prog. Mater. Sci., 30 (1986) 81.
2. L. Schultz, Mater. Sci. Eng., 97, (1988) 15.
3. R. Bormann, F. Gärtner and K. Zöltner, J. less-common Metals, 145 (1988) 19.
4. G. Cocco, I. Soletta, S. Enzo, L. Battezzati and M. Baricco, Phil Mag. b, (1990), May issue.
5. R. B. Schwartz, P. Nash and D. Turnbull, J. Mater. Res., 2 (1987) 456.
6. G. Cocco, L. Schiffini, I. Soletta, M. Magini and N. Cowlam, J. de Phys. in press.
7. T. B. Massalski and C. G. Woychik, Acta Met., 33 (1985) 1873.
8. L. Battezzati, M. Baricco, G. Riontino and I. Soletta, J. de Phys., in press.
9. L. Battezzati and A. L. Greer, Intern. J. Rapid Solidif., 3 (1987).
10. L. Battezzati, Phil. Mag. B, (1990) May issue.

262

12. R. R. Hultgren, P. D. Desai, D. P. Hawkins, M. Gleiser, K. K. Kelley and D. D. Wagman, Selected Values of the Thermodynamic Properties of the Elements, Amer, Soc. for Metals, Metals Park, Ohio, 1973.

13. H. Fenglai, N. Cowlam, G. E. Carr and J. B. Suck, Phys. Chem. Liq., 16 (1986) 99.

14. F. Sommer, K. -H. Klappert, I. Arpshofen and B. Predel, Z. Metallknde, 73 (1982) 581.

15. N. Saunders, CALPHAD, 9 (1987) 297.

16. O. J. Kleppa and S. Watanabe, Met. Trans., 13B (1982) 391.

17. A. W. Weeber and H. Bakker, Physica B, 153 (1988) 93.

18. M. Arita, R. Kinaka and M. Someno, Metall. Trans., 10A (1979) 529.

19. H. J. Fecht, G. Han, Z. Fu and W. L. Johnson, J. Appl. Phys., 67 (1990) 1744.

20. J. L. Murray, Metall. Trans. A, 19 (1988) 243.

21. L. Battezzati and M. Baricco, J. less-common Met., 145 (1988) 31.

22. T. Egami and Y. Waseda, J. non-crystalline Solids, 64 (1984) 113.

23. A. R. Miedema and A. K. Niessen, Physica, 114B (1982) 367.

24. J. A. Graves, J. H. Perepezko, C. H. Ward and F. H. Froes, Scripta Met., 21 (1987) 567.

25. L. Battezzati and M. Baricco, Phil. Mag. B, 56 (1987) 139.

26. Present Authors, submitted to 7th Int. Conf. on Rapidly Quenched Materials, Stockholm, Aug 1990.

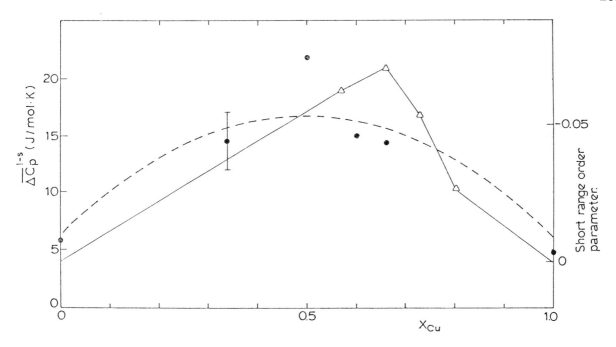

Fig. 1. Average specific heat difference between liquid and solid phases in Cu-Ti as a function of composition (black points -left scale). Short range order parameter in liquid Cu-Ti alloys (triangles-right scale), after [13].

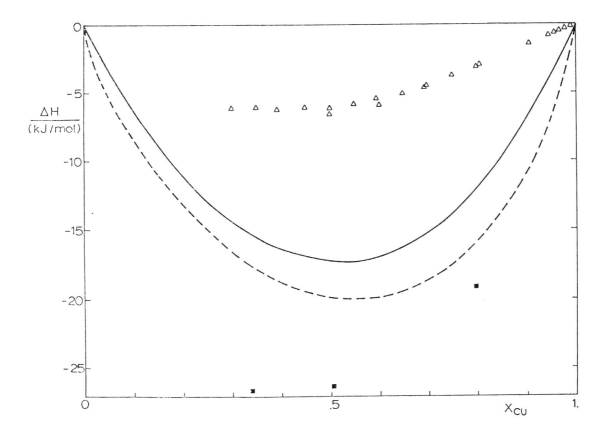

Fig. 2. Enthalpy of mixing in Cu-Ti. Triangles: experimental values determined at 1373 K [16] and reassessed in [8]. Squares: heat of formation of intermetallic compounds at 0 K [18]. Full line: calculated according to the Miedema model at 0 K [17]. Dashed line: calculated in the present work at 0 K.

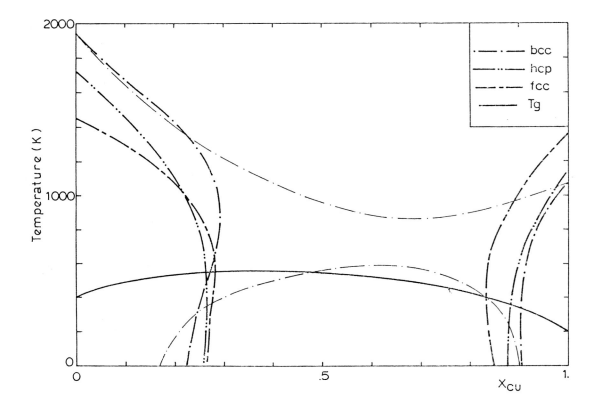

Fig. 3 . T_0 curves for the Cu-Ti system. Full line: T_g trend calculated by means of eq. (2). Thin lines: T_0 for the bcc phase calculated using a regular solution model for the liquid with the addition detailed in the text. Thick lines: T_0 curves for solution phases (see legend) calculated in the present work.

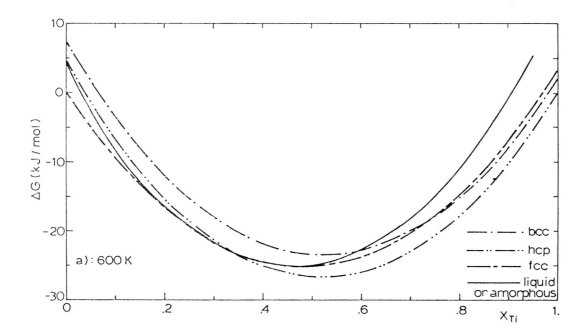

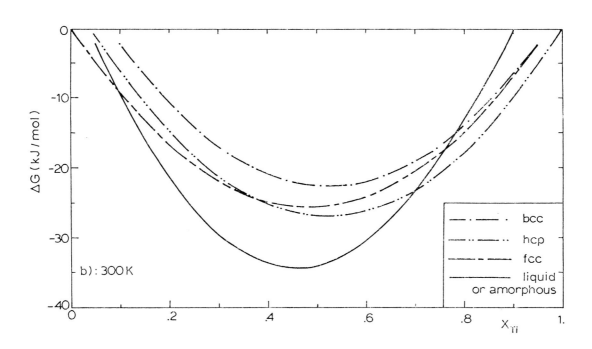

Fig. 4 . Free enthalpy curves for solution phases (see legend) in Cu-Ti at 600 K (a) and 300 K (b).

266

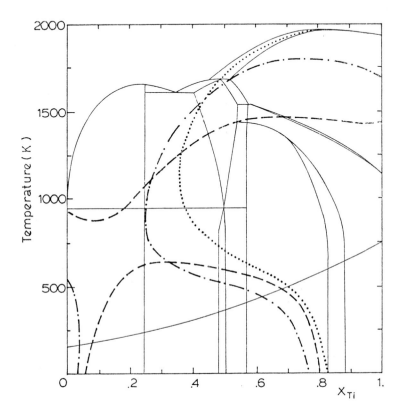

Fig. 5. To curves for Al-Ti calculated as described in the text and superimposed to the phase diagram assessed in [20]. Dotted line: bcc, dashed line: fcc, dotted-dashed line: hcp.

31

INTERACTIVE EXPERIMENTAL MEASUREMENTS AND COMPUTER CALCULATIONS OF PHASE DIAGRAMS, FOR DETERMINING A REAL THERMODYNAMIC DESCRIPTION OF BINARY SYSTEMS: APPLICATIONS TO TRANSITION ALLOYS

J. CHARLES, J. C. GACHON, M. NOTIN, M. RAHMANE AND J. HERTZ

Laboratory of Metallurgical Thermodynamics, University of Nancy I, U.R.A. CNRS 1108, B.P. 239, F 54506, Vandoeuvre-lès-Nancy - Cedex, France

ABSTRACT

The thermochemical properties of a multicomponent system are considered to be well known if a perfect consistency between experimental and calculated thermodynamic functions exists. We describe in the following the interactions between experiments and computations which take place in the normal process of studying a binary system in this laboratory.

INTRODUCTION

Experimental techniques such as calorimetry, DTA, EMF, or vapour pressure measurements, give information about the thermodynamic functions of a binary system. Our laboratory is equiped to determine enthalpies of formation of alloys by direct reaction calorimetry between components, up to 1800 K, or by indirect methods, such as dissolution in a metallic bath. Also, emf measurements allow us to study the activities of very oxidable components. But, in many cases, the data obtained are incomplete and it is not an easy task to verify the consistency between all the available experimental results.

NANCYUN is a computer program which can take into account all the known experimental data and sets of points (x, T) read from experimental phase diagram.

The Gibbs energy of formation of each phase is represented with a mathematical formalism,

$H^f - T.S^f$ for stoichiometric phases,

$$x (1 - x) \sum (a_i + b_i T) L_i(x) + R T (x \ln (x) + (1 - x) \ln (1 - x))$$

for liquid or solid solutions; $L_i(x)$ is the Legendre polynomial of degree i.

Starting from various experimental measurements: enthalpies and entropies of formation and melting partial or integral Gibbs energies and tie-lines of phase diagram, the computer is able to determine a coherent set of thermodynamic functions which permits an estimation of the thermodynamic quantities not yet experimentally obtained. The values of the thermodynamic functions given by the computer are compared with the corresponding experimental data. In order to obtain good consistency between experimental and calculated phase diagram, the user can modify, for each equation, a weighing factor chosen in respect of the information reliability and of its effect on the computed results.

SYSTEM Mg-Ga

Experimental information:
The primary information used in the study of the MgGa system is:

- H^{melt} and S^{melt} of Mg and Ga[1]
- Experimental phase diagram by Clark et al.[2]

Formation enthalpies of the five intermetallic compounds (Mg Ga is assumed to be stoichiometric)[3]
- 7 partial excess Gibbs energy of Mg in the liquid phase at T=800 K from $X_{Ga} = 0.3$ to $X_{Ga} = 0.9$[3]
- Entropies of melting of the five compounds[4]

All this information gives 34 equations and the 6 phases are represented through 22 unknown parameters: - 5 formation enthalpies and entropies of the intermetallic compounds $G^{form} = H^{form} - TS^{form}$
- 12 adjustable coefficients to represent the excess Gibbs energy of the liquid:

$$G^E = x (1 - x) \sum (a_i + b_i T) L_i (x), \text{ for } i = 0 \text{ to } 5.$$
(the solubility of Ga in the primary Mg solution is taken to be zero)

Calculation:
The results of the calculation are very satisfactory:

- the computed phase diagram is very close to the experimental one. However, as can be seen, MgGa has a congruent melting point as in the Predel and Stein diagram[5] and not a peritectic decomposition,

- the differences between excess enthalpies and entropies of the liquid phase, obtained experimentally[3] and calculated are less than 600 J/mole and 0.5 J/mole.K respectively,

- the comparison between the experimental data and the computed values for the 5 compounds shows good

Coefficients of the liquid phase*		Intermetallic compounds* refered to Mg and Ga solid						
			exp[3] -H^{form}	calc -H^{form}	exp[3] S^{form}	calc S^{form}	exp[4] S_{melt}	calc S_{melt}
a_0=-43603	b_0=11.78							
a_1= 7697	b_1=-8.34	Mg$_5$Ga$_2$	10900	10789	1.57	1.85	12.0	12.3
a_2= 9147	b_2=-6.81	Mg$_2$Ga	11700	11746	2.29	2.37	12.1	12.2
a_3= 1269	b_3=-3.67	MgGa	13000	13072	2.76	3.48	12.6	12.5
a_4= - 998	b_4= 0.74	MgGa$_2$	11400	11324	2.59	3.64	14.1	14.3
a_5= - 286	b_5= 0.71	Mg$_2$Ga$_5$	9900	10711	2.07	2.22	14.6	16.4

	Liquid phase refered to Mg and Ga liquid T=800K*					
X_{Ga}	exp[2,3] G^E_{Ga}	calc G^E_{Ga}	exp[2,3] H^E	calc H^E	exp[2,3] S^E	calc S^E
0.1		-116	-4341	-4099	-1.47	-1.42
0.2		-849	-7815	-7526	-2.64	-2.40
0.3	-2798	-2759	-10241	-10146	-3.41	-3.15
0.4	-6198	-5964	-11519	-11750	-3.74	-3.69
0.5	-10098	-10061	-11633	-12138	-3.62	-3.87
0.6	-14198	-14392	-10650	-11224	-3.11	-3.49
0.7	-18498	-18463	-8721	-9120	-2.33	-2.51
0.8	-22388	-22270	-6078	-6181	-1.44	-1.19
0.9	-26298	-26336	-3038	-2952	-0.55	-0.11

* All numerical data are given in J/mole of at. or J/(K. mole of at.).

agreement as well as for the formation and melting quantities.

SYSTEM Hf-Ni

Experimental information:
- Enthalpies of melting and melting points of Hf and Ni[1].
- Phase diagram: from the experimental phase diagram, established by P. and A. Nash[6], we use the information for the 10 invariant solid-liquid equilibria, the melting temperatures of Hf_2Ni_7[7] and of HfNi, and 5 points of the liquidus.
- Stoichiometric compounds: the enthalpies formation of 5 among the 9 compounds have been determined by a very high temperature calorimetric study[7]. In order to have a sufficient number of equations it was necessary to introduce into the computer estimates of the entropies of melting entropies for the 9 compounds. A small weighing factor has been affected to these entropy values.
- Liquid phase: only one integral enthalpy value is known[7].

Calculation:
The excess Gibbs energy of the liquid is represented with a polynomial expansion of order n = 5:

$$G^E = x\,(1-x) \sum (a_i + b_i T)\, L_i(x), \text{ for } i = 0 \text{ to } 5.$$

Coefficients			Liquid phase refered to liquid pure elements					
i	a_i	b_i	x_{Hf}	G^{mix}	H^{mix}	S^{mix}	$G^{mix}{}_{Ni}$	$G^{mix}{}_{Hf}$
0	-229416	34.25	0.1	-22210	-23574	-0.783	-2203	-202277
1	34923	-1.02	0.2	-39777	-46859	-4.064	-10439	-157127
2	-44999	21.20	0.3	-50821	-60623	-5.624	-29142	-101402
3	-115966	44.25	0.4	-54270	-60981	-3.850	-54663	-53683
4	4909	-2.31	0.5	-51387	-51269	0.068	-77298	-25477
5	20003	-7.73	0.6	-44905	-39089	3.337	-88913	-15567
			0.7	-37331	-31051	3.603	-91224	-14235
			0.8	-29255	-27772	0.851	-99899	-11594
			0.9	-18464	-21653	-1.829	-140322	- 4925
				(1743K)			(1743K)	(1743K)

The values of the thermodynamic functions concerning the stoichiometric compounds refered to both solid pure cubic elements are reported in the next table:

	H^{form}[7]	H^{form}(calc)	S^{form}(calc)	S^{melt}(calc)	T^{melt}
$Ni_{0.83}Hf_{0.17}$	-38000	-37946	-3.99	11.05	1528
$Ni_{0.78}Hf_{0.22}$	-52118	-52107	-6.44	11.79	1702
$Ni_{0.75}Hf_{0.25}$		-57408	-8.14	12.90	1639
$Ni_{0.72}Hf_{0.28}$		-60839	-9.14	13.58	1590
$Ni_{0.70}Hf_{0.30}$		-62670	-9.48	13.82	1553
$Ni_{0.59}Hf_{0.41}$		-63347	-8.34	14.81	1573
$Ni_{0.55}Hf_{0.45}$	-57277	-57338	-4.68	12.56	1655
$Ni_{0.50}Hf_{0.50}$	-51036	-51048	-1.45	11.38	1803
$Ni_{0.33}Hf_{0.67}$	-40695	-40616	-4.88	18.57	1579

CONCLUSIONS

The computed diagram seems to be similar to the experimental one, if we except a great instability of the liquid phase near $x_{Hf}= 0.65$. However this point appears also in the experimental diagram established by Svechnikov[8].

The calculated and experimental enthalpies formation of the compounds are very close to each other. But the computer program is not able to obtain the low temperature decomposition of $Ni_{0.72}Hf_{0.28}$ and $Ni_{0.70}Hf_{0.30}$ respectively at 1450K and 1290K.

The calculation of this system gives useful information which is consistent with the experimental data to the experimenter. But, with the use of the computer, it can be seen that small fluctuations the primary data induce a large modification of the entropy coefficients of all phases. In order to obtain the best representation of the thermodynamic functions, it will be desirable to have more reliable experimental data about the development of the system with temperature.

SYSTEM MoGa

This system has been studied previously, the available information is:

- Diagram of Bornand, Siemens and Oden[9] exhibiting a large solubility of molybdenum in liquid gallium,
- Diagram of Brewer and Lamoreaux[10] which has narrow solubility domain,
- The structure of some intermediate phases (GaMo, Ga_2Mo) being established by Yvon [11]. (these data were not taken into account because there is no equilibrium with the liquid),
- Calorimetric determination of the enthalpies of formation of Ga_5Mo and $GaMo_3$ [4].

In the optimization of the system, several difficulties were experienced.
(a) Difficulties appeared in drawing the liquidus line on the gallium side: a new experimental determination of the enthalpy of formation of Ga_5Mo was performed; it confirmed the first results.
(b) The optimization was not possible with the liquidus of[9]; the diagram of[10] was adopted.
(c) No evident reason exists for a demixing or a non-demixing curve of the liquidus line near $x_{Ga} = 0.70$. Until now, experimental determinations are missing and the numerical optimization is not able to distinguish between these.

Coefficients of Legendre polynomials for the liquid phase :	Coefficients of Legendre polynomials for the solid solution
$G^E = x\,(1-x)\Sigma(a_i + b_i\,T)L_i(x)$	$G^E = x\,(1-x)(a_0 + b_0\,T) + x\,(h^{tr} - Ts^{tr})$
$a_0 = -3036 \qquad b_0 = -2.86$	$a_0 = 61098 \qquad b_0 = -68.42$
$a_1 = 15572 \qquad b_1 = 9.00$	
$a_2 = -4.10^{-4} \qquad b_2 = -1.10^{-6}$	$h^{tr} = -26938 \qquad\qquad s^{tr} = -35.40$
$a_3 = -737 \qquad b_3 = -0.43$	(transition of Ga solid to Mo structure)

Formation and melting function values for stoichiometric compounds refered to Mo solid and Ga liquid					
Phase	$H^{form}[4]$	$H^{form}(calc)$	S^{form}	$S^{form}(calc)$	$S^{melt}(calc)$ $T^{melt}(calc)$
Mo_3Ga	-19000	-21000	-7.29	-6.24	20.8 2086
Mo_8Ga_{41}	-25000	-24480		-19.25	23.3 1266

SYSTEM HfPd

In this system it is impossible to obtain good compatibility between all of the experimental data and the numerical modelling. In fact, to begin with the phase diagram, two different versions have been proposed: one by Savitsky[12] and the other by Shurin and Pet'kov[13]. Several questions arise:

(a) Temperature stability of the compound $Hf_{0.33}Pd_{0.67}$ ($HfPd_2$). The enthalpy results we obtained do not permit a eutectoid decomposition[14] at low temperature as shown in Shurin and Pet'kov[13]. But the same compound is very likely to present a higher temperature of peritectoid decomposition (or melting point) than the one given by Savitsky.

(b) Solid solubility of hafnium in palladium. The solid solubility limit of hafnium in palladium is in question. The computation cannot reproduce the solubility limit at about 20 at.% and gives less than 10 at.%. The equilibrium of this solution with the liquid is given in the Shurin and Pet'kov diagram as a single two phase field while the computation gives a congruent point as shown in the figure. There is also a congruence in Savitsky's diagram, but the temperature reported is lower than the one of the nearest peritectic instead of being higher, as in our computation. Finally, on the hafnium side of the diagram, preliminary tests with Lukas' program[14] show that difficulties arise when dealing with β hafnium.

As a conclusion here we can say that further experimental work is required on this system.

Coefficients of Legendre polynomials for the liquid phase :	Coefficients of Legendre polynomials for the solid solution
$G^E = x(1-x)\Sigma(a_i + b_i T)L_i(x)$	$G^E = x(1-x)(a_0 + b_0 T) + x(h^{tr} - Ts^{tr})$
$a_0 = -268863.842 \qquad b_0 = -23.598$	$a_0 = -247893.939 \qquad b_0 = 52.2093$
$a_1 = -147162.477 \qquad b_1 = 20.73293$	$a_1 = -211772.762 \qquad b_1 = 4.9079$
	transition Hf (cc) → Hf (cfc)
	$h^{tr} = 3340 \qquad s^{tr} = 0 \qquad [15]$
References are solid Pd and β Hf at their melting points	

Formation and melting function values for stoichiometric compounds refered to solid Pd and β Hf

Phase	$H^{form}(exp)[4]$	$H^{form}(calc)$	$S^{form}(est)$	$S^{form}(calc)$	$S^{melt}(calc)$	T^{melt}
$Pd_{0.33}Hf_{0.67}$	-55000	-55024	-6	-5.372	16.584	1695
$Pd_{0.50}Hf_{0.50}$	-69000	-74235	-12	-4.320	13.809	2015
$Pd_{0.67}Hf_{0.33}$	-101450	-101760	-14	-12.691	20.827	2441
$Pd_{0.75}Hf_{0.25}$	-102220	-98815	-15	-15.276	23.208	2318

GENERAL CONCLUSION

By means of these examples it can be seen that the use of the NANCYUN program permits us to criticically assess the accepted experimental data. Finally, it gives a complete numerical representation of the binary system under study. Therefore, a sufficient amount of experimental data is available, it is possible to have reasonable estimates for the thermodynamic values which have not yet been determined experimentally and to check the consistency of what is already known.

REFERENCES

1. R. HULTGREN, P. DESAI, D. HAWKINS, M. GLEISER, K. KELLEY, Selected Values of the Thermodynamic Properties of Binary Alloys, American Society for Metals, Ohio, 1973.
2. A.A. NAYEB-HASHEMI, J.B . CLARK, Bull. of Alloy Phase Diagrams, 6 (5), 434, (1985).
3. Z. MOSER, E. KAWECKA, F. SOMMER, B. PREDEL, Met. Trans., 13B, 71, (1982).
4. D. BELBACHA, Thesis, Nancy, 1989.
5. B. PREDEL, D.W. STEIN, J. Less-Com. Met., 18, 202 (1979).
6. P. NASH, A. NASH, Bull. of Alloy Phase Diagrams, 4 (3)(1983), 250.
7. N. SELHAOUI,Thesis, Nancy, 1990.
8. V.N. SVECHNIKOV, A.K. SHURIN, G.P. DMITRIYEVA, Russ. Met. (Met), 6, 95-96, (1967).
9. J.D. BORNAND, R.E. SIEMENS and L.L. ODEN, J. Less-Com. Met., 30, (1973) 205.
10. L. BREWER and R.H. LAMOREAUX, Molybdenum: physicochemical properties of its compounds and alloys, IAEA, Vienna (1980), Atomic energy review, Special issue no.7.
11. K. YVON, Acta cryst., B30, (1974) 853; B31(1975) 117.
12. E. SAVITSKY, V. POLYAKOVA, N. GORINA, N. ROSHAU, Physical Metallurgy of Platinum Metals, MIR publishers, Moscow,1978; English translation, Pergamon Press.
13. A.K. SHURIN, V.V. PET'KOV, Russian Metallurgy, 2 (1972) 122.
14. H.L. LUKAS, Max Planck Institute, Stuttgart, RFA, private communication.
15. L. KAUFMAN, H. BERNSTEIN, Computer calculation of Phases Diagrams, Academic Press, New-York, 1970.

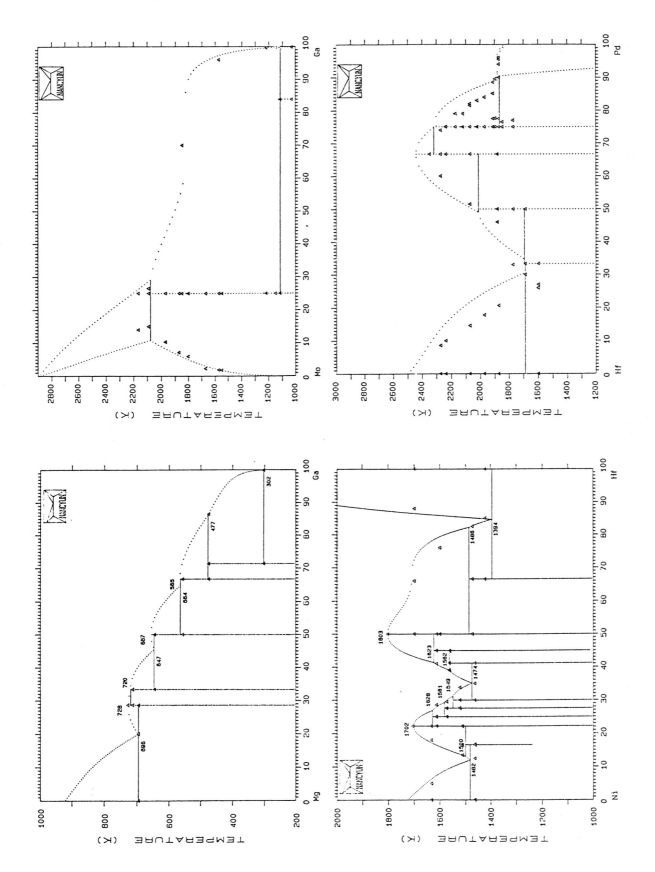

Fig. 1 Comparison between experimental points (Δ) and computed phase diagrams.

32
INFLUENCE OF Dy ON CONSTITUTION AND MAGNETIC PROPERTIES OF Fe-Nd-B BASED MAGNET COMPOSITIONS

K. Fritz, B. Grieb, E. -Th. Henig and G. Petzow

Max-Planck-Institut für Metallforschung, Institut für Werkstoffwissenschaft, PML, Heisenbergstrabe 5, 7000 Stuttgart 80, Germany

INTRODUCTION

One possible way of improving the intrinsic magnetic properties of Fe-Nd-B based magnets is to replace the Nd with Dy. The Φ phase ($Fe_{14}RE_2B$) formed from Fe, Dy and B has more than three times the coercivity of Φ formed from Fe, Nd and B [85Bol]. On the other hand the magnetization is lowered considerably by the heavy rare-earth element Dy [85Bol]. This improvement in coercivity is caused by the increased anisotropy. On the other hand changes in the microstructure are observed. The influence of additives, contaminations, and phases in the grain boundaries is known to be of major importance [84Sag]. The ternary boundary systems were investigated by Schneider et al. [86Sch] and Grieb et al. [89Gri1]. In addition Grieb has investigated the effect of substituting Nd with Dy in the Φ phase [89Gri2].

EXPERIMENTAL DETAILS

The technical alloy composition with 75 at.% Fe, 18.5 at.% RE and 6.5 at.% B was chosen. Nd was replaced by Dy in steps of 10 % or less referred to the RE content.

The alloys were prepared from materials of the following purities in mass%: Fe 99.9, Dy 99.9, Nd 99.9 and a prealloy of 18.5B, balance Fe. The specimens were melted in an arc furnace under argon. Samples were annealed at different temperatures or were investigated by DTA reaching different maximum temperatures. The formation mechanism of the Φ phase depends on the maximum temperature reached by the sample. The stable formation of Φ as a primary phase is observed only in samples with less than 60 % replacement of Nd by Dy and when the maximum temperature of the melt is kept near the formation temperature of Φ. At higher Dy contents $Fe_{17}RE_2$ is the primary phase. Superheating leads to metastable behaviour with $Fe_{17}RE_2$ as the primary phase in all alloys with subsequent formation of Φ at a temperature about 100 K lower [87Sch, 89Gri1, 89Gri2].

Special methods to prepare metallographical samples (lapping) had to be used before optical analysis and phase determination by EPMA was possible.

RESULTS

Microstructure of Annealed and DTA Samples
The Figs. 1 to 3 show the phases which appear in the DTA samples and in the annealed samples. The Φ phase always occurs. In the DTA samples with primary $Fe_{17}Nd_2$ (60 % Dy and more) Fe precipitates are found inside the Φ grains. In these samples the $Fe_{17}RE_2$ phase was formed primarily followed by transformation to Φ + Fe.

Microstructres of the Grainboundary Regions
The main phase in the grain boundary region between the Φ grains is an Nd-rich RE(Nd) phase. It contains about 97 at.% Nd, 2 at.% Dy and 1 at.% Fe. This phase separates the Φ grains from each other. In addition to the RE(Nd) phase several other phases are present:

Fe_2RE:
Two different Fe_2RE-phases can be distinguished and specified in the DTA samples.

Both phases are stoichiometric with a constant Fe:RE ratio of 2:1, first, $Fe_2(Nd)$ with a composition of 66.7 at.% Fe, 32 at.% Nd and 1.3 at.% Dy and second, Fe_2RE with the same iron content but 9.6 at.% Nd and 23.7 at.% Dy. Annealing (10 d, 1000°C) indicated $Fe_2(Nd)$ to be metastable (sample consists of $Fe_{17}RE_2$ and RE(Nd)). Annealed samples of the Fe_2RE composition are almost homogeneous.

Fe₃RE:

Fe_3RE:

In the DTA samples Fe_3RE is observed at lower Dy contents than in the annealed samples. The DTA samples are not fully in equilibrium because the grain boundary region is enriched in Dy (Fig. 4).

$Fe_{23}RE_6$:

Only in the samples with 80 to 100 % Dy does the $Fe_{23}RE_6$ phase appear.

Constitution Along Fe75 (Nd100-xDyx)18.5 B6.5

Nd and Dy are elements with great similarities. Therefore, similar phase relations and equilibria conditions can be expected. Contrary to this expectation, the results of DTA and metallographic analyses along $Fe75(Nd100-xDyx)18.5B6.5$ in the quaternary systems showed that the section has to be split into two parts: The first part, 0 to 40 % Dy (Fig. 5) can be described as a pseudoternary system but the second one, ranging from 50 to 100 % Dy is a true quaternary one (Fig. 6). In addition, in both parts stable (Figs. 5 and 6) and metastable (Figs. 7 and 8) systems have to be distinguished due to large undercooling effects in the metastable behaviour.

The phases occurring only at higher Dy amounts (Fe_2RE, Fe_3RE, $Fe_{23}RE_6$) are known from the binary Fe-Dy diagram.

Some of the phase fields are missing (such as the four phase field between $L + Fe_{17}RE_2 + Fe_{23}RE_6$ and $L + \Phi + Fe_{23}RE_6$) due to the threshold sensitivity of the DTA equipment.

SINTERED MAGNETS BY POWDER METALLURGICAL PROCESS

Magnets containing from 0 to 60 % Dy were prepared from the alloys described above by the standard methods of milling, sieving, sintering (1 h, 1050°C), and annealing (1 h, 600°C).

Microstructure

Magnets with 0 to 30 % Dy consists of Φ and a grain boundary region of RE(Nd) and $Fe_2RE(Nd)$. The volume fraction of $Fe_2RE(Nd)$ in the grain boundary region decreases with increasing Dy content.

In the magnets with 40 to 60 % Dy the Fe_2RE phase is observed in increasing amounts. The amounts of $Fe_2RE(Nd)$ and Fe_2RE in the sample with 35 % Dy are rather low and negligible. Microstructures of magnets with 10 and 50 % Dy are shown in Figs. 9a and 9b.

Magnetic Properties

In Fig. 10 the coercivity $\mu_o H_c$ is plotted as a function of the Dy amount. At 35 % Dy the coercivity shows a maximum of 3.2 T. This is combined with a high value of B_R of 0.8 T (Fig.11). The $\mu_o H_c$ vs. T relation that is characteristic of a material with 35 % Dy can be seen in Fig. 12. $\mu_o H_c$ decreases linearly with increasing temperature and therefore the temperature coefficient is improved compared to the Fe-Nd-B material [89Sch].

SUMMARY

Magnets based on Fe-Nd-B show improved coercivities combined with a decrease in remanence due to Dy additions. Highest values of coercivity are found in magnets with 30 to 40 % Dy. The coercivity maximum is found at about 35 % Dy and coincides with the occurrence of the Fe_2RE and the $Fe_2RE(Nd)$ phases. These phases are magnetically soft and act as nucleation centres in the material. Temperature dependence of coercivity is linear .

REFERENCES

[85Bol] E. B. Boltich, E. Oswald, M. Q. Huang, S. Hirosawa and W. A. Wallace, J. Appl. Phys. 57 (1985) 4106.
[84Sag] M. Sagawa, S. Fujimura, N. Togawa, H. Yamamoto, Y. Matsuura, J. Appl. Phys. 55 (1984) 2083.
[86Sch] G. Schneider, E. -Th. Henig, G. Petzow and H. H. Stadelmaier, Z. Metallkde. 77 (1986) 755.
[89Gri1] B. Grieb, E. -Th. Henig, G. Schneider and G. Petzow, Z. Metallkde. 80 (1989) 95.
[89Gri2] B. Grieb, E. -Th. Henig, G. Schneider and G. Petzow, Z. Metallkde. 80 (1989) 515.
[89Sch] G. Schneider, E. -Th. Henig, B. Grieb und G. Knoch, in "Concerted European Action on Magnets (CEAM)", edited by I. V. Mitchell, J. M. D. Coey, D. Givord, I. R. Harris and R. Hanitsch, Elsevier Science Publishers Ltd., Essex, UK (1989) 335.

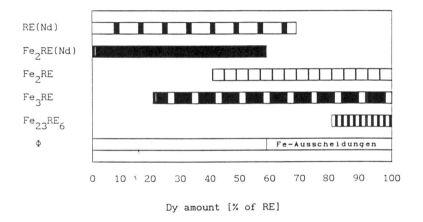

Fig. 1. Phases appearing in DTA samples.

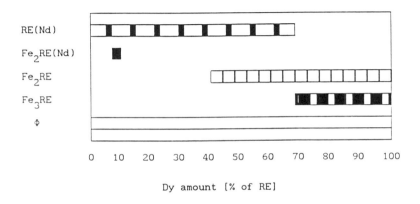

Fig. 2. Phases appearing in annealed samples (950 °C, 9 d).

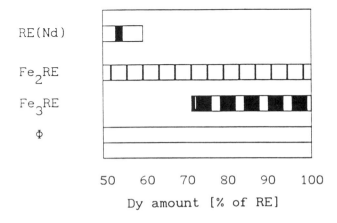

Fig. 3. Phases appearing in annealed samples (800 °C, 14 d).

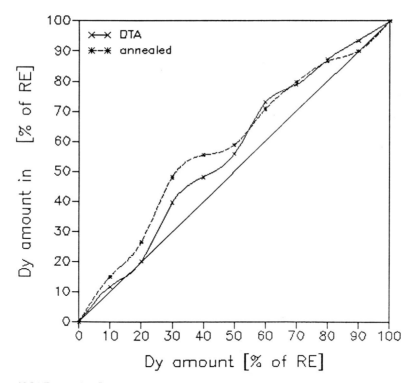

Fig. 4. Ratio of Nd:Dy in the Φ phase in dependence on the percentage amount of Dy in the sample.

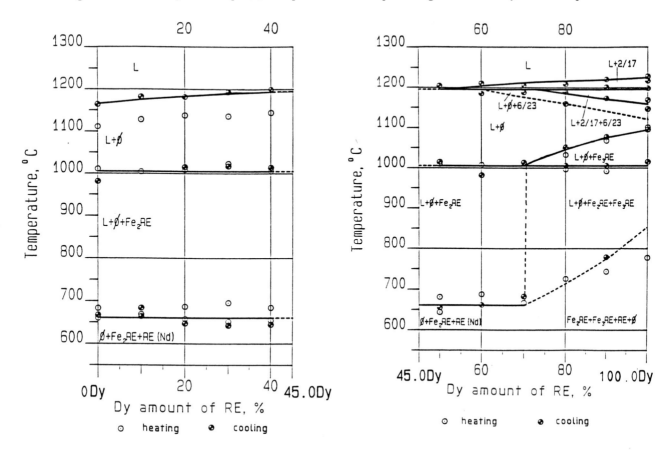

Fig. 5. Temperature-concentration diagram, concentration range 0 to 40 % Dy, equilibrium state.

Fig. 6. Temperature-concentration diagram, concentration range 50 to 100 % Dy, equilibrium state.

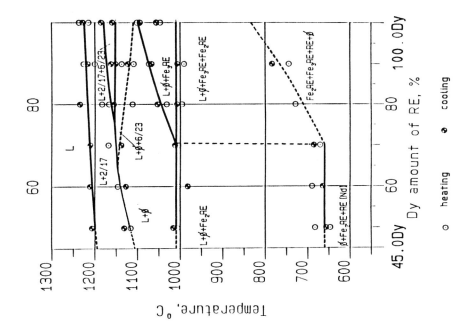

Fig. 8. Temperature-concentration diagram, concentration range 50 to 100 % Dy, metastable state.

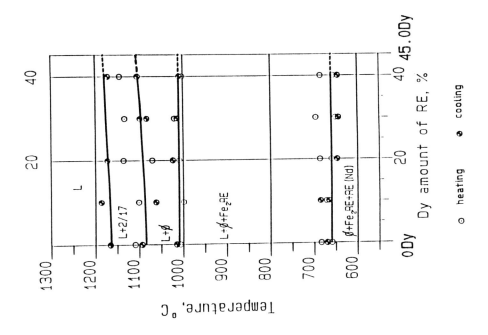

Fig. 7. Temperature-concentration diagram, concentration range 0 to 40 % Dy, metastable state.

278

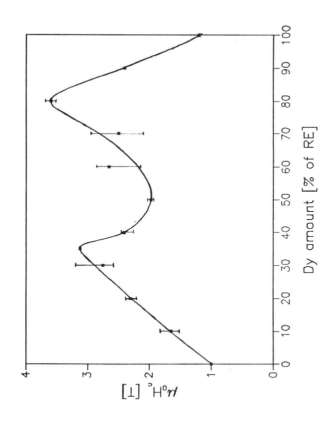

Fig. 10. Dependence on $\mu_0 H_c$ on the Dy amount.

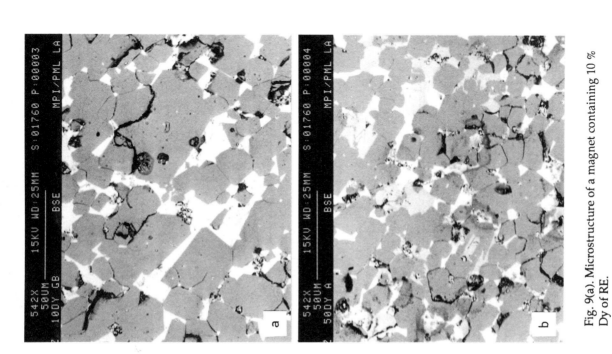

Fig. 9(a). Microstructure of a magnet containing 10 % Dy of RE.

Fig. 9(b). Microstructure of a magnet containing 50 % Dy of RE.

279

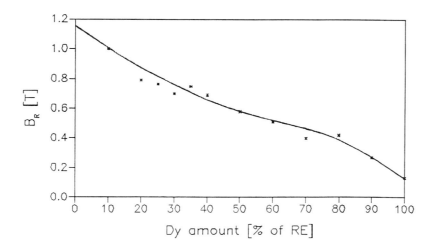

Fig. 11. Dependence of B_R on the temperature.

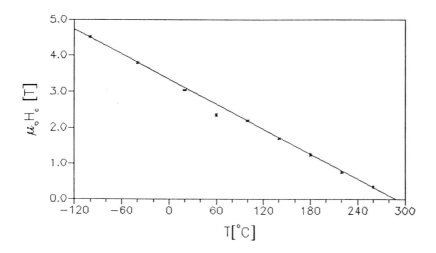

Fig.12. Dependence of $\mu_o H_c$ on the temperature.

33
ASSESSMENT OF THE Al-Zn-Sn SYSTEM

S. G. Fries, H. L. Lukas, S. Kuang and G. Effenberg*

Max-Planck-Institut für Metallforschung, Institut für Werkstoffwissenschaft, PML,
Heisenbergstr. 5, D-7000 Stuttgart 80, Germany
*Materials Science International Services, GmbH, Data and Knowledge Base for Alloy Development,
Heisenbergstr. 5, D-7000 Stuttgart 80, Germany

ABSTRACT

The ternary Al-Zn-Sn phase diagram was first extrapolated from the already calculated binary systems. It was optimised taking into account the ternary experimental data found in the literature. The ternary phase diagram extrapolated from the binaries is already a good approximation of the optimised diagram.

1 INTRODUCTION

The thermodynamic characteristics of alloys containing aluminium have both theoretical and practical importance because their knowledge enables the optimisation of the steps necessary for light alloys production and design. Al-Zn alloys have many industrial aplications and their phase diagram was determined after many controversies. The sensitivity of Al-Zn alloys to impurities such as Sn makes the Al-Zn-Sn ternary system interesting.

The computer is a powerful tool for the optimisation of thermodynamic descriptions. In this laboratory software has been developed [77Luk], which takes into account all the experimental data found in the literature, phase diagram data as well as calorimetric, e.m.f. or vapour pressure measurements. Model descriptions are chosen for the thermodynamic properties of all phases involved and their adjustable coefficients are determined by the least squares method.

One important feature of this procedure is, that the thermodynamic descriptions of multicomponent systems can be extrapolated from those already calculated for their subsystems.

Secondly it enables one to estimate safely equilibria e. g. solubilities by extrapolation into temperature ranges, where diffusion is too slow for equilibrium to be reached. Finally, the method enables checking of the consistency between the different kinds of experimental data. Therefore this optimisation procedure must start with the binary systems.

The Al-Zn binary was optimised by [92anM], and the Zn-Sn and Al-Sn systems were optimised by [92Fri].

2 LITERATURE EXPERIMENTAL DATA

2.1 *Phase diagram data*

The ternary phase diagram was initially investigated by thermal analysis of the liquidus surface in the whole concentration triangle by Plumbridge [11Plu]. A ternary eutectic point was reported at 198°C and 3.0 at.% Al, 10.5 at.% Zn, 86.5 at.% Sn. [23Cre], [23Los], [68Pro] and [75Nay] again studied the ternary liquidus surface by thermal analysis. The shape of the liquidus isotherms in these investigations is in general agreement although slight differences do exist. Regarding the concentration of the eutectic melt [23Cre] found primary crystals of the Al solid solution in an alloy, where [11Plu] gave the composition of the ternary eutectic and therefore assumed a much lower Sn content of the ternary eutectic liquid. [69Pro], however, precisely determined the composition of the eutectic melt by zone melting to be in mass-%: 0.6 Al, 7.7 Zn and 91.7 Sn, which is in mole-%: 2.4 Al, 12.9 Zn and 84.7 Sn and thus is very near to the composition given by [11Plu]. [75Nay] agrees with the Sn concentration to be 2.5 mole %, but reports 15 mole % Zn.

Solid solubilities were not mentioned by the cited authors. [80Vin], [81Vinl], [81Vin2] and [87Seb] reported a large solubility of Sn in the Zn rich aluminium solid solution, (Al'). In a vertical section from $Al_{28}Zn_{72}$ to Sn the boundary between the (Al')+(Al") two phase field and the L+(Al')+(Al") three phase field is drawn in a way that the derivative $dT/d\left(x_{Sn}^{Al'}\right)$ is nearly equal to zero, which is very doubtful as explained in section 3.2.

2.2 *Gibbs energy data*

[65Tik] measured the e.m.f. of the reversible galvanic cell: ternary alloy against pure Al in a molten salt of equimolar amounts of NaCl and KCl with an addition of 2 % $AlCl_3$. Twenty-four alloys were investigated in the temperature range 973 to 1073 K.

3 STRATEGY OF THE CALCULATION

3.1 Models

The pure solid elements in their states stable at 298.15 K were chosen as the reference states of the system. (Stable Element Reference, SER). The Gibbs energies of the elements as functions of the temperature were compiled by Dinsdale [89Din]. The equations appear in the form:

$$G^{\phi} - H^{SER} = a + b \cdot t + c \cdot T \cdot \ln T + d \ T^2 + e/T + f \cdot T^3 + i \cdot T^7 + j \cdot T^9$$

where in different temperature ranges different sets of the coefficients a to j may be used.

The Gibbs energies of the phases as functions of the concentration were represented by the Redlich-Kister-Muggianu formalism:

$$G^{\phi} - H^{SER} = {}^{ref}G + {}^{ideal}G^{\phi} + {}^{Ebin}Gf + {}^{Eter}G^{\phi}$$

with

$$
{}^{ref}G = \left[{}^{o}G_{Al}^{ref\ 1}\ (T) - H_{Al}^{SER}\ (298.15\ K)\right] \cdot x_{Al} +
$$
$$
\left[{}^{o}G_{Zn}^{ref\ 2}\ (T) - H_{Sn}^{SER}\ (298.15\ K)\right] \cdot x_{Zn} +
$$
$$
\left[{}^{o}G_{Sn}^{ref\ 3}\ (T) - H_{Sn}^{SER}\ (298.15\ K)\right] \cdot x_{Sn}
$$

$$
{}^{ideal}G^{\phi} = R \cdot T \left[x_{Al} \cdot \ln (x_{Al}) + x_{Zn} \cdot \ln (x_{Zn}) + x_{Sn} \cdot \ln (x_{Sn})\right]
$$

$$
{}^{Ebin}G^{\phi} = \sum_{i=1}^{2} \sum_{j=i+1}^{3} x_i \cdot x_j \cdot \sum_{v=0}^{mij} K_{i\ j\ v} (x_i - x_j)^v
$$

$$
{}^{Eter}G^{\phi} = x_{Al} \cdot x_{Zn} \cdot x_{Sn} \cdot \left[K_{Al} \cdot x_{Al} + K_{Zn} \cdot x_{Zn} + K_{Sn} \cdot x_{Sn}\right]
$$

The variables in the above equations mean:

G^{ϕ} molar Gibbs energy of phase ϕ.

x_{Al}, x_{Zn}, x_{Sn} gross mole fractions of the phase.

H_i^{SER} (298.15 K) enthalpy of the pure element i in its stable state at the reference temperature 298.15 K

H^{SER} is an abbrevation of H_{Al}^{SER} (298.15 K)] $\cdot x_{Al} + H_{Zn}^{SER}$ (298.15 K)] $\cdot x_{Zn} + H_{Sn}^{SER}$ (298.15 K)] $\cdot x_{Sn}$

${}_{o}G_{Al}^{ref\ 1}$ (T) Gibbs energy of pure Al at temperature T in the state of phase ref 1.

${}_{o}G_{Zn}^{ref\ 2}$ (T) Gibbs energy of pure Zn at temperature T in the state of phase ref 2.

${}_{o}G_{Sn}^{ref\ 3}$ (T) Gibbs energy of pure Sn at temperature T in the state of phase ref 3.

m_{ij} degree of the Redlich-Kister polynomial of the binary i-j-system (i, j = 1, 2, 3 correspond to Al, Zn, Sn)

$K_{ij\nu}$ binary Redlich-Kister parameter, given as linear function of the temperature $K_{ij\nu} = a_{ij\nu} + b_{ij\nu} \cdot T + c_{ij\nu} \cdot T \cdot \ln T$

K_{Al}, K_{Zn}, K_{Sn} ternary parameters, given as linear functions of T ($K_i = a_i + b_i \cdot T$)

3.2 Van't Hoff- and Scheil-Clausius-Clapeyron Equations

[80Vin] concluded from their results that a large solubility of Sn exists in the Zn-rich (Al,Zn) solid solution above the eutectoid temperature. [11Plu] and [23Cre] found the temperature of the ternary four phase equilibrium between liquid, (Al'), (Al") and (Zn) near 265°C, which is nearly identical to the binary eutectoid temperature, in the Al-Zn system. Scheil [49Sch] deduced an equation, which he called by the generalized Clausius-Clapeyron equation, for the slope of the boundary lines of a three phase equilibriun in a ternary system. This equation is formally identical to van't Hoff's equation for the freezing point depression of a dilute solution [81Goo]:

$$\frac{dT}{dx_{Sn}^{Al'}} = \frac{R \cdot T^2}{\Delta H}$$

where ΔH is the enthalpy change during the eutectoid reaction. It implies that if the temperature of the four phase equilibrium is near to that of the binary three phase equilibrium, then the Sn concentration of (Al') cannot be significantly larger than those of the (Al") and (Zn) phases, which are known to be small from the binary Al-Sn and Zn-Sn systems.

[80Vin] tried to combine a large Sn solubility above the eutectoid temperature with a negligible Sn solubility at the eutectoid temperature by an extremely retrograde solubility of Sn in the (Al,Zn') phase. The strongly retrograde solubility, however, seems to be impossible, because when its lower part is translated to an Arrhenius plot (ln(x) vs. 1/T) a line extremely far from a straight line is obtained.

Therefore in the present optimisation the values of [80Vin] were ignored and the Sn solubility in the whole (Al,Zn) solid solution was extrapolated from the binary Al-Sn and Zn-Sn systems by providing no ternary excess term (^{Eter}G in eq. (1)) for this phase.

3.3 Selection of the Adjustable Parameters

If the only experimental data for a phase are Gibbs energies in a narrow range of temperatures, then it is not possible to independently adjust the enthalpy and entropy parameters. Liquidus temperatures are roughly equivalent to partial Gibbs energies.

In the present system the temperature, where [65Tik] measured μ_{Al} values, is not far enough from the liquidus to safely determine a temperature dependence of the Gibbs energy. Therefore the parameters K_i of the term ^{Eter}G in eq. (1) cannot be split into two independent coefficients a_i and b_i. During the optimisation therefore a constraint between a_i and b_i must be applied to have virtually one coefficient only.

For this constraint an estimate of Kubaschewski [79Kub], which recently was refined by Tanaka et al. [90Tan], is valid. This estimate gives a relationship between the maximal values of the excess enthalpy and excess entropy of a liquid. Kubaschewski assumed proportionality between both values with a universal factor of about 3000 K, whereas after [90Tan] this factor is system dependent and for binary systems is approximated by $14 \cdot (1/T_1 + 1/T_2)^{-1}$, where T_1 and T_2 are the melting temperatures of the two elements. The logical extension of this factor into ternary systems is: $21 \cdot (1/T_1 + 1/T_2 + 1/T_3)$. For Al-Zn-Sn with the melting temperatures 934, 693 and 505 K the factor is 4673 K.

For the optimisation the ternary term ^{Eter}G was modelled with K_{Al}, K_{Zn} and K_{Sn} given as linear functions of the temperature T, $K_i = a_i + b_i \cdot T$, but with the constraints $a_i = b_i \cdot 4673$ K.

If the work of [80Vin] is neglected, as discussed in section 3.2, then the solid phases have small solubilities in the ternary. Therefore in their thermodynamic descriptions a ternary term ^{Eter}G is insignificant since in the common factor $x_{Al} \cdot x_{Zn} \cdot x_{Sn}$ always one of the x_i is very small.

For the fcc (Al) solid solution the contribution to ^{Ebin}G from the Zn-Sn binary system is missing, because there is no fcc Zn-Sn phase. As an estimate the missing term was replaced by the excess term of the binary hcp Zn-Sn phase. For the (Zn) and (Sn) solid solutions the Al-Sn and Al-Zn terms respectively are missing. Since the two phases have small solubilities for both of the other elements, in the whole ranges of homogeneity in eq. (1) these terms are multiplied with the product of two very small factors $x_i \cdot x_j$. Therefore they are insignificant and may be neglected.

4 RESULTS

The three independent coefficients selected for the ternary term ^{Eter}G were adjusted by the least squares method using the liquidus temperatures of [11Plu], [23Cre], [23Los] and [68Pro], the eutectic concentrations of [69Pro] and the μ_{Al}^{liq} values of [65Tik]. The data of [80Vin] were not used for the reasons discussed in section 3.2. The values of [75Nay] could not be used as he gave only a liquidus surface plot, from which the temperatures of the experimental points cannot be read satisfactorily. Plotting the data showed, that the temperatures given by [23Cre] are significantly lower than those of the other authors. Therefore the calculation was repeated without the data of [23Cre].

Tables 1 and 2 present the unary and binary coefficient sets, respectively, used in the calculation. Table 3 presents the set of optimised ternary excess coefficients.

Figure 1 shows the optimised liquidus surface. The lines calculated from the description without the ternary term ^{ter}G are drawn by dashed lines. In the vertical sections, Figures 2 to 4 some of the experimental points are compared with the calculated diagram. Figure 5 shows the partial Gibbs energy of Al in liquid at two different temperatures with experimental values. In Fig. 6 Scheil's reaction scheme is drawn.

The deviation of the optimised ternary phase diagram from that thermodynamically extrapolated from the binary subsystems is nearly within the scatter of the experiments.

ACKNOWLEDGEMENTS

One of the authors (S.G.F) thanks the International Scientific Cooperation EC - Brazil and the CNPq agency for financial support during her postdoctoral fellowship at the Max-Planck-Institut at Stuttgart.

Financial support by the BMFT (Bundesministerium für Forschung und Technologie) in the COST 507 project (03K07093) is gratefully acknowledged.

REFERENCES

[11Plu] D. V. Plumbridge. "On the Binary and Ternary Alloys of Al, Zn, Cd and Sn" (In German), thesis, University of Munchen, Germany (1911).

[23Cre] E. Crepaz, "The Ternary Al-Zn-Sn System" (In Italian), Giorn. Chim. ind. appl. 6, 285-290 (1923).

[23Los] L. Losana and E. Carozzi, (In Italian), Gazz. chim. ital. 53, 546-554 (1923).

[49Sch] E. Scheil, "The Law of Freezing Point Depression of Binary Eutectic Melts by Small Additions of a Third Element" (In German), Z. Metallkde. 40, 246-248 (1949).

[65Tik] A. A. Tikhomirov and I. T. Sryvalin, "Investigations of the Properties of Liquid Alloys of the Al-Zn-Sn System by the E.M.F. Method" (In Russian), Sb. Nauch. Tr., Perm. Politekh. Inst No. 18, 193-198 (1965).

[68Pro] S. Prowans and M. Bohatyrewicz, "The Al-Sn-Zn System" (In Polish), Arch. Hutn. 13, 217-233 (1968).

[69Pro] S. Provans and M. Bogatyrevich, "Determining Eutectic Compositions by the Zone Melting Method", translated from Zavod. Lab., (1) 62-63 (1969).

[75Nay] A. K. Nayak, "Constitution of Ternary Al-Zn-Sn Alloys determined by an Iso-Peribol Calorimeter", Trans. Indian Inst. Met. 28, 148-153 (1975).

[77Luk] H. L. Lukas, E.-Th. Henig and B. Zimmermann, "Optimization of Phase Diagrams by a Least Squares Method using Simultaneously Different Types of Data", Calphad 1, 225-236 (1977).

[79Kub] O. Kubaschewski and C. B. Alcock, "Metallurgical Thermochemistry", Pergamon Press, New York (1979) 55.

[81Goo] D. Goodman, J. W. Cahn and L. Bennet, "The Centennial of the Gibbs-Konovalov Rule", Bull. Alloy Phase Diagram 2, 29-34 (1981).

[80Vin] D. Vincent, "Contribution to the Study of the Ternary Al-Zn-Sn System"(In French), Thesis No. 84, Université Claude Bernard, Lyon I, France, 1980.

[81Vin1] D. Vincent and A. Sebaoun "Study of the Al-Zn-Sn Ternary System" (In French), J. Therm. Anal. 20, 419-433 (1981).

[81Vin2] D. Vincent and A. Sebaoun "Interpretation of the Al-Zn-Sn Phase Diagram: Mechanism of Degeneration of Zn-Al Alloys by Sn" (In French), Mém. Etud. Sci. Rev. Métall. 78, 165-172 (1981).

[87Seb] A. Sebaoun, D. Vincent and D. Tréheux, "Al-Zn-Sn Phase Diagram — Isothermal Diffusion in Ternary Systems", Mat. Sci, Tech. 3, 241-248 (1987).

[89Din] A. T. Dinsdale, "SGTE Data for Pure Elements", NPL Report DMA(A) 195 (1989).

[90Tan] T. Tanaka, N. A. Gokcen and Z. Morita, "Relationship between Partial Enthalpy of Mixing and Partial Excess Entropy of Solute Elements in Infinitely Dilute Solutions of Liquid Binary Alloys", Z. Metallkde. 81, 349-353 (1990).

[92anM] S. an Mey, to be published.

[92Fri] S. G. Fries and H. L. Lukas, to be submitted to Calphad.

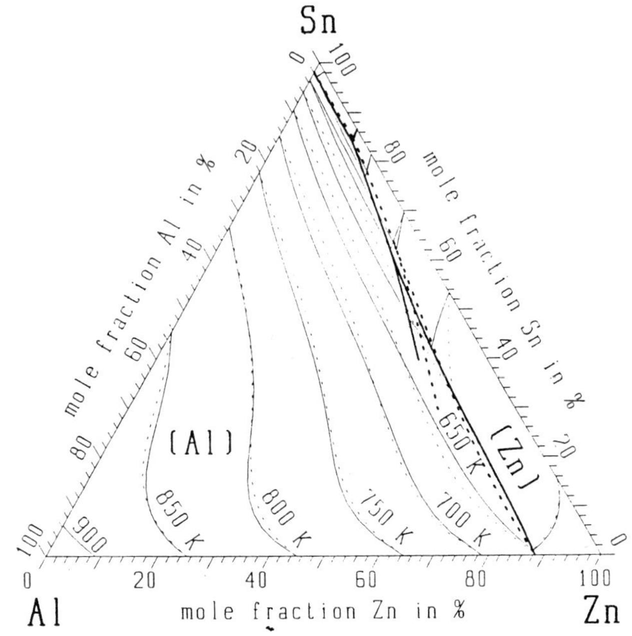

284

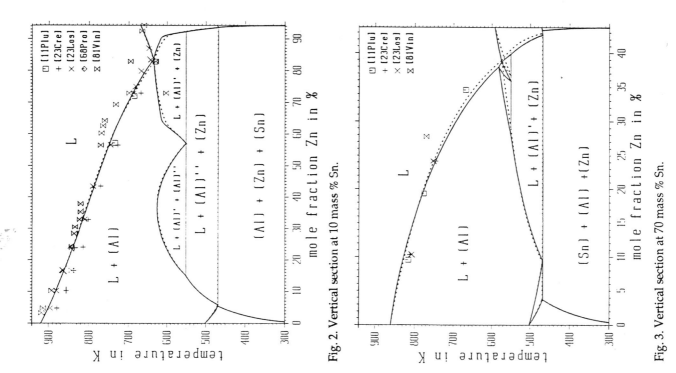

Fig. 2. Vertical section at 10 mass % Sn.

Fig. 3. Vertical section at 70 mass % Sn.

Fig. 1. Liquidus surface. Full lines optimised. Dashed lines extrapolated (without term EterG).

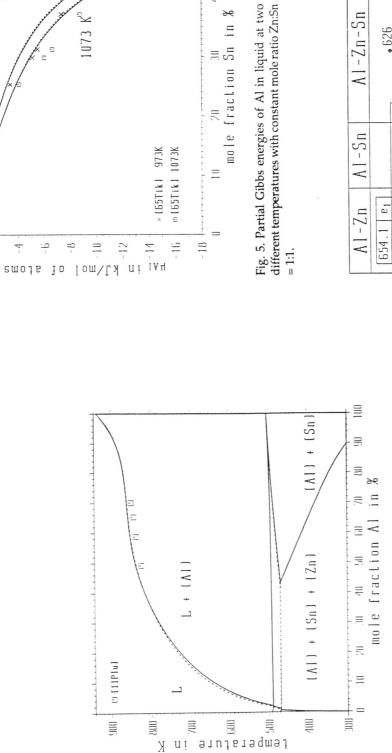

Fig. 5. Partial Gibbs energies of Al in liquid at two different temperatures with constant mole ratio Zn:Sn = 1:1.

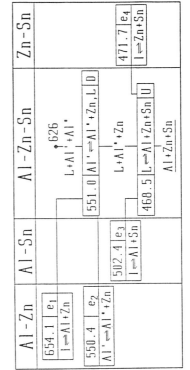

Fig. 6. Scheil's reaction scheme.

Fig. 4. Vertical section at constant mole ratio Zn:Sn = 51:949.

Table 1 Gibbs energies of the unary phases [89Din], given in J/mol of atoms

$$G^\phi(T) - H^{SER} = a + b\cdot T + c\cdot T\cdot \ln T + d\cdot T^2 + e/T + f\cdot T^3 + i\cdot T^7 + j\cdot T^{-9}$$

	a	b	c	$d\cdot10^3$	e	$f\cdot10^6$	$i\cdot10^{20}$	$j\cdot10^{-26}$
Al(fcc)								
298 to 700 K	-7976.15	137.07154	-24.367193	-1.884662	74092.	-0.877664		
700 to 933 K	-11276.24	223.02695	-38.584430	18.531982	74092.	-5.764227		
933 to 2900 K	-11277.68	188.66199	-31.748192					-123.4
Al(liq)								
298 to 700 K	3029.40	125.23067	-24.367198	-1.884662	74092.	-0.877664	7.940	
700 to 933 K	-270.69	211.18608	-38.584430	18.531982	74092.	-5.764227	7.940	
933 to 2900 K	-795.71	177.40997	-31.748192					
Al								
(hcp) - (fcc)	5481.00	-1.80000						
(bcc) - (fcc)	10083.00	-4.81300						
Zn(hcp)								
273 to 693 K	-7285.79	118.46927	-23.701314	-1.712034		-1.264963		
above 693 K	-11070.60	172.34491	-31.380000					4.70657
Zn(liq)								
273 to 693 K	-128.52	108.17693	-23.701314	-1.712034		-1.264963	-35.8652	
above 693 K	-3620.47	161.60868	-31.380000					
Zn								
(bcc) - (hcp)	2886.96	-2.51040						
(fcc) - (hcp)	2969.82	-1.56968						
Sn (bct)								
250 to 550 K	-5855.14	65.42789	-15.961000	-18.870000	-61960.	3.121167		
550 to 800 K	2524.72	3.98985	-8.259049	-16.814429	-1081244.	2.623131		
above 800 K	-8256.96	138.98146	-28.451200					-0.12307
Sn (liq)								
250 to 550 K	1247.96	51.34012	-15.961000	-18.870000	-61960.	3.121167	147.031	
550 to 800 K	9496.31	-9.82454	-8.259049	-16.814429	-1081244.	2.623131	147.031	
above 800 K	-1285.37	125.16707	-28.451200					
Sn								
(fcc) - (bct)	4150.00	-5.20000						
(hcp) - (bct)	3900.00	-4.40000						

Table 2 Adjusted excess parameters for the binary systems, given in J/mol of atoms

$$K_{ij\nu} = a_{ij\nu} + b_{ij\nu}\cdot T + c_{ij\nu}\cdot T\cdot \ln T$$

Phase	System $i - j$	ν	$a_{ij\nu}$	$b_{ij\nu}$	$c_{ij\nu}$	Reference
Liquid	Al-Zn	0	10465.55	-3.39259		[92anM]
	Al-Sn	0	16329.85	-4.98306		[92Fri]
		1	4111.97	-1.15145		
		2	1765.43	-0.57390		
	Zn-Sn	0	19314.64	-75.89949	8.751396	[92Fri]
		1	5696.28	-4.20198		
		2	1037.22	0.98362		
(Al)(fcc)	Al-Zn	0	7297.48	0.47512		[92anM]
		1	6612.88	-4.59110		
		2	-3097.19	3.30635		
	Al-Sn	0	45297.84	8.39814		[92Fri]
	Zn-Sn	0	33433.94	-11.14466		as hcp
(Zn)(hcp)	Al-Zn	0	18820.95	-8.95255		[92anM]
		3	-702.79			
	Zn-Sn	0	33433.94	-11.14466		[92Fri]
(Sn)(bct)	Al-Sn	0	14136.95	-4.71231		[92Fri]
	Zn-Sn	0	6514.76	25.70957		[92Fri]

Table 3 Coefficient set for Al-Zn-Sn system, given in J/mol of atoms

Term ^{Eter}G of eq. (1)

$$^{Eter}G^\phi = x_{Al}\cdot x_{Zn}\cdot x_{Sn}\cdot [(a_{Al}+b_{Al}\cdot T)\cdot x_{Al} + (a_{Zn}+b_{Zn}\cdot T)\cdot x_{Zn} + (a_{Sn}+b_{Sn}\cdot T)\cdot x_{Sn}]$$

Phase	Index	a	b
Liquid	Al	-2777.03	.59427
	Zn	15225.63	-3.25821
	Sn	-16198.13	3.46632

34

ChemSage : A COMPUTER PROGRAM FOR THERMOCHEMICAL APPLICATION CALCULATIONS

G. Eriksson and K. Hack*

CRCT, Ecole Polytechnique, Montreal, Canada
*Theoretische Huettenkunde, RWTH Aachen, Germany

INTRODUCTION

ChemSage is derived from the widely renowned SOLGASMIX[1,2]. It is a powerful and flexible computer program, which enables phase equilibria and energy changes associated with a wide variety of material problems to be investigated using thermodynamic calculation techniques. It is designed to be user friendly and provide research and industry with a highly versatile tool for daily use in practical applications of thermochemistry.

FIELDS OF APPLICATION

ChemSage is applied in such varied fields as, for example:

- alloy development (steels, superalloys, light-metal alloys)
- semiconductor and superconductor production
- environmental pollution studies
- toxic and non-toxic waste disposal
- recycling operations
- impurity distribution between liquid metal and slag
- stable and metastable coating deposition
- geochemical phenomena

It provides considerable savings in time and costs generally associated with exclusive use of experimental investigations to resolve these problems.

ChemSage is also used by industry to ensure minimum wastage of materials and energy in optimizing existing production techniques and in the development of new processes to manufacture products of a desired composition and purity.

FEATURES OF THE PROGRAM

Features of ChemSage include a library of models for the description of

- non-ideal gases
- dilute alloy solutions
- concentrated alloys (both solid and liquid)
- non-stoichiometric phases
- liquid slags
- concentrated aqueous solutions
- magnetic transformations

By selecting a particular model for a specific phase, the user has the capability to

- calculate phase equilibria rapidly in multiphase, multicomponent systems
- calculate equilibrium or non-equilibrium thermodynamic function values for specified components in a defined phase of a complex system
- simulate a multi-stage reactor with the possibility of defining energy and materials flow between stages.

REQUIREMENTS FOR USE OF ChemSage

ChemSage is available for main-frame computers or IBM compatible PCs.

Hardware requirements for use on PCs are:

> At least 500 kBytes of free CPU space
> A mathematical co-processor
> About 3 MBytes free space on harddisk
> One floppy drive; 5.25 inch/1.2 MBytes or 3.5 inch/720 kBytes
> An extended memory of 4 MBytes CPU will improve the processing speed significantly.

ChemSage is compatible with SGTE and other databases, as well as customised subsets and datafiles compiled to customer requirements, which can be supplied additionally with either version. In addition, the user can create and use his own datafiles with ChemSage.

For implementation, a set of files is supplied made up of three distinct groups.

1. Files related to SAGE itself
Main-frame : FORTRAN77 source files
PC : executable code

2. Files containing data examples

3. Files containing explanatory texts

for   Installation of ChemSage on to a PC consists of loading the executable code into the subdirectory Sage; a main-frame, it consists of loading, compiling and linking the source files to generate an executable module. After installation, the data and text files serve to check proper performance of the program. Additionally the data files serve as masterfiles to help create user data files based on self-generated thermodynamic data. The text files also serve as a user manual.

THE GENERAL STRUCTURE OF CHEMSAGE

In terms of the calculation capabilities of ChemSage, three major modules may be distinguished. These comprise the calculation of thermodynamic function values, complex heterogeneous phase equilibria, and material flow in simple multistage reactors. Figure 1 shows schematically the different program-blocks of ChemSage.

MAIN Program
The MAIN program serves primarily as the input module to feed the thermodynamic data for any system under consideration into the program, either by reading a ready made datafile or by assembling data manually, using a dialogue oriented routine. A datafile for further use can thus be created. For actual calculations, the user branches off into the appropriate modules.

THERMODYNAMIC FUNCTIONS Module
Many thermodynamic properties of solutions can be derived from the Gibbs energy by taking the appropriate temperature and composition derivatives. In the routines for thermodynamic function calculations the following quantities are calculated: integral and partial values of specific heat, enthalpy, total or excess entropy, and total or excess Gibbs energy for single solution phases and specific heat capacity, enthalpy, entropy, and Gibbs energy for stoichiometric condensed phases.

PHASE EQUILIBRIUM Module
The phase equilibrium routines enable a user to predict the chemical equilibrium state of a system which has been uniquely defined with respect to temperature, pressure or volume, and total amounts and/or equilibrium activities for independent system components. A calculation can also be controlled by the change of an extensive property of the system, e.g. enthalpy to calculate an adiabatic flame temperature or volume to calculate a vapour pressure.

The computational efficiency and convergence reliability of ChemSage are enhanced through a routine which provides close starting estimates of the equilibrium mole numbers. This facilitates the calculation of equilibria for which the number of dominant compounds is less than the number of elements.

REACTOR MODEL Module
A set of routines, originally developed by Eriksson and Johansson[3,4] is devoted to the modelling of processes. This module permits steady-state calculations for dynamic chemical reactors with con- or counter-current material flow and exchange of energy between different stages of the reactor or with the environment. These features are typical for many metallurgical processes, e.g. rotary kilns, blast furnaces, electrothermal furnaces, reverberatory furnaces,

etc., but a process like the gradual solidification of a liquid alloy according to the Scheil-Gulliver scheme may also be described by this approach.

It should be noted that for the REACTOR calculations two different data files are required: one containing the thermodynamic information on the system and the second describing the features of the actual reactor.

Non-ideal solution models
There is direct access from the three calculational modules to the thermodynamic model routines which contain the descriptions of the nonideal mixture phases. Table I below gives a list of the models that are currently available in ChemSage.

Characteristic APPLICATIONS

Calculation of an adiabatic flame temperature
In this application the adiabatic flame temperature for the combustion of CO with O_2 is calculated. The heat balance is fundamental to the equilibrium calculation. It is used as a constraint to calculate the temperature by entering the value of the heat balance as 0. The system is fed with 1 mol of CO/Gas/ and 1 mol of O_2/Gas/ at room temperature and 1 bar pressure. The thermochemical data for this application were taken from the thermodynamic databank system THERDAS, RWTH Aachen.

Calculation of a slag equilibrium for the system CaO-SiO$_2$
This system exhibits liquid immiscibility. Note the components of the liquid phase derived from the stoichiometric formula (Ca2+)P(O2-,SiO4(4-),SiO3(3-),SiO2)Q with P and Q variable stoichiometric coefficients of the respective sublattices according to the charge balance constraint. The composition is silica rich, the temperature above the melting point of silica. At the bottom of the table the compositions of the two liquids are given with respect to the system components CaO and SiO_2. The thermodynamic data were given by Sundman[23].

CALCULATION OF A GEOCHEMICAL SYSTEM

An aqueous system with high concentrations (= molalities). Halite and Nahcolite are co-precipitated. Note that the gas phase is not stable as only CO_2(G) has been considered as a gas species the equilibrium partial pressure of which is less than one. It can yet be calculated as the potentials of oxygen and carbon are known from the calculated equilibrium state between the condensed phases. Ten moles of each Halite and Nahcolite were fed at ambient temperature into one kilogram (i.e. 55.508 moles) of water. The thermodynamic data were taken from Harvie et al.[18].

REFERENCES

1. G. Eriksson and E. Rosen: Chem. Scr. 1973, vol. 4, pp. 193-194.
2. G. Eriksson: Chem. Scr. 1975, vol. 8, pp. 100-103.
3. G. Eriksson and T. Johansson: Scand. J. Metall. 1978, vol. 7, pp. 264-270.
4. T. Johansson and G. Eriksson: J. Electrochem. Soc. 1984, vol. 131, pp. 365-370.
5. O. Redlich and A.T. Kister: Ind. Eng. Chem. 1948, vol. 40, pp. 345-348.
6. Y.M. Muggianu, M. Gambino, and J.P. Bros: J. Chim. Phys. Phys.-Chim. c Biol. 1975, vol. 72, pp. 83-88.
7. L. Kaufman and H. Nesor: CALPHAD, 1978, vol. 2, pp. 35-53.
8. F. Kohler: Monatsh. Chem. 1960, vol. 91, pp. 738-740.
9. M. Margules: Sitzungsber. Wien Akad. 1895, vol. 104, pp. 1243-1278.
10. B. Sundman and J. Agren: J. Phys. Chem. Solids 1981, vol. 42, pp. 297-301
11. M. Hillert, B. Jansson, B. Sundman, and J. Agren: Met. Trans. A 1985, vol. 16A, pp. 261-266.
12. A.D. Pelton: CALPHAD, 1988, vol. 12, pp. 127-142.
13. H. Gaye and J. Welfringer: Metall. Slags Fluxes, Int. Symp., Proc., 2nd 1984, pp. 357-375.
14. M.L. Kapoor and M.G. Frohberg: Chem. Metall. Iron Steel, Proc. Int. Symp. Metall. Chem. — Appl. Ferrous Metall. 1971, pp. 17-22.
15. A.D. Pelton and M. Blander: Metall. Trans. B 1986, vol. 17B, pp. 805-815.
16. G.W. Toop: Trans. AIME 1965, vol. 233, pp. 850-855.
17. A.D. Pelton and C.W. Bale: Metall. Trans. A 1986, vol. 17A, pp. 1211-1215.
18. C.E. Harvie, N. Moeller, and J.H. Weare: Geochim. Cosmochim. Acta 1984, vol. 48, pp. 723-751.
19. K.S. Pitzer: J. Solution Chem. 1975, vol. 4, pp. 249-265.
20. R. Holub and P. Vonka: The Chemical Equilibrium of Gaseous Systems, D. Reidl, Dordrecht, Holland, 1976, pp. 162-195.
21. M. Hillert and M. Jarl: CALPHAD, 1978, vol. 2, pp. 227-238.
22. G. Inden: Project Meeting CALPHAD V, Max-Planck-Institut für Eisenforschung, Duesseldorf, Germany, 1976, pp. IV.1-1-IV.1-35.
23. B. Sundman, KTH, Stockholm, Sweden, private communication, 1988.

```
     T = 2871.81 K
     P = 1.0000E+00 BAR
     V = 3.9743E+02 DM3

     REACTANTS:          AMOUNT/MOL    TEMPERATURE/K    PRESSURE/BAR
     CO/GAS/             1.0000E+00      298.15         1.0000E+00
     O2/GAS/             1.0000E+00      298.15         1.0000E+00

                        EQUIL AMOUNT     PRESSURE        FUGACITY
     PHASE: GAS              MOL           BAR             BAR
     CO2                 7.4277E-01     4.4625E-01      4.4625E-01
     O2                  5.9278E-01     3.5614E-01      3.5614E-01
     CO                  2.5724E-01     1.5455E-01      1.5455E-01
     O                   7.1668E-02     4.3058E-02      4.3058E-02
     O3                  2.7112E-07     1.6289E-07      1.6289E-07
     C                   1.9014E-12     1.1424E-12      1.1424E-12
     C2                  7.4851E-20     4.4970E-20      4.4970E-20
     C3                  3.4990E-26     2.1022E-26      2.1022E-26
     TOTAL:              1.6645E+00     1.0000E+00
                            MOL                          ACTIVITY
     C 0.0000E+00  8.0155E-08
********************************************************************
  DELTA H/J DELTA S/J.K-1 DELTA G/J   DELTA U/J   DELTA A/J  DELTA V/DM3
********************************************************************
  0.0000E+00   1.0107E+02 -1.3566E+06 -3.4785E+04 -1.3914E+06   3.4785E+02
```

Printout 1

```
T = 2000.00 K
P = 1.0000E+00 BAR

REACTANTS:          AMOUNT/MOL
CaO                 1.0000E-01
SiO2(quartz)        9.0000E-01


                    EQUIL AMOUNT   MOLE FRACTION      ACTIVITY
PHASE: LIQUID          MOL
Ca2O2               1.1376E-05     9.3238E-05     4.6745E-08
Si2O4               7.7908E-02     6.3851E-01     9.7226E-01
Ca4(SiO4)2          1.2083E-04     9.9032E-04     5.4969E-05
Ca2(SiO3)2          4.3974E-02     3.6040E-01     9.1254E-02
TOTAL:              1.2201E-01     1.0000E+00
PHASE: LIQUID          MOL        MOLE FRACTION      ACTIVITY
Ca2O2               3.5308E-06     1.0765E-05     4.6745E-08
Si2O4               3.2223E-01     9.8242E-01     9.7226E-01
Ca4(SiO4)2          5.3031E-06     1.6168E-05     5.4969E-05
Ca2(SiO3)2          5.7588E-03     1.7557E-02     9.1254E-02
TOTAL:              3.2800E-01     1.0000E+00
                       MOL                            ACTIVITY
SiO2(crist)         0.0000E+00                     9.8501E-01
SiO2(trid)          0.0000E+00                     9.8018E-01
SiO2(quartz)        0.0000E+00                     9.1135E-01
Pseudo-woll         0.0000E+00                     3.1494E-01
Wollastonite        0.0000E+00                     2.5864E-01
@-Ca2SiO4           0.0000E+00                     1.4837E-02
Bredigite           0.0000E+00                     1.4119E-02
Larnite             0.0000E+00                     7.7639E-03
Olivine             0.0000E+00                     4.6113E-03
Rankinite           0.0000E+00                     1.9234E-03
CaO                 0.0000E+00                     1.0552E-03
Hatrurite           0.0000E+00                     1.7918E-05


MOLE FRACTIONS OF THE SYSTEM COMPONENTS IN THE LIQUID PHASE:
CaO                 2.6606E-01
SiO2                7.3394E-01


MOLE FRACTIONS OF THE SYSTEM COMPONENTS IN THE LIQUID PHASE:
CaO                 1.7296E-02
SiO2                9.3270E-01
```

Printout 2

```
T =   298.15 K
P = 1.0000E+00 BAR
V = 0.0000E+00 DM3

REACTANTS:              AMOUNT/MOL
NaHCO3                  1.0000E+01
NaCl                    1.0000E+01
H2O/AQUEOUS/            5.5508E+01

                   EQUIL AMOUNT      PRESSURE          FUGACITY
PHASE: GAS             MOL             BAR               BAR
CO2                0.0000E+00       1.3306E-01        1.3306E-01
TOTAL:             0.0000E+00       1.3306E-01
PHASE: AQUEOUS         MOL         MOLE FRACTION       ACTIVITY
H2O                5.5509E+01       8.1602E-01        7.5069E-01
H(+)               5.1138E-09       7.5176E-11        2.2784E-08
Na(+)              6.2577E+00       9.1991E-02        6.0265E+00
OH(-)              6.0371E-07       8.8748E-09        3.3203E-07
HCO3(-)            1.9314E-01       2.8392E-03        6.5601E-02
CO3(2-)            1.3156E-03       1.9340E-05        1.3184E-04
CO2(AQ)            1.3162E-03       1.9349E-05        4.3303E-03
Cl(-)              6.0619E+00       8.9113E-02        6.1708E+00
TOTAL:             6.8025E+01       1.0000E+00
                       MOL                            ACTIVITY
NaHCO3             9.8042E+00                        1.0000E+00
NaCl               3.9381E+00                        1.0000E+00
Na3H(CO3)2         0.0000E+00                        1.1775E-02
Na2CO3 7H2O        0.0000E+00                        1.8557E-03
Na2CO3 10H2O       0.0000E+00                        1.8175E-03
Na2CO3 H2O         0.0000E+00                        1.1852E-03

MOLE FRACTIONS OF THE SYSTEM COMPONENTS IN THE AQUEOUS PHASE:
Na(+)              9.1729E-02
CO3(2-)            2.8697E-03
H(+)               2.8697E-03
H2O                8.1367E-01
Cl(-)              8.8859E-02
```

Printout 3

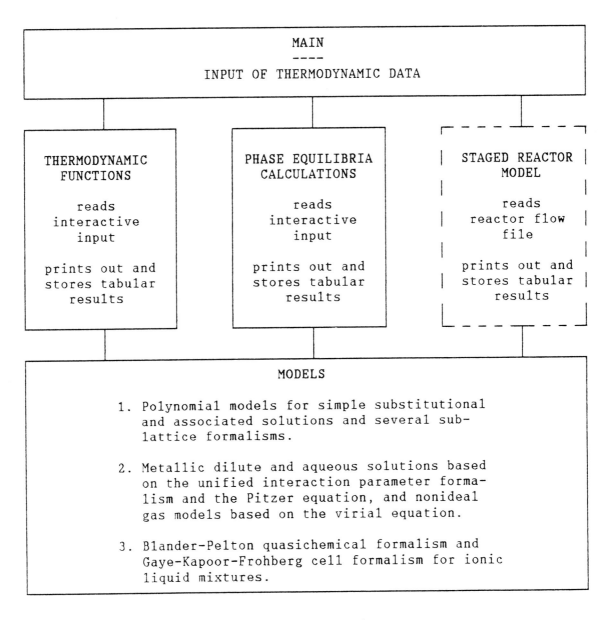

The solid lines indicate the standard features of **ChemSage**; the staged reactor module is optional (**ChemSage-REACTOR**). In conjunction with the main-frame version of **ChemSage** the staged reactor module would be integrated into the source as indicated above; in the PC version, it is supplied as an individual module linked to the others by common thermodynamic datafiles.

Fig. 1. Schematic illustration of the major modules of ChemSage.

Table 1 Models currently available

IDMX: Ideal mixture phase.

RKMP: Simple substitutional and associate models with the Redlich-
Kister-Muggianu equation [5,6].

RKXP: As for RKMP but for pseudobinary mixtures (the stoichiometry of
the sublattice where mixing takes place is an additional input
parameter).

KKOP: Kaufman-Kohler equation as for RKMP [7,8].

KKXP: Kaufman-Kohler equation as for RKXP.

MARP: Four-suffix Margules equation as for RKMP [9].

SUBL: Multisublattice formalism. Ideal behaviour is assumed for a
mixture phase with more than two sublattices [10].

SUBI: Two-sublattice ionic formalism for varible stoichiometries [11].

SUBM: Two-sublattice equivalence fraction formalism for molten
salts [12].

GAYE: Gaye-Kapoor-Frohberg cell formalism for ionic oxidic liquid
mixtures [13,14].

QUAS: Blander-Pelton quasichemical formalism for ionic liquid
mixtures [15].

QKTO: Kohler/Toop general polynomial formalism [8,16].

WAGN: Metallic dilute solutions with a modified Wagner formalism [17].

PITZ: Aqueous solutions with the Pitzer equation [18,19].

VIRN: Virial expansion, no estimates [20].

VIRA: Virial expansion, arithmetic means [20].

VIRG: Virial expansion, geometric means [20].

VIRL: Virial expansion, Lorentz means [20].

Magnetic contributions will be accepted as an additional term by
giving an M as the fifth character. These terms are treated according
to the formalism proposed by Hillert and Jarl [21] which is based on a
method originally given by Inden [22].

35
A CONTRIBUTION TO Au-Ge-X AND Au-Pb-X SYSTEMS

M.T.Z Butt, C. Bodsworth and A. Prince

Department of Materials Technology, Brunel, The University of West London, Uxbridge, Middlesex, UB8 3PH, UK

ABSTRACT

The partial constitutions of Au-Ge-X and Au-Pb-X ternary alloys have been investigated, where X is a metallic element, [selected from the B sub-groups of period II and III of the periodic table (In, Ga, Zn, or Cd)], which forms one or more stable compounds with gold, but which forms no stable compound with Ge and Pb. The AuX-Ge sections were found to be pseudobinary eutectic systems. However, AuX-Pb sections have not pseudobinary characteristics.

EXPERIMENTAL TECHNIQUES

The S+mith Thermal Analysis Method, supplemented by metallographic and X-ray techniques, was used to determine the constitutions of the ternary systems. Specimens for Thermal Analysis were prepared from high purity metals; Au, Ge, In, Zn, Pb were 99.9999% pure, while the purity of Ga and Cd was 99.99% and 99.9998% respectively.

The arrest temperatures for the solidus line were obtained from the heating and cooling curves; arrest temperatures for the liquidus were taken only from the cooling curves. The reproducibility of liquidus temperature are considered to be within $\pm 1°$ C and for solidus temperature $\pm 3°$C, while the overall accuracy was of the order of $\pm 1°$C.

RESULTS

Au-Ge-X Systems

Prior to this investigation, the only published work on the phase relationships in Au-Ge-X ternary systems was an examination of the Au-rich corner of the Au-Ge-In system [1]. In the present investigation eutectiferous, quasibinary systems were found between Ge and the stable compounds AuX (AuIn, AuIn$_2$, AuGa, AuGa$_2$, AuZn and AuCd). The liquidus and solidus curves in all of the systems were experimentally determined up to 50 or 60 at.% Ge; expansion on solidification shatters the silica crucibles at higher Ge concentrations. Both liquidus and solidus curves were extrapolated, from 50 (or 60 at.% Ge) to pure germanium, adopting a melting point of 938.5°C for Ge [2].

Calculations of the liquidus curves for most of the AuX-Ge alloys using the equation:

$$\ln\left(1 - X_{Ge}\right) = \frac{\left[\dfrac{1}{T_{AuX}}\right] - \left[\dfrac{1}{T_{Eu}}\right] H_{AuX}^{F}}{R}$$

and a value from the literature [3,4] for the enthalpy of fusion for AuX, predicted the eutectic composition in good agreement with the experimental values.

The solubility of Ge in the AuX compounds was not determined directly. However, it was 1.3 at.% Ge for Zn and Cd containing alloys and less than 1.0 at.% Ge for In and Ga containing alloys at the eutectic temperatures. All alloys having more than this limited concentration of Ge, showed one liquidus and one solidus arrest at the eutectic temperature. All Au-Ge-X systems show complete miscibility in the liquid state for all compositions. The eutectic temperatures, eutectic compositions and solubilities of the various systems which were examined are summarized in Figs. 1-6 and in Table 1.

AuIn-Ge-AuIn$_2$ Partial Ternary Section (Isopleth of Au$_{41.5}$In$_{58.5}$→Ge Section)

Since the AuIn-Ge and AuIn$_2$-Ge joins were found to be eutectic systems and AuIn and AuIn2 also form a eutectic system [10,11], it was assumed that the AuIn-Ge-AuIn$_2$ partial section would contain a ternary eutectic. To determine the ternary eutectic composition and temperature, STA was performed on the isopleth from Au$_{41.5}$In$_{58.5}$ to Ge. The thermal analysis results on this isopleth gave the phase diagram shown in Fig. 7(a). A magnified portion of this isopleth for alloys containing less than 8 at.% Ge is shown in Fig. 7(b). The monovariant curve from the AuIn$_2$-Ge eutectic e$_3$ to the ternary eutectic E, Fig. 7(c) is intersected at 4.0 at.% Ge at 488°C. The tie line between the ternary eutectic E and AuIn$_2$ is intersected at 2.75 at.% Ge at 471°C. The ternary eutectic composition, corresponding to the reaction L$_E$ ⇔ AuIn$_2$ + AuIn + Ge, is estimated to be 3.5 at.% Ge, 43.25 at.% Au at 471°C.

AuIn$_2$-Ge-In Partial Ternary Section (Isopleth of Au$_{20}$In$_{80}$:→Ge Section)

A ternary eutectic was also anticipated in this section, so a Au$_{20}$In$_{80}$ to Ge isopleth was determined by thermal analysis. The derived phase diagram from these results is shown in Fig. 8(a), and the magnified portion of this diagram, with Ge contents less than 7 at.% is given in Fig. 8(b). The presence of a degenerate ternary eutectic is evident, located very close to the In corner at a temperature of 156°C.

AuGa-Ge-AuGa$_2$ Partial Ternary Section (Isopleth of Au$_{40}$Ga$_{60}$→Ge section)

The AuGa-Ge and AuGa$_2$-Ge joins were found to be eutectic experimentally, while AuGa and AuGa$_2$ also form a eutectic system with each other [12]. It was assumed that the AuGa-Ge-AuGa$_2$ partial system would contain a ternary eutectic. To determine the ternary eutectic composition and temperature, STA was performed on the isopleth from Au$_{40}$Ga$_{60}$ to Ge. The resulting phase diagram is shown in Fig. 9(a). The magnified portion of this diagram for alloys containing less than 8 at.% Ge is shown in Fig. 9(b). The monovariant curve from the AuGa$_2$-Ge eutectic e$_3$ to the ternary eutectic E, Fig. 9(c) is intersected at 3.9 at.% Ge at 466°C. The tie line between the ternary eutectic E and AuGa$_2$ is intersected at 2.25 at.% Ge at 438°C. The ternary eutectic composition, corresponding to the reaction L$_E$ ⇔ AuGa$_2$ + AuGa + Ge, is estimated to be 3.2 at.% Ge, 41.8 at.% Au at 438°C.

Au-Pb-X SYSTEMS

The partial constitutions of Au-Pb-X ternary alloys has also been investigated, where X = Ga, Zn and Cd. Again a combination of quantitative Thermal Analysis and metallographic techniques, were used. The investigation of AuX-Pb sections in Au-Pb-X ternary system, showed extensive liquid immiscibility over a wide range of compositions; immiscibility was easily seen on metallographic sections. The solubilities of AuX in Pb or vice versa are less than 2 at.%.

Au-Pb-Ga SYSTEM

Prior to the current investigation of this system, no published data was available. Both the AuGa-Pb and AuGa$_2$-Pb sections were determined; neither have pseudobinary characteristics. Liquid immiscibility was detected at 438°C and at 481°C in the AuGa-Pb and AuGa$_2$-Pb sections respectively, extending at least over the range from 5.0 to 80.0 at.% Pb for both sections. In each section two further arrests were observed. The AuGa-Pb section showed an isothermal reaction at 317°C and at about 202°C; the AuGa$_2$-Pb section showed reactions at 325°C and 317°C. No interpretation of this data is possible without further investigation.

Au-Pb-Zn SYSTEM

The only data on the phase relationships in the Au-Pb-Zn system is restricted to the lead-rich corner [13]. In the present study the AuZn-Pb section was examined; it has not pseudobinary characteristics. Liquid immiscibility was detected at 750°C in the Au-rich portion but this temperature decreased with increasing Pb contents to 712°C at 90 at.% Pb. An isothermal reaction at 211°C occurs throughout the section and one additional arrest was found varying from 322°C at 95 at.% Pb to 281°C at 10 at.% Pb. Further work is required before this data can be interpreted.

Au-Pb-Cd SYSTEM

Before this study, the only published data for the Au-Pb-Cd system was an examination of Pb-rich alloys [14]. The present investigation showed that the AuCd-Pb section is complex and certainly not a pseudobinary section. Liquid immiscibility was found for alloys containing up to 70 at.% Pb, associated with temperatures varying from 595°C to 584°C; and a number of arrests occurred at temperatures below the melting point of Pb for each alloy composition. No interpretation of this data is possible without further investigation.

Au-Pb-In SYSTEM

Previous work [15] has shown that the AuIn-Pb and AuIn$_2$-Pb are pseudobinary sections. The AuIn-Pb section contains a monotectic reaction and a eutectic reaction. The AuIn$_2$-Pb section was reported to be a simple eutectic system. However, according to Prince (16), the AuIn-Pb section is a pseudobinary whereas the AuIn$_2$-Pb section is not.

ACKNOWLEDGEMENTS

M. T. Z. Butt gratefully acknowledges a Government of Pakistan scholarship award. The authors would like to thank Dr. D. E. J. Talbot for his kind advice on experimental techniques and Prof. B. Ralph, Head of the Department of Materials Technology, Brunel University, U.K., for provision of facilities. The authors also thank Englehard Limited for the loan of gold and GEC Hirst Research Centre, Wembley, for the supply of germanium.

REFERENCES

1. C. R. M. Grovenor, Thin Solid Films, 104, 409, (1983).
2. Private Communication from T. G. Chart, National Physical Laboratory, U.K (SGTE Pure Component Transition Data, March, (1988).
3. P. C. Wallbrecht, R. Blachnik and K. C. Mills, Thermochimica Acta., 48, 69, (1981).
4. E. A. Brandes, Smithells Metal Reference Book (6th ed.), Butterworth and Co. Ltd., London (1983).
5. M. T. Z. Butt and C. Bodsworth, Mater. Sci. Tech., 6(2), 134, (1990).
6. M. T. Z. Butt, C. Bodsworth and A. Prince, J Less-Common Metals, 154, 229, (1989).
7. M. T. Z. Butt, C. Bodsworth and A. Prince, Scripta Met., 24(3), 481, (1990).
8. M. T. Z. Butt, C. Bodsworth and A. Prince, Scripta Met., 23(7), 1105, (1989).
9. M. T. Z. Butt, C. Bodsworth and A. Prince, Unpublished data.
10. V. K. Nikitina, A. A. Babitsyna and Yu. K. Lobanova, Izv. Akad. Nauk. SSSR, Neorg. Mater., 7(3), 421, (1971) in Russian; TR: Inorg. Mater., 7(3), 371, (1971).
11. S. E. R. Hiscocks and W. Hume-Rothery, Proc. Roy. Soc. (London), A 282, 318, (1964).
12. C. J. Cooke and W. Hume-Rothery, J Less-Common Metals, 10, 42, (1966).
13. R. D. Pehlke and K. Okajima, Trans. Metall. Soc. AIME, 239, 1354, (1967).
14. C. T. Heycock and F. H. Neville, J. Chem. Soc., 65, 65, (1894).
15. M. M. Karnowsky and F. G. Yost, Metall. Trans. A, 7A, 1149, (1976).
16. A. Prince, G. V. Raynor and D. S. Evans, "Phase Diagrams of Ternary Gold Alloys", Institute of Metals, London, (1990).

298

Table 1 Summary of the AuX-Ge pseudobinary systems

Section	Eutectic Composition (at.% Ge)	Eutectic Temperature (°C)	Solubility of Ge in AuX	Phase Diagrams Figures	References
AuIn-Ge	2.0	488	< 1 at.% Ge	1	5
AuIn$_2$-Ge	4.1	522	< 1 at.% Ge	2	5
AuGa-Ge	5.5	446	< 1 at.% Ge	3	6
AuGa$_2$-Ge	5.0	476	< 1 at.% Ge	4	7
AuZn-Ge	12.2	673	1.3 at.% Ge	5	8
AuCd-Ge	10.0	555	1.3 at.% Ge	6	9

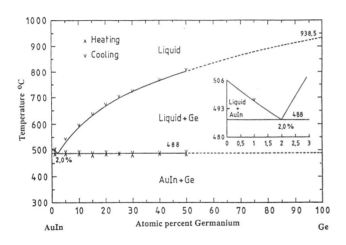

Fig 1. The pseudobinary eutectic system of AuIn-Ge section.

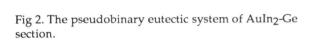

Fig 2. The pseudobinary eutectic system of AuIn₂-Ge section.

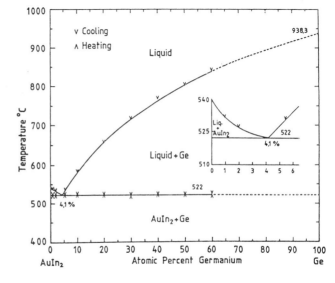

Fig 3. The pseudobinary eutectic system of AuGa-Ge section.

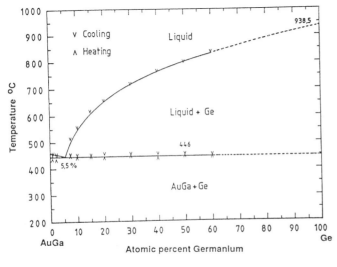

300

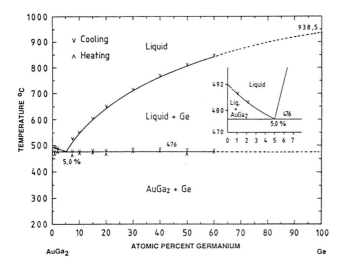

Fig 4. The pseudobinary eutectic system of AuGa₂-Ge section.

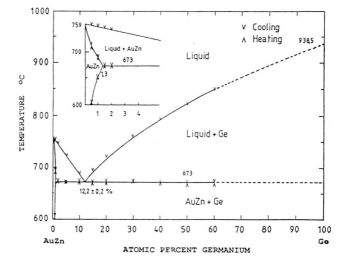

Fig 5. The pseudobinary eutectic system of AuZn-Ge section.

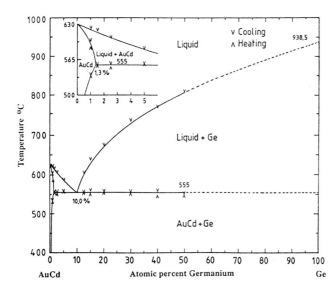

Fig 6. The pseudobinary eutectic system of AuCd-Ge section.

301

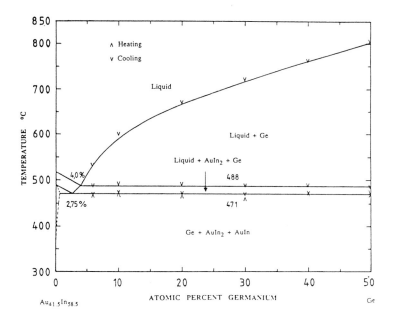

Fig. 7(a). Isopleth from Au$_{41.5}$In$_{58.5}$ to Ge.

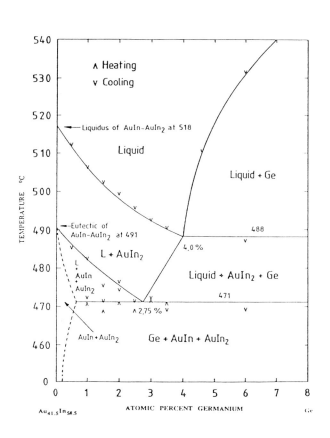

Fig. 7(b). Magnified portion of Fig. 7(a) for alloys containing less than 8 at.% Ge.

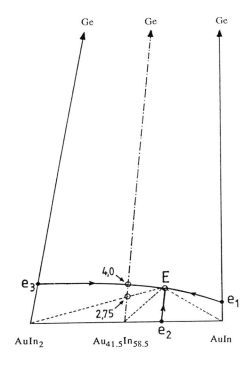

----- = Isopleth Studied, Au$_{41.5}$In$_{58.5}$ → Ge.

e$_1$ = Eutectic between AuIn and Ge at 2.0 at.% Ge and 488°C.

e$_2$ = Eutectic between AuIn and AuIn$_2$ at 56.5 at.% In and 491°C.

e$_3$ = Eutectic between AuIn$_2$ and Ge at 4.1 at.% Ge and 522°C.

E = Ternary Eutectic Point at 3.5 at.% Ge, 43.25 at.% Au and 53.25 at.% In and 471°C.

Fig. 7(c). Liquidus projection of the partial ternary system AuIn-AuIn$_2$-Ge.

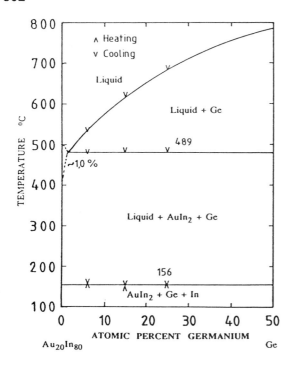

Fig.8(a). Isopleth from Au$_{20}$In$_{80}$ to Ge.

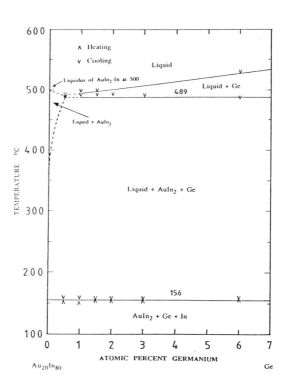

Fig.8(b). Magnified portion of Fig. 8(a) for alloys containing less than 7 at.% Ge.

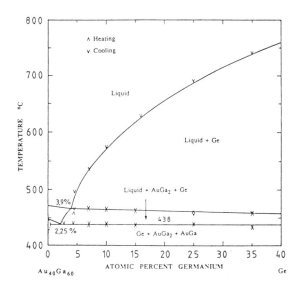

Fig. 9(a). Isopleth from $Au_{40}Ga_{60}$ to Ge.

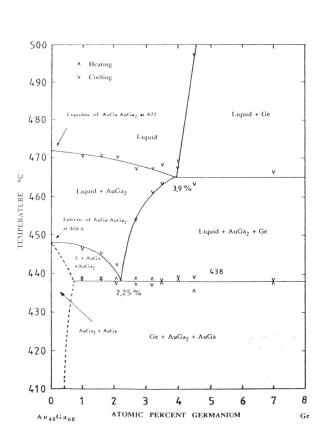

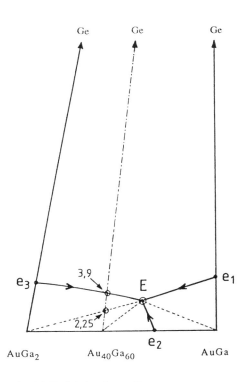

---·- = Isopleth Studied, $Au_{40}Ga_{60} \rightarrow$ Ge.

e_1 = Eutectic between AuGa and Ge at 5.5 at.% Ge and 446°C.

e_2 = Eutectic between AuGa and $AuGa_2$ at 55.4 at.% Ga and 448.6°C.

e_3 = Eutectic between $AuGa_2$ and Ge at 5.0 at.% Ge and 476°C.

E = Ternary Eutectic Point at 3.2 at.% Ge, 41.8 at.% Au and 55.0 at.% Ga and 438°C.

Fig. 9(b). Magnified portion of Fig. 9(a) for alloys containing less than 8 at.% Ge.

Fig. 9(c). Liquidus projection of the partial ternary system AuGa-$AuGa_2$-Ge.